Lecture Notes in Artificial Intelligence 10558

Subseries of Lecture Notes in Computer Science

More information about this series at http://www.springer.com/series/1244

Akihiro Yamamoto · Takuya Kida
Takeaki Uno · Tetsuji Kuboyama (Eds.)

Discovery Science

20th International Conference, DS 2017
Kyoto, Japan, October 15–17, 2017
Proceedings

 Springer

Editors
Akihiro Yamamoto
Kyoto University
Kyoto
Japan

Takuya Kida
Hokkaido University
Sapporo
Japan

Takeaki Uno
National Institute of Informatics
Tokyo
Japan

Tetsuji Kuboyama
Gakushuin University
Tokyo
Japan

ISSN 0302-9743   ISSN 1611-3349 (electronic)
Lecture Notes in Artificial Intelligence
ISBN 978-3-319-67785-9   ISBN 978-3-319-67786-6 (eBook)
DOI 10.1007/978-3-319-67786-6

Library of Congress Control Number: 2017953423

LNCS Sublibrary: SL7 – Artificial Intelligence

Printed on acid-free paper

This Springer imprint is published by Springer Nature
The registered company is Springer International Publishing AG
The registered company address is: Gewerbestrasse 11, 6330 Cham, Switzerland

Preface

The 20th International Conference on Discovery Science (DS 2017) was held in Kyoto, Japan, during October 15–17, 2017. As in previous years, the conference was co-located with the International Conference on Algorithmic Learning Theory (ALT 2017), which was already in its 28th year. First held in 2001, ALT/DS has been one of the longest running series of co-located events in computer science. The unique combination of recent advances in the development and analysis of methods for automatic scientific knowledge discovery, machine learning, intelligent data analysis, and their application to knowledge discovery on the one hand, and theoretical and algorithmic advances in machine learning on the other hand, makes every instance of this joint event unique and attractive.

This volume contains all the papers presented at the 20th International Conference on Discovery Science, while the papers of the 28th International Conference on Algorithmic Learning Theory are published as a volume in the JMLR Workshop and Conference Proceedings series. The 20th Discovery Science conference received 42 international submissions. Each submission was reviewed by at least three Program Committee members. The co-chairs eventually decided to accept 18 papers as regular papers and 6 papers as short papers. A special issue on the topics of Discovery Science has also been scheduled for the Springer journal Machine Learning, thus offering the option of publishing in this prestigious journal an extended and reworked version of papers presented at Discovery Science 2017.

The program included 4 invited talks. In the joint DS/ALT invited talk, Masashi Sugiyama from RIKEN, the University of Tokyo, gave a presentation on "Machine Learning from Weak Supervision – Towards Accurate Classification with Low Labeling Costs."The DS invited talk by Koji Tsuda from the University of Tokyo was on "Automatic Design of Functional Molecules and Materials." DS participants also had the opportunity to attend the ALT invited talks, which were given by Adam Kalai from Microsoft Reasearch New England and by Alexander Rakhlin from the University of Pennsylvania. Abstracts of the joint invited talk and the DS invited talk are included in this volume. The program also included one special session organized by Takeaki Uno from National Institute of Informatics, Japan.

We would like to thank all the authors of submitted papers, the Program Committee members, and the additional reviewers for their efforts in evaluating the submitted papers, as well as the invited speakers. We are grateful to Steve Hanneke and Lev Reyzin for ensuring the smooth coordination with ALT. We are grateful to the people behind EasyChair, too. It was an essential tool in the paper submission and evaluation process, as well as in the preparation of the Springer proceedings. We are also grateful to Springer for their continuing support of Discovery Science and for publishing the conference proceedings.

We would like to thank the local arrangement chairs, Yasuaki Kobayashi and Matthew de Brecht from Kyoto University, and all their team. Also, we wish to express our thanks to Kaori Deguchi for helping us with various affairs. All of them worked very hard to make both conferences a great success.

Finally, we gratefully appreciate the financial support of JST CREST (Data Particlization for Next Generation Data Mining).

August 2017

Akihiro Yamamoto
Takuya Kida
Tetsuji Kuboyama
Takeaki Uno

Organization

Program Committee

Annalisa Appice	University of Bari Aldo Moro, Italy
Hiroki Arimura	Hokkaido University, Japan
Yukino Baba	Kyoto University, Japan
Hideo Bannai	Kyushu University, Japan
Michelangelo Ceci	University of Bari Aldo Moro, Italy
Bruno Cremilleux	Université de Caen, France
Ivica Dimitrovski	Ss. Cyril and Methodius University in Skopje, Macedonia
Sašo Džeroski	Jožef Stefan Institute, Slovenia
Floriana Esposito	University of Bari Aldo Moro, Italy
Daiji Fukagawa	Doshisha University, Japan
Johannes Fürnkranz	TU Darmstadt, Germany
Mohamed Gaber	Birmingham City University, UK
João Gama	INESC TEC, University of Porto, Portugal
Dragan Gamberger	Rudjer Boskovic Institute, Croatia
Kouichi Hirata	Kyushu Institute of Technology, Japan
Jaakko Hollmén	Aalto University School of Science, Finland
Kimihito Ito	Hokkaido University, Japan
Alipio M. Jorge	FCUP, Univ. do Porto/LIAAD, INESC Porto L.A., Portugal
Takuya Kida	Hokkaido University, Japan
Dragi Kocev	Jožef Stefan Institute, Slovenia
Stefan Kramer	Johannes Gutenberg University Mainz, Germany
Tetsuji Kuboyama	Gakushuin University, Japan
Nada Lavrač	Jožef Stefan Institute, Slovenia
Philippe Lenca	IMT Atlantique, France
Gjorgji Madjarov	Ss. Cyril and Methodius University in Skopje, Macedonia
Donato Malerba	University of Bari Aldo Moro, Italy
Giuseppe Manco	ICAR-CNR, Italy
Elio Masciari	ICAR-CNR, Italy
Robert Mercer	The University of Western Ontario, Canada
Tetsuhiro Miyahara	Hiroshima City University, Japan
Anna Monreale	University of Pisa, Italy
Masaaki Nishino	NTT Communication Science Laboratories, Japan
Pance Panov	Jožef Stefan Institute, Slovenia
Dino Pedreschi	University of Pisa, Italy
Ruggero G. Pensa	University of Turin, Italy
Bernhard Pfahringer	University of Waikato, New Zealand
Gianvito Pio	University of Bari Aldo Moro, Italy
Jan Ramon	Inria, France

Chedy Raïssi	Inria, France
Chiara Renso	ISTI-CNR, Italy
Kazumi Saito	Univesity of Shizuoka, Japan
Hiroshi Sakamoto	Kyushu Institute of Technology, Japan
Marina Sokolova	University of Ottawa and Institute for Big Data Analytics, Canada
Jerzy Stefanowski	Poznan University of Technology, Poland
Mahito Sugiyama	National Institute of Informatics, Japan
Yasuo Tabei	RIKEN Center for Advanced Intelligence Project, Japan
Takeaki Uno	National Institute of Informatics, Japan
Herna Viktor	University of Ottawa, Canada
Ryo Yoshinaka	Tohoku University, Japan
Blaž Zupan	University of Ljubljana, Slovenia
Tomislav Šmuc	Rudjer Boskovic Institute, Croatia

Additional Reviewers

Bioglio, Livio	Petković, Matej
Gerow, Aaron	Sousa Lima, Wesllen
Haidar, Diana	Sousa, Ricardo
Jafer, Yasser	Zopf, Markus
Mignone, Paolo	

Abstracts of Invited Talks

Machine Learning from Weak Supervision — Towards Accurate Classification with Low Labeling Costs

Masashi Sugiyama

[1] RIKEN, 1-4-1 Nihonbashi, Chuo-ku, Tokyo 103-0027, Japan
[2] The University of Tokyo, 5-1-5 Kashiwanoha, Kashiwa-shi,
Chiba 277-8561, Japan
sugi@k.u-tokyo.ac.jp

Abstract. Machine learning from big training data is achieving great success. However, there are various application domains that prohibit the use of massive labeled data. In this talk, I will introduce our recent advances in classification from weak supervision, including classification from two sets of unlabeled data, classification from positive and unlabeled data, a novel approach to semi-supervised classification, and classification from complementary labels. Finally, I will briefly introduce the activities of RIKEN Center for Advanced Intelligence Project.

Automatic Design of Functional Molecules and Materials

Koji Tsuda

Graduate School of Frontier Sciences, University of Tokyo, 5-1-5
Kashiwa-no-ha, Kashiwa-shi, Chiba-ken 277-8561, Japan
tsuda@k.u-tokyo.ac.jp

Abstract. The scientific process of discovering new knowledge is often characterized as search from a space of candidates, and machine learning can accelerate the search by properly modeling the data and suggesting which candidates to apply experiments on. In many cases, experiments can be substituted by first principles calculation. I review two basic machine learning techniques called Bayesian optimization and Monte Carlo tree search. I also show successful case studies including Si-Ge nanostructure design, optimization of grain boundary structures and discovery of low-thermal-conductivity compounds from a database.

Contents

Bioinformatics

Knowledge Discovery

Online Learning

Context-Based Abrupt Change Detection and Adaptation for Categorical Data Streams

Sarah D'Ettorre[1]([✉]), Herna L. Viktor[1], and Eric Paquet[1,2]

[1] School of Electrical Engineering and Computer Science,
University of Ottawa, Ottawa K1N 6N5, Canada
{sdett026,hviktor}@uottawa.ca, eric.paquet@nrc-cnrc.gc.ca
[2] National Research Council of Canada, Ottawa, ON K1A 0R6, Canada

Abstract. The identification of changes in data distributions associated with data streams is critical in understanding the mechanics of data generating processes and ensuring that data models remain representative through time. To this end, concept drift detection methods often utilize statistical techniques that take numerical data as input. However, many applications produce data streams containing categorical attributes, where numerical statistical methods are not applicable. In this setting, common solutions use error monitoring, assuming that fluctuations in the error measures of a learning system correspond to concept drift. Context-based concept drift detection techniques for categorical streams, which observe changes in the actual data distribution, have received limited attention. Such context-based change detection is arguably more informative as it is data-driven and directly applicable in an unsupervised setting. This paper introduces a novel context-based algorithm for categorical data, namely **FG-CDCStream**. In this unsupervised method, multiple drift detection tracks are maintained and their votes are combined in order to determine whether a real change has occurred. In this way, change detections are rapid and accurate, while the number of false alarms remains low. Our experimental evaluation against synthetic data streams shows that **FG-CDCStream** outperforms the state-of-the art. Our analysis further indicates that **FG-CDCStream** produces highly accurate and representative post-change models.

Keywords: Data streams · Categorical data · Concept drift · Context-based change detection · Unsupervised learning · Ensembles · Online learning

1 Introduction

Data streams, that are characterized by a continuous flow of high speed data, require learning methods that incrementally update models as new data become available. Further, such algorithms should be able to adapt appropriately as underlying concepts in the data evolve over time. Explicit detection of this evolution, known as concept drift, is beneficial in ensuring that models remain

© Springer International Publishing AG 2017
A. Yamamoto et al. (Eds.): DS 2017, LNAI 10558, pp. 3–17, 2017.
DOI: 10.1007/978-3-319-67786-6_1

accurate and provide insights into the mechanics of the data generating process [6]. Thus, concept drift detection has been of continuous interest to machine learning researchers.

The majority of concept drift detection algorithms utilize statistical methods requiring numerical input [5]. However, real world data attributes are often categorical [9]. For example, point-of-sales streams include categorical attributes such as colour {red, green, blue} or size {small, medium, large, x-large}. Environmental data attributes might contain categorical attributes like predators {eagle, owl, fox} or land cover {desert, forest, tundra}. This prevalence of categorical data poses a challenge for change detection researchers. Currently, the majority of research on change detection in categorical data streams utilize error changes in the learning system as an indicator of concept drift [9]. While these techniques have proven to be reasonably successful, it remains that fluctuations in error measures cannot be definitively attributed to concept drift alone. Relatively few studies in the literature have examined context-based change detection in categorical data streams [9]. In this case, concept drift is detected when changes in the actual data distribution are observed, providing more precise information about particular changes. This opens the door to unsupervised change detection [9], a non-trivial task [4]. Since class information is not always available, facilitating unsupervised change detection broadens the spectrum of categorical data which may be analyzed.

To this end, this paper focuses on improving the quality of knowledge extraction from evolving streams of categorical data through the use of context-based change detection and adaptation strategies. This paper introduced the **FG-CDCStream** technique, which extends the **CDCStream** algorithm [9], in order to rapidly detect abrupt changes, using a fine-grained drift detection technique. The **FG-CDCStream** method employs a voting-based algorithm in order to track the evolution of the data as the stream evolves. Adaptation is improved by ensuring that a post-change classifier is trained on a reduced batch of highly relevant data. This ensures that the evolving classification models are more representative of the post-change concept. Our experimental evaluation confirms that our algorithm is highly suitable for abrupt change detection.

This paper is organized as follows. Section 2 introduces background work. In Sect. 3, we detail the **FG-CDCStream** algorithm. Section 4 discusses our experimental evaluation and results. Finally, Sect. 5 presents our conclusion and highlights future work.

2   Background

This section discusses related works in terms of measuring the similarity of categorical data and context-based drift detection.

2.1   DILCA Context-Based Similarity Measure

Categorical variables are abundant in real-world data and this fact has lead to a large and diverse collection of proposed distance measures spanning various fields of study, many of which arising in the context of categorical data clustering. Similarity measures can be context-free or context-sensitive, supervised or unsupervised. Context-free measures do not depend on relationships between instances while context-sensitive measures do. Based on [4,5], distance measures for categorical values may be classified into six groups that are not necessarily mutually exclusive: simple matching approaches, frequency-based approaches, learning approaches, co-occurrence approaches, probabilistic approaches, information theoretic approaches and rough set theory approaches.

DILCA is a recent state-of-the-art similarity measure that is purely context-based, that makes no assumptions about data distribution and that does not depend on heuristics to determine inter-attribute relationships [8]. These properties make **DILCA** attractive as it minimizes bias and can be applied to a great range of data sets with a wide range of characteristics. The **DILCA** similarity measure is computed in two steps, namely context selection and distance computation. Informative context selection is non-trivial, especially for data sets with many attributes. Consider the following set of m categorical attributes: $F = \{X_1, X_2, ..., X_m\}$. The context of the target attribute Y is defined as a subset of the remaining attributes: $context(Y) \subseteq F/Y$. **DILCA** uses the Symmetric Uncertainty (SU) feature selection method [10] to select a relevant, non-redundant set of attributes which are correlated to the target concept in the context.

$$SU(Y,X) = 2\frac{IG(Y;X)}{H(Y) + H(X)} \tag{1}$$

DILCA selects a set of attributes with high SU with respect to the target Y. A strength of the SU measure, as defined in Eq. 1, is that it is not biased towards features of greater cardinality and is normalized on $[0,1]$. Once the context has been extracted, the distances between attribute values of the target attribute are computed using Eq. 2:

$$d(y_i, y_j) = \sqrt{\frac{\Sigma_{X \in context(Y)} \Sigma_{x_k \in X} (P(y_i|x_k) - P(y_j|x_k))^2}{\Sigma_{X \in context(Y)} |X|}} \tag{2}$$

For each value of each context attribute, the Euclidean distance between the conditional probabilities for both values of Y is computed and summed. This value is then normalized by the total number of values in X. The pairwise distances computation between each of the values of Y results in a symmetric and hollow (where diagonal entries are all equal to 0) matrix $M_i = |Y| \times |Y|$.

2.2 CDCStream Categorical Drift Detector

The **CDCStream** algorithm, as created by [9], utilizes the above-mentioned **DILCA** method [9] in order to detect drift in categorical streams, as follows.

Consider an infinite data stream S where each attribute X is categorical. A buffer is used to segment the stream into batches: $S = \{S_1, S_2, ..., S_n, ...\}$. (Note that, if the class information is present in the stream, it is removed during segmentation, thus creating an unsupervised change detection context.) The set of distance matrices $M = \{M_1, M_2, ..., M_s\}$ produced by **DILCA** is aggregated to numerically summarize the data distribution of each batch in a single statistic using Eq. 3. The resulting statistic, in $[0, 1]$, represents both intra- and inter-attribute distributions of a batch.

$$extractSummary(M) = \frac{\Sigma_{M_l \in M} \frac{2 * \sqrt{\Sigma_{i=0}^{|X_l|} \Sigma_{j=i+1}^{|X_l|} M_{X_l}(i,j)^2}}{|X_l| * (|X_l| - 1)}}{|F|} \tag{3}$$

Data are analyzed in batches, by using a dynamic sliding window method. The historical window L consists of all batch summaries, except the most recent batch which constitutes the current window. The dynamic window grows and shrinks appropriately based on the change status, by forgetting the historical window and/or absorbing the most recent batch. In periods of stability, the historical window continues to grow summary by summary. When a change is detected, abrupt forgetting is employed and the historical window is dropped and replaced with the current window.

As noted in [9], a two-tailed statistical test, which does not assume a specific data distribution is required for context-based, unsupervised change detection. To this end, Chebychev's inequality was adopted. Formally, Chebychev's Inequality states that if X is a random variable with mean μ_X and standard deviation σ_X, for any positive real number k:

$$Pr(|X - \mu_X| \geq k\sigma_X) \leq \frac{1}{k^2} \tag{4}$$

In this equation, $\frac{1}{k^2}$ is the maximum number of the values that may be beyond k standard deviations from the mean. **CDCStream** utilizes this property in order to warn for, and subsequently detect, changes in the distributions. Values of k representing warning and change thresholds, are denoted by k_w and k_c, respectively. These values were empirically set to $k_w = 2$ and $k_c = 3$. That is, in order for a warning to be flagged, at least 75% of the values in the distribution must be within two standard deviations of the mean. For a change to be flagged, at least 88.89 % of the values must be within three standard deviations of the mean. The **CDCStream** adaptation method employs global model replacement. When a warning is detected (k_w), a new background classifier is created and updated alongside the current working model. Once a change is detected (k_c), the current model is entirely replaced with the background model.

Limitations of CDCStream for Abrupt Drift Detection. A recent study showed that **CDCStream** performs competitively in terms of accuracy and adaptability, when compared to two state-of-the art algorithms with similar goals but different structures [9]. In this prior study, the main focus was on detecting gradual drifts. **CDCStream** does, however, have a number of limitations, notably when aiming to address abrupt drift. Firstly, Chebychev's Inequality is conservative, leading to high detection delay, due to the fact that the grain of the bound must be quite coarse. Secondly, aggregation into a single summary statistic may have a diluting effect, depending on factors such as change magnitude, duration and location, which may result in increased missed detections and detection delays. In an abrupt drift setting, a main goal is to offer a fast response to change. Lastly, temporal bounds on the post-change replacement classifier training data are unnecessarily broad, potentially leading to warnings caused by noise or minor fluctuations. This broadness could severely effect any post-change classifications. Collectively, these limitations have the greatest effect on the detection of abrupt concept drifts. An increased detection delay due to the coarseness of Chebychev's inequality or aggregation dilution would be most detrimental in the case of abrupt drift. Since the distributions change so rapidly, predictions based on the previous distribution would be quite erroneous. This is more so than in the gradual change case, where the stream still contains some instances of the previous distribution. Aggregation dilution would also be more likely to miss changes altogether if the transition period was shorter. Finally, introducing a new classifier that remains partially representative of the previous distribution, is unlikely to be effective in classifying the data of an abrupt change. In the next section, we present our **FG-CDCStream** algorithm that extends **CDCStream** to address the abrupt drift scenario.

3   FG-CDCStream Algorithm

The aims of the **FG-CDCStream** approach, as depicted in Algorithm 1, are to improve detection delay and reduce missed detections of the batch-based scenario and its associated aggregation. Ergo, **FG-CDCStream** overlays a series of batches, or *tracks*, each shifted by one instance from the previous track in order to simulate an instance by instance analysis. A detailed technical explanation follows.

Our **FG-CDCStream** technique uses a dynamic list L of i contiguous batches S of fixed size n to detect a change between the current batch and the previous batches remaining in the sliding window. To solve the grain size problem, we employ overlapped dynamic lists deemed tracks. More formally, **FG-CDCStream** uses a series of n tracks, each overlapping the previous track's most recent $n - 1$ instances. This allows a change test to be performed as each instance is received, while incorporating the use of batches. Note that there is a single initial delay of $2n$ before the first two batches may be compared.

Figure 1 displays the batch and track construction process as instances arrive. For simplicity, this example shows only the structure of batches and tracks and does not include responses to concept drift. Let us assume a batch size of three instances (a batch size much too small in reality, but sufficient to understand track construction). Until the first three instances are collected, no batches exist. Once the third instance arrives, the first batch, represented by the purple rectangle, is complete. From this point on, a batch is completed upon the arrival of a new instance. This is demonstrated with the creation of the blue and the green batches. Upon the arrival of the sixth instance $(2n)$, a second batch is added to

---

Algorithm 1. FG-CDCStream

Input: S: stream of instances, W: window size

1: Initialize $trackPointer = 0$, $k = 0$, $k_w = 2$, $k_c = 3$, $votes = 0$, $votes_w = 1$,
   $votes_c = 15$, $inst = null$
2: **for** $count = 0 \rightarrow W$ **do**
3:   $tracks.add$(new change detector)
4: **while** $hasMoreInstances(S)$ **do**
5:   $inst = S.nextInstance$
6:   $buffer.add$(new batch container)
7:   **for** $count = 0 \rightarrow buffer.size$ **do**
8:     $buffer.get(count).add(inst)$
9:   **if** $buffer.size == W$ **then**
10:     $k = tracks.get(trackPointer).getKValue(buffer.get(0))$
11:     **if** $k \geq k_c$ **then**
12:       $votes + +$
13:       **if** $votes == votes_w$ **then**
14:         $warningPeriod = true$; $initiateBackgroundClassifier(buffer.get(0))$
15:       **if** $votes == votes_c$ **then**
16:         $changePeriod = true$; $warningPeriod = false$
17:         replace classifier with background classifier; nullify background classifier
18:     **else if** $k < k_w$ **then**
19:       $votes = 0$
20:       **if** $warningPeriod$ **then**
21:         $warningPeriod = false$; nullify background classifier
22:       **if** $changePeriod$ **then**
23:         $changePeriod = false$
24:   $updateClassifier(inst)$; $updateBackgroundClassifier(inst)$
25:   **if** $buffer.size == W$ **then**
26:     $buffer.remove(0)$;
27:     update track pointer

---

the original purple track. Note that the track with the newest batch is shown at the front and the least recently updated track in the rear. This process continues until the end of the stream (or to infinity).

It should be noted that, in reality, instances are stored explicitly only until a batch is complete. A buffer with a container for each track stores instances until n have been collected. When a batch is complete, it is input to the corresponding track's change detector object, which maintains summary statistics, as described in Eq. 3. That track's buffer is then cleared and the main algorithm begins building its next batch upon the arrival of the next instance. The appearance of the next instance fills a batch belonging to the next track and the process continues.

The algorithm requires a data stream and a user-defined window size parameter. An integer variable *trackPointer* keeps an account of which track the current batch (the batch completed by the current instance) belongs to. The k variable refers to the value calculated by Chebychev's inequality, which is initially zero. The *votes* variable stores the accumulated votes of the tracks seen so

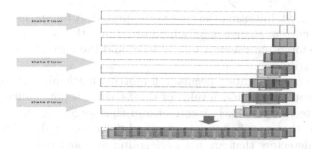

Fig. 1. Track and batch building visualization

far. As each instance arrives, a buffer, which is a list of instance containers (or batches), adds a new batch object to the list. A copy of the current instance is then added to each of the existing batches. Note that upon the arrival of the first instance, only one batch container exists. When the second instance arrives, a new batch is added to the buffer and that instance is added to both batches. At this point, the first batch contains the first and second instances, and the second batch contains only the second instance. This produces contiguous batches that contain data that are shifted by one instance. This process continues until the buffer contains n batches. At this point the first batch is complete, i.e. it contains n instances. The complete batch is processed, as described below. After the batch is processed and summarized, it is removed from the buffer.

To process a completed batch, the batch is sent to the change detector associated with the current track. The change detector returns the value of k for this batch, as calculated using Chebychev's Inequality. Recall that, in **CDCStream**, if this value is equal to two (k_w), a warning occurs. Similarly, a value of three (k_c) indicates that a change is detected by this track. Otherwise, a value of zero is returned.

The **FG-CDCStream** algorithm differs from **CDCStream** in two aspects. Firstly, it does not immediately flag a warning when a track encounters $k = k_w = 2$, but only maintains this statistic. Secondly, a value of $k = k_c = 3$ initiates a warning period, rather than reporting a change, if this is the first change detection in a series. This launches the creation of the background classifier which is built from the current batch. Subsequent reports of k_c in this warning period increment the *votes* count. Reported values of k_w are permitted within a warning period, but do not terminate it nor increment the vote count. If a value of less than k_w is reported by a track during the warning period, the warning period is terminated, the *votes* count is reset to zero and the background classifier is removed. This initiates a static period, until the next warning occurs.

If at least $votes_c$ tracks confirm a value of k_c, the system acknowledges the change. (It follows that the value of the $votes_c$ parameter is domain-dependent and determined through experimentation.) A confirmed change triggers adaptation by replacing the current classifier with the background classifier. The change period remains, whereby no new warnings or changes may occur, until a value of $k < 2$ is reported, initiating the next static period. A fall in the value of k signifies that the current change is fully integrated into the system, i.e. the change detectors have forgotten the past distribution and the current model represents the current distribution.

Whether or not adaptation occurs, a forgetting mechanism is applied to any change detector that produces a value of k_c. If change is not confirmed, the original change detection was likely incorrect and thus forgetting the corresponding information for that specific track is assumed to be reasonable. This effectively resets change detectors that are not performing well and permits outlier information to be discarded.

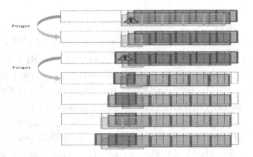

Fig. 2. Forgetting mechanism example

Figure 2 illustrates the system's forgetting mechanics. Firstly, the green track's most recent batch detects a change (represented by the exclamation mark in the red triangle). It then forgets all of its past batches retaining only the current one. The next batch to arrive, belonging to the purple track, also detects the change and forgets its past batches. The next batch, belonging to the blue track does not detect the change so it forgets nothing. The next green batch also does not detect change, so track building (or remembering) proceeds. The same is true for the next purple batch. This process occurs regardless of the state of the system: in-control, warning or out-of-control. In this example, the green and purple tracks may have detected a change due to outlier interference. It is beneficial for these two tracks to forget this information. On the other hand, the blue track, whose summary statistics may not have been as affected by the outlier(s) due the shifted sample, would retains this information.

4   Experimentation

Experimentation was conducted using the **MOA** framework for data stream mining [1], an open source software closely related to its offline counterpart **WEKA** [7]. We used both synthetic and real data streams in our evaluation. Due to space restrictions, we are only reporting the results against the synthetic data streams. Five of **MOA**'s synthetic data set generators, summarized in Table 1, were used in various configurations to produce the synthetic data. Varying degrees of noise were tested using synthetic data in order to test and compare the algorithms' robustness to noise. Each synthetic data set was injected with 0, 1, 2, 3, 4, 5, 10, 15, 20 and 25% noise using the **WEKA** "addNoise" filter. This noise was applied to every attribute but the class attribute, since **CDC-Stream** and **FG-CDCStream** are unsupervised change detectors. Experimentation was performed on a machine with an Intel i7-4770 processor, 16GB of memory, using the Windows 10 Pro x64 Operating System. The original **CDC-Stream** study [9] tested its strategy using only the Naïve Bayes classifier. For a more comprehensive understanding of the behaviour of both **CDCStream** and **FG-CDCStream**, we employed the Naïve Bayes classifier, the Hoeffding tree incremental learning, and the K-NN lazy learning strategy. Each of the classifiers used for experimentation are available in **MOA**.

Four measures [2] were used to evaluate change detection strategies, as follows. The mean time between false alarms (MTFA) describes the average distance between changes detected by the detector that do not correspond to true changes in the data. It follows that a high MTFA value is desirable. The mean time to detection (MTD) describes how quickly the change detector detects a true change and a low MTD value is sought. Further, the missed detection rate (MDR) gives the probability that the change detector will not detect a true change when it occurs and it follows that a low value is preferred. Finally, the calculated mean time ratio (MTR) describes the compromise between fast detection and false alarms, as shown in the equation, and a higher MTR value is required.

$$MTR = \frac{MTFA}{MTD}(1 - MDR) \tag{5}$$

These performance measures allow for a detailed examination of a change detector's effectiveness in detecting true changes quickly while remaining robust to noise and issuing few false alarms, thus providing researchers with a way to directly assess a change detector's performance [2].

A more indirect way of evaluating change detection methods, and the most common in the literature, is the measuring of accuracy-type performance measures. We considered the classification accuracy, κ and κ^+ accuracy-type measures focusing on the κ^+ results due to their comprehensiveness. The κ statistic considers chance agreements, and the κ^+ statistic [3] considers the temporal dependence often present in data streams.(Interested readers are referred to [3] for a detailed discussion on the evaluation of data stream classification algorithms.) We considered the progression of accuracy-type statistics throughout the stream, not only the final values, in order to gain more insights into change detector performance. For instance, the steepness of the drop in accuracy-type performance at a change point, and the swiftness of recovery provides more information than a single value representative of an overall accuracy.

Abrupt changes were injected and streams of one, four and seven changes were studied in various orders. Different stream sample sizes, change widths, patterns and distances between changes (as well as magnitudes and orderings) were studied on account of comprehensiveness. For single change scenarios, changes

Table 1. Synthetic data (basic characteristics)

Dataset	Classes	Features	Categorical	Numerical
LED	10	24	24	0
Stagger	2	3	3	0
Mixed	2	4	2	2
Agrawal	2	9	3	6
ConceptDriftStream	Varies	Varies	Varies	Varies

were injected half way through the stream in order to observe system behaviour well before and well after the change. For multiple abrupt changes, four different distances between changes were studied, namely 500, 1000, 2500 and 5000 instances. This was done in order to assess the change detectors' abilities to detect changes in succession and to compare recovery times.

4.1 Results

Table 2 shows the average performance of the classifiers in abrupt drift scenarios. The table shows that the Hoeffding Tree incremental learner performs the best, overall, in all cases. Further, the **FG-CDCStream** algorithm generally outperforms **CDCStream**. The greater gap between the performance in Hoeffding Tree and Naïve Bayes change detectors in **CDCStream** compared to **FG-CDCStream**, is likely a consequence of the Hoeffding Tree's superior ability to conform to streaming data naturally, through data acquisition and without explicit concept drift detection.

Table 2. Average algorithm performance comparison by classifier

Change type	Algorithm	Classifier	κ^+	κ	Accuracy (%)
Abrupt	FG-CDC	HT	**93.74**	93.67	95.72
		NB	91.42	91.45	94.55
		IbK	79.93	79.92	84.55
	CDC	HT	71.70	76.28	83.81
		NB	57.58	60.05	71.65
		IbK	60.33	69.34	80.72

Next, we focus on our experimental results against single abrupt change data streams generated using the LED data set, as shown in Figs. 3 and 4. The value of $votes_c$ was set to 15, by inspection. Note that the information for MTFA, MTD and MTR performance measures are not available for **CDCStream**. This is because **CDCStream** was unsuccessful in detecting changes in the case of the

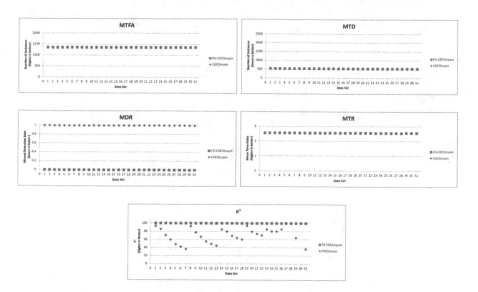

Fig. 3. A graphical comparison of performance statistics of FG-CDCStream and CDC-Stream on 31 different data streams containing a single abrupt change each.

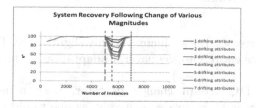

Fig. 4. System recovery following a single abrupt change. The long dashed line indicates the true change, the medium dashed line the approximate change detection and the short dashed line the full classifier recovery.

single abrupt drifting data streams. In contrast, **FG-CDCStream** produces good change detection statistics in the single abrupt change scenario. These measures remain consistent across all of the streams. Specifically, our algorithm issues false alarms at a low rate (with MTFA values around 1400) and successfully detects the true change in every case.

Next, we consider the multiple change scenario. Recall that multiple changes were injected into streams at varying distances from one another, namely 500, 1000, 2500 and 5000 instances. The averaged results and the corresponding trends may be observed in Fig. 5. Similar to the single abrupt change scenario, **CDCStream** failed to detect concept drift in the multiple abrupt change scenario. (Note that our evaluation confirms that its accuracy-type statistics are equivalent to those of a regular incremental Naïve Bayes classifier with no explicit change detection functionality.) The κ^+ graph in Fig. 5 shows that as distance

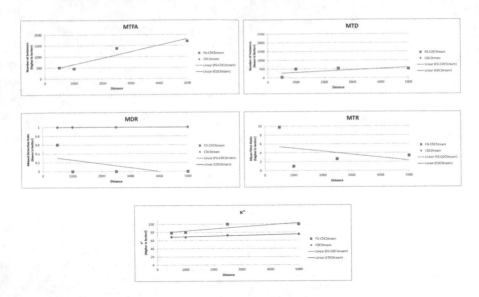

Fig. 5. A graphical comparison of averaged performance statistics of FG-CDCStream and CDCStream for multiple abrupt drifts and the associated trends as distance between the changes increases.

between changes increases, the κ^+ increases only slightly for **CDCStream**. This slight increase is due entirely to the natural adaptation of the incremental classifier over time.

For the **FG-CDCStream** change detector, performance and distance between changes generally correlate positively, as one would expect. If there is a longer period between the changes, **FG-CDCStream** has more information to detect and recover from change. This leads to a more streamlined representation of the current concept by growing its more accurate tracks' windows of the current concept and pruning the less accurate ones. A clear increase in MTFA occurs with increasing distance between changes. This trend corresponds to the systems' ability to forget more of previous concepts (discard more of the less accurate tracks) when more time is available between changes. When less time is available, it is likely that more tracks still contain the information of previous concepts. Nevertheless, the **FG-CDCStream** is able to detect concept drift fast, while maintaining a low false detection rate.

The effect of concept drift injection magnitudes on the κ^+ performance measure throughout the stream is shown in Fig. 6. The particular stream represented was injected with seven changes from low to high change magnitudes. At shorter distances, change points are less defined and the system has a more difficulty to recover, especially from higher magnitude change. This is due to less reliable change detection, as discussed above and therefore less representative models.

Finally, we turn our attention to the case when increasing levels of noise were injected into streams with multiple abrupt drifts. The averaged results and

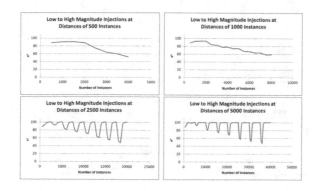

Fig. 6. FG-CDCStream κ^+ performance on LED data stream with varying distances between change injection points.

corresponding trends may be observed in Fig. 7. The results indicate that **FG-CDCStream** is robust to noise, especially when multiple changes are present and the distances between those changes are large. It follows that, the level of acceptable noise would depend on the particular application. In general, though, since a value greater than 15% noise in every attribute is a rather high noise ratio to be occurring in real data, it is likely that **FG-CDCStream** would perform satisfactorily on most streams containing abrupt drifts.

4.2 Discussion

The above-mentioned experimental results confirm that the **FG-CDCStream** algorithm leads to improved abrupt change detection and adaptation.

Improving Abrupt Change Detection. FG-CDCStream was designed with the intention of retaining the appealing qualities of **CDCStream** while improving change detection elements where opportunities exist. It essentially uses an ensemble of change detectors, each containing slightly shifted information, that vote on whether or not a change has occurred. Only the change detectors in close proximity in the forward direction to the detector that first flags change submits a vote. This is appropriate, due to the temporal nature of the data and the desire to detect a change quickly. The voting system decreases the chances of completely losing information due to aggregation. This not only increases the probability of detecting the change (reducing MDR) and detecting it quickly (reducing MTD), but provides another advantage of decreasing the probability of false detections (increasing MTFA). Since a vote is required to confirm a change, slight variations in distribution that would flag a false change in **CDCStream** would, in **FG-CDCStream**, be outvoted and therefore not flag a false change. This further increases the algorithm's robustness. In summary, **FG-CDCStream** overcomes the issues that **CDCStream** has with regard to the batch scenario and dilution due to aggregation, improving its change detection capabilities.

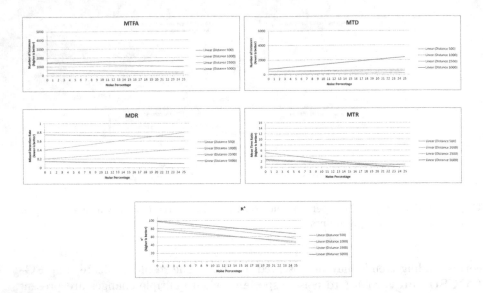

Fig. 7. A graphical comparison of linear trends of the averaged performance statistics of FG-CDCStream on six data sets containing a multiple abrupt changes at varying distances from one another over increasing noise levels.

Improving Adaptation. A major limitation of **CDCStream** is its potential for a very unrepresentative replacement classifier during adaptation. This is caused by the termination requirement of a warning period being only a change period regardless of how closely or distantly that change period occurs. This requirement potentially results in a background classifier that is non-representative of the new distribution following a change detection. **FG-CDCStream** reconciles this by permitting a warning period to terminate if it is followed by either a static period or a change. In **FG-CDCStream**, if a warning period is followed by a static period, the background classifier initiated by the warning is ignored rather than continuing to build until the next change occurs. The only background classifier that may be used as an actual replacement classifier in **FG-CDCStream** is one that is initiated by a warning immediately prior (i.e. there is no static period in between) a detected change. If, by chance, a change occurs without a preceding warning, the replacement classifier is created from the most recent batch only. This follows, since a change occurring absent of any warning is likely to be abrupt, and therefore a short historical window is appropriate.

5   Conclusion

This paper introduces the **FG-CDCStream** algorithm, an unsupervised and context-based change detection algorithm for streaming categorical data. This algorithm provides users with more precise information than that of learning-based methods about the changing distributions in categorical data streams and provides context-based change detection capabilities for categorical data, previously undocumented in machine learning research. The experimental results show that **FG-CDCStream** method is able to detect abrupt drift fast, while maintaining both lower false alarm rates and lower detection misses.

This research would benefit from further exploration of the effects of different types of data streams on algorithm performance. For instance, a comprehensive study on real data might provide further insight into algorithm behaviour. Additional research on no-change data streams would also be beneficial. Further, studying how attribute cardinality effects the algorithm would be useful as well.

References

1. Bifet, A., Holmes, G., Kirkby, R., Pfahringer, B.: MOA: massive online analysis. J. Mach. Learn. Res. **11**, 1601–1604 (2010)
2. Bifet, A., Read, J., Pfahringer, B., Holmes, G., Žliobaitė, I.: CD-MOA: change detection framework for massive online analysis. In: Tucker, A., Höppner, F., Siebes, A., Swift, S. (eds.) IDA 2013. LNCS, vol. 8207, pp. 92–103. Springer, Heidelberg (2013). doi:10.1007/978-3-642-41398-8_9
3. Bifet, A., Read, J., Žliobaitė, I., Pfahringer, B., Holmes, G.: Pitfalls in benchmarking data stream classification and how to avoid them. In: Blockeel, H., Kersting, K., Nijssen, S., Železný, F. (eds.) ECML PKDD 2013. LNCS, vol. 8188, pp. 465–479. Springer, Heidelberg (2013). doi:10.1007/978-3-642-40988-2_30
4. Boriah, S., Chandola, V., Kumar, V.: Similarity measures for categorical data: a comparative evaluation. In Proceedings of the 2008 SIAM International Conference on Data Mining, pp. 243–254 (2008)
5. Cao, F., Zhexue Huang, J., Liang, J.: Trend analysis of categorical data streams with a concept change method. Inf. Sci. **276**, 160–173 (2014)
6. Gama, J., Zliobaite, I., Bifet, A., Pechenizkiy, M., Bouchachia, A.: A survey on concept drift adaptation. ACM Comput. Surv. **46**(4), 1–37 (2014)
7. Hall, M., Frank, E., Holmes, G., Pfahringer, B., Reutemann, P., Witten, I.H.: The WEKA data mining software: an update. ACM SIGKDD Explor. **11**(1), 10–18 (2009)
8. Ienco, D., Pensa, R.G., Meo, R.L.: From context to distance: learning dissimilarity for categorical data clustering. ACM Trans. Knowl. Discov. Data **6**(1), 1–25 (2012)
9. Ienco, D., Bifet, A., Pfahringer, B., Poncelet, P.: Change detection in categorical evolving data streams. In: Proceedings of the 29th Annual ACM Symposium on Applied Computing (SAC 2014), pp. 274–279 (2014)
10. Yu, L., Liu, H.: Feature selection for high-dimensional data: a fast correlation-based filter solution. In: Proceedings of Twentieth International Conference on Machine Learning, vol. 2, pp. 856–863 (2003)

A New Adaptive Learning Algorithm
and Its Application to Online Malware Detection

Ngoc Anh Huynh[1,2](✉), Wee Keong Ng[1,2], and Kanishka Ariyapala[1,2]

[1] Nanyang Technological University, Singapore, Singapore
hu0001nh@e.ntu.edu.sg, wkn@pmail.ntu.edu.sg,
kanishka.ariyapala@math.unifi.it
[2] University of Padua, Padua, Italy

Abstract. Nowadays, the number of new malware samples discovered every day is in millions, which undermines the effectiveness of the traditional signature-based approach towards malware detection. To address this problem, machine learning methods have become an attractive and almost imperative solution. In most of the previous work, the application of machine learning to this problem is batch learning. Due to its fixed setting during the learning phase, batch learning often results in low detection accuracy when encountered zero-day samples with obfuscated appearance or unseen behavior. Therefore, in this paper, we propose the FTRL-DP online algorithm to address the problem of malware detection under concept drift when the behavior of malware changes over time. The experimental results show that online learning outperforms batch learning in all settings, either with or without retrainings.

Keywords: Malware detection · Batch learning · Online learning

1 Introduction

VirusTotal.com is an online service which analyzes files and urls for malicious content such as virus, worm and trojan by leveraging on an array of 52 commercial antivirus solutions for the detection of malicious signatures. On record, VirusTotal receives and analyzes nearly 2 million files every day. However, only a fraction of this amount (15%) can be identified as malicious by at least one antivirus solution. Given the fact that it is fairly easy nowadays to obfuscate a malware executable [23], it is rather reasonable to believe that a sheer number of the unknown files are actually obfuscated malware samples. In principle, the rest of the unknown cases should be manually reverse engineered to invent new signatures, but this is infeasible due to the large number of files to be analyzed. Therefore, looking for an automated way to address this problem is imperative and has attracted a lot of research effort, especially in the direction of using machine learning which has gained a lot of successes in various domains of pattern recognition such as face analysis [22] and sentiment analysis [4].

In a recent paper, Saxe et al. [20] train a 3-layer neural network to distinguish between malicious and benign executables. In the first experiment, the

© Springer International Publishing AG 2017
A. Yamamoto et al. (Eds.): DS 2017, LNAI 10558, pp. 18–32, 2017.
DOI: 10.1007/978-3-319-67786-6_2

author randomly splits the whole malware collection into train set and test set. The trained network can achieve a relatively high detection accuracy of 95.2% at 0.1% false positive rate. It is noted that the first experiment disregards the release time of the executables, which is an important dimension due to the adversarial nature of malware detection practice since malware authors are known to regularly change their tactics in order to stay ahead of the game [7]. In the second experiment, the author uses a timestamp to divide the whole malware collection into train set and test set. The results obtained show that the detection accuracy drops to 67.7% at the same false positive rate. We hypothesize that the reasons for this result are two-fold: the change in behavior of malware over time and the poor adaptation of neural network trained under batch mode to the behavioral changes of malware. In addition, another recent study also reports similar findings [3].

The working mechanism of batch learning is the assumption that, the samples are independently and identically drawn from the same distribution (iid assumption). This assumption may be true in domains such as face recognition and sentiment analysis where the underlying concept of interest hardly changes over time. However, in various domains of computer security such as spam detection and malware detection, this assumption may not hold [2] due to the inherently adversarial nature of cyber attackers, who may constantly change their strategy so as to maximize the gains. To address this problem of concept drift, we believe online learning is a more appropriate solution than batch learning. The reason is that, online algorithms are derived from a theoretical framework [11] which does not impose the iid assumption on the data, and hence can work well under concept drift or adversarial settings [5]. Motivated by this knowledge, we propose the Follow-the-Regularized-Leader with Decaying Proximal (FTRL-DP) algorithm – a variant of the proximal Follow-the-Regularized-Leader (FTRL-Proximal) algorithm [12] – to address the problem of malware detection.

To be specific, the contributions of this paper are as follow:

- A new online algorithm (FTRL-DP) to address the problem of concept drift in Windows malware detection. Our main claim is that online learning is superior to batch learning in the task of malware detection. This claim is substantiated in Sect. 7 by analyzing the accuracy as well as the running time of FTRL-DP, FTRL-Proximal and Logistic Regression (LR). The choices of the algorithms are clarified in Sect. 4.
- An extensive data collection of more than 100k malware samples using the state-of-the-art open source malware analyzer, Cuckoo sandbox [6], for the evaluation. The collected data comprises many types such as system calls, file operations, and others, from which we were able to extract 482 features. The experiment setup for data collection and the feature extraction process are described in Sect. 5.

For LR, the samples in one month constitutes the test set and the samples in a number of preceding months are used to form the train set; for FTRL-DP, a sample is used for training right after it is tested. The detailed procedure is presented in Sect. 6. In Sect. 2, we review the previous works related to the problem of malware detection using machine learning. We formulate the problem of malware detection as a regression problem in Sect. 3. Lastly, in Sect. 8, we conclude the paper and discuss the future work.

2   Related Work

2.1   Batch Learning in Malware Detection

The analysis of suspicious executables for extracting the features used for auto-mated classification can be broadly divided into two types: static analysis and dynamic analysis. In static analysis, the executables are executed and only features extracted directly from the executables are used for classification such as file size, readable strings, binary images, n-gram words, etc. [10,19].

On the other hand, dynamic analysis requires the execution of the executables to collect generated artifacts for feature extraction. Dynamic features can be extracted from host-based artifacts such as system call traces, dropped files and modified registries [17]. Dynamic features can also be extracted from network traffic such as the frequency of TCP packets or UDP packets [16,18]. While static analysis is highly vulnerable to obfuscation attacks, dynamic analysis is more robust to binary obscuration techniques.

To improve detection accuracy, the consideration of a variety of dynamic features of different types is recently gaining attention due to the emergence of highly effective automated malware analysis sandboxes. In their work, Mohaisen et al. [14] studied the classification of malware samples into malware families by leveraging on a wide set of features (network, file system, registry, etc.) provided by the proprietary AutoMal sandbox. In a similar spirit, Korkmaz et al. [9] used the open-source Cuckoo sandbox to obtain a bigger set of features aiming to classify between traditional malware and non-traditional (Advanced Persistent Threat) malware. The general conclusion in these papers is that combining dynamic features of different types tends to improve the detection accuracy.

In light of these considerations, we decided to use the Cuchoo sandbox to execute and extract the behavioral data generated by a collection of more than 100k suspicious executables. Our feature set (Sect. 5.3) is similar to that of Korkmaz et al. although we address a different problem: regressing the risk levels of the executables. Furthermore, our work differs from previous work in the aspect that we additionally approach this problem as an online learning problem rather than just a batch learning one.

2.2   Online Learning in Malware Detection

The importance of malware's release time has been extensively studied in the domain of Android malware detection. In their paper [2], Allix et al. studied the effect of history on the biased results of existing works on the application of machine learning to the problem of malware detection. The author notes that most existing works evaluate their methodology by randomly picking the samples for the train set and the test set. The conclusion is that, this procedure usually leads to much higher accuracy than the cases when the train set and the test set are historically coherent. The author argues that this result is misleading as it is not useful for a detection approach to be able to identify randomly picked samples but fail to identify zero-day or new ones.

To address this problem of history relevance, Narayanan et al. [15] have developed an online detection system, called DroidOL, which is based on the online Passive Aggressive (PA) algorithm. The novelty of this work is the use of an online algorithm with the ability to adapt to the change in behavior of malware in order to improve detection accuracy. The obtained result shows that, the online PA algorithm results in a much higher accuracy (20%) compared to the typical setting of batch learning and 3% improvement in the settings when the batch model is frequently retrained.

3   Problem Statement

Given 52 antivirus solutions, we address the problem of predicting the percentage of solutions that would flag an executable file as malicious. Formally, this is a regression problem of predicting the output $y \in [0, 1]$ based on the input $x \in R^n$ which is the set of 482 hand-crafted features extracted from the reports provided by the Cuckoo sandbox (Sect. 5.3). The semantic of the output defined in this way can be thought of as the risk level of an executable. We augment the input with a constant feature which always has the value 1 to simulate the effect of a bias. In total, we have 483 features for each malware sample.

In this case, we have framed the problem of malware detection as the regression problem of predicting the risk level of an executable. We rely on the labels provided by all 52 antivirus solutions and do not follow the labels provided by any single one as different antivirus solutions are known to report inconsistent labels [8]. In addition, we also do not use two different thresholds to separate the executables into two classes, malicious and benign, as in [20] since it would discard the hard cases where it is difficult to determine the nature of the executables, which may be of high value in practice.

4   Methodology

To allow a fair comparison between batch learning and online learning, we use the models of the same linear form, represented by a weight vector w, in both cases. The sigmoid function ($\frac{1}{1+e^{-z}}$) is then used to map the dot product (biased by the introduction of a constant feature) between the weight vector and the input, $w^\top x$, to the [0, 1] interval of possible risk levels. Additionally, in both cases, we optimize the same objective function, which is the sum of logistic loss (log loss – the summation term in Eq. 1). In the batch learning setting, the sum of log loss is optimized in a batch manner in which each training example is visited multiple times in minimizing the objective function. The resultant algorithm is usually referred to as Logistic Regression with log loss (Sect. 4.1). On the other hand, in the online setting, we optimize the sum of log loss in an online manner, in which each training sample is only seen once. The resultant algorithm is the proposed FTRL-DP algorithm (Sect. 4.2).

4.1 Batch Learning – LR

Given a set of n training examples $\{(x_i, y_i)\}_{i=1}^n$, Logistic Regression with log loss corresponds to the following optimization problem:

$$\underset{w}{\operatorname{argmin}} \left\{ -\sum_{i=1}^n \left(y_i \log(p_i) + (1 - y_i) \log(1 - p_i) \right) + \lambda_1 \|w\|_1 + \frac{1}{2}\lambda_2 \|w\|_2^2 \right\} \tag{1}$$

$$\text{in which } p_i = \operatorname{sigmoid}(w^\top x_i)$$

The objective function of Logistic Regression (Eq. 1) is a convex function with respect to w as it is the sum of three convex terms. The first term is the sum of log losses associated with all training samples (within a time window). The last two terms are the L1–norm regularizer and the L2–norm regularizer. The L1 regularizer is a non–smoothed function used to introduce sparsity into the solution weight w. On the other hand, the L2 regularizer is a smooth function used to favor low variance models that have small weight.

4.2 Online Learning – FTRL-DP

Online Convex Optimization. The general framework of online convex optimization can be formulated as follows [21]. We need to design an algorithm that can make a series of optimal predictions, each at one time step. At time step t, the algorithm makes a prediction, which is a weight vector w_t. A convex loss function $l_t(w)$ is then exposed to the algorithm after the prediction. Finally, the algorithm suffers a loss of $l_t(w_t)$ at the end of time step t (Algorithm 1). The algorithm should be able to learn from the losses in the past so as to make better and better decisions over time.

Algorithm 1. Online Algorithm

1: **for** t = 1,2,... **do**
2:      Make a prediction w_t
3:      Receive the lost function $l_t(w)$
4:      Suffer the lost $l_t(w_t)$

The objective of online convex optimization is to minimize the regret with respect to the best classifier in hindsight (Eq. 2). The meaning of Eq. 2 is that we would like to minimize the total loss incurred up to time t with respect to the supposed loss incurred by the best possible prediction in hindsight, w^*.

$$\mathbf{Regret}_t = \sum_{s=1}^{t} l_s(w_s) - \sum_{s=1}^{t} l_s(w^*) \qquad (2)$$

Since the future loss functions are unknown, the best guess or the greedy approach to achieve the objective of minimizing the regret is to use the prediction that incurs the least total loss on all past rounds. This approach is called Follow-the-Leader (FTL), in which the leader is the best prediction that incurs the least total loss with respect to all the past loss functions. In some cases, this simple formulation may result in algorithms with undesirable properties such as rapid change in the prediction [21], which lead to overall high regret. To fix this problem, some regularization function is usually added to regularize the prediction. The second approach is called Follow-the-Regularized-Leader (FTRL), which is formalized in Eq. 3.

$$w_{t+1} = \operatorname*{argmin}_{w} \left\{ \sum_{s=1}^{t} l_s(w) + r(w) \right\} \qquad (3)$$

It is notable to see that the FTRL framework is formulated in a rather general sense and performs learning without relying on the iid assumption. This property makes it more suitable to adversarial settings or settings in which the concept drift problem is present.

The Proposed FTRL-DP Algorithm. In the context of FTRL-DP, an online classification or an online regression problem can be cast as an online convex optimization problem as follows. At time t, the algorithm receives input x_t and makes prediction w_t. The true value y_t is then revealed to the algorithm after the prediction. The loss function $l_t(w)$ associated with time t is defined in terms of x_t and y_t (Eq. 4). Finally, the cost incurred at the end of time t is $l_t(w_t)$. The underlying optimization problem of FTRL-DP is shown in Eq. 5.

$$l_t(w) = -y_t \log(p) - (1 - y_t) \log(1 - p)$$
$$\text{in which } p = \text{sigmoid}(w^\top x_t) \qquad (4)$$

Compared with Eq. 3, Eq. 5 has the actual loss function $l_t(w)$ replaced by its linear approximation at w_t, which is $l_t(w_t) + \nabla l_t(w_t)^\top (w - w_t) = g_t^\top w + l_t(w_t) - g_t^\top w_t$ (in which $g_t = \nabla l_t$). The constant term $(l_t(w_t) - g_t^\top w_t)$ is omitted in the final equation without affecting the optimization problem. This approximation is to allow the derivation of a closed–form solution to the optimization problem at each time step, which is not possible with the original problem in Eq. 3.

$$w_{t+1} = \operatorname*{argmin}_{w} \left\{ g_{1:t}^\top w + \lambda_1 \|w\|_1 + \frac{1}{2}\lambda_2 \|w\|_2^2 + \frac{1}{2}\lambda_p \sum_{s=1}^{t} \sigma_{t,s} \|w - w_s\|_2^2 \right\}$$
$$\text{in which } g_{1:t}^\top = \sum_{i=1}^{t} g_t^\top \qquad (5)$$

FTRL-DP utilizes 3 different regularizers to serve 3 different purposes. The first two regularizers of L1–norm and L2–norm serve the same purpose as in the case of Logistic Regression introduced in Sect. 4.1. The third regularization function is the proximal term used to ensure that the current solution does not deviate too much from past solutions with more influence given to most recent ones by using an exponential decaying function $\left(\sigma_{t,s} = \gamma^{t-s} \text{ with } 1 > \gamma > 0\right)$. This is our main difference from the original FTRL-Proximal algorithm [12]. The replacement of the per coordinate learning rate schedule by the decaying function proves to improve the prediction accuracy in the face of concept drift (discussed in Sect. 7). The solution to the objective function of FTRL-DP is stated in Theorem 1, whose proof is presented in Appendix A.

Theorem 1. *The optimization problem in Eq. 5 can be solved in the following closed form:*

$$w_{t+1,i} = \begin{cases} 0 & \text{if } \|z_{t,i}\|_1 \leq \lambda_1 \\ -\dfrac{z_{t,i} - \lambda_1 \text{sign}(z_{t,i})}{\lambda_2 + \lambda_p \frac{1-\gamma^t}{1-\gamma}} & \text{otherwise.} \end{cases} \tag{6}$$

$$\text{in which } z_t = g_{1:t} - \lambda_p \sum_{s=1}^{t} \sigma_{t,s} w_s$$

Regret Analysis of FTRL-DP. In Theorem 2, we prove a result that bounds the regret of FTRL-DP. The bound is dependent on the decaying rate γ.

Theorem 2. *Suppose that $\|w_t\|_2 \leq R$ and $\|g_t\|_2 \leq G$. With $\lambda_1 = \lambda_2 = 0$ and $\lambda_p = 1$, we have the following regret bound for FTRL-DP:*

$$Regret(w^*) \leq 2R^2 \frac{1}{1-\gamma} + \frac{G^2}{2} \frac{1 + \ln T}{\gamma^T} \tag{7}$$

Due to space constraint, the proof of Theorem 2 will be provided in an extended version of the paper.

In summary, we aim to compare between the performance of batch learning and online learning on the problem of malware detection. To make all things equal, we use the models of the same linear form and optimize the same log loss function, which lead to the LR algorithm in the batch learning case and the FTRL-DP algorithm in the online learning case. For LR, only the samples within a certain time window contribute to the objective function (Eq. 1). On the other hand, the losses associated with all previous samples equally contribute to the objective function of FTRL-DP (Eq. 5). This difference is critical as it leads to the gains in the performance of FTRL-DP over LR, which is discussed in Sect. 7.

5   Data Collection

5.1   Malware Collection

We used more than 1 million files collected in the duration from March 2016 to Apr 2016 by VirusShare.com for the experiments. VirusShare is an online

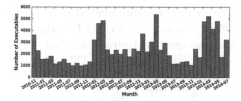

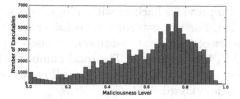

Fig. 1. Distribution of executables   **Fig. 2.** Distribution of risk level.

malware analyzing service that allows Internet users to scan arbitrary files against an array of 52 antivirus solutions (the scan results are actually provided by VirusTotal). In this study, we are only interested in executable files and able to separate out more than 100k executables from the 1 million files downloaded.

Figure 1 shows the distribution of the executables with respect to executables' compile time. The horizontal axis of Fig. 1 shows the months during the 4 years from Nov/2010 until Jul/2014, which is the period of most concentration of executables and chosen for the study. The vertical axis of Fig. 1 indicates the number of executables compiled during the corresponding month. Figure 2, instead, shows the maliciousness distribution of the executables. The horizontal axis indicates the maliciousness measure and the vertical axis the number of corresponding executables.

5.2 Malware Execution

We make use of the facility provided by DeterLab [13] as the testbed for the execution of the executables. DeterLab is a flexible online experimental lab for computer security, which provides researchers with a host of physical machines to carry out experiments. In our setup, we use 25 physical machines with each physical machine running 5 virtual machines for executing the executables. Each executable is allowed to run for 1 min. The experiment ran for more than 20 days and collected the behavioral data of roughly 100k executables.

5.3 Feature Extraction

In this paper, we mostly consider dynamic features of the following 4 categories for regression: file system category, registry category, system call category, and the category of other miscellaneous features.

API Call Category. API (Application Programming Interface) calls are the functions provided by the operating system to grant application programs the access to basic functionality such as disk read and process control. Although these calls may ease the process of manipulating the resources of the machine, it also provides hackers with a lot of opportunities to obtain confidential information. For this category, we consider the invoking frequencies of the API calls as a set of features. In addition, we also extract as features the frequencies that the API files are linked. The total number of features in this category is 353 and the complete set of API calls as well as the set of API files are available at https://git.io/vDywd.

Registry Category. In Windows environments, the registry is a hierarchical database that holds the global configuration of operating system. Ordinary programs often use the registry to store information such as program location and program settings. Therefore, the registry system is like a gold mine of information for malicious programs, which may refer to it for information such as the location of the local browsers or the version of the host operating system. Malicious program may also add keys to the registry so as to be able to survive multiple system restarts. We extract the following 4 registry related features: the number of registries being written, opened, read and deleted.

File System Category. File system is the organization of the data that an operating system manages. It includes two basic components: file and directory. File system-related features are an important set of features to consider since malware has to deal with the file system in one way or another in order to cause harm to the system or to steal confidential information. We consider the following file-related features: the number of files being opened, written, in existence, moved, read, deleted, failed and copied. In addition, we also consider the following 3 directory related features: the number of directories being enumerated, created and removed. In total, we were able to extract 11 features in this category.

Miscellaneous Category. In addition to out-of-the-box functionalities, Cuckoo sandbox is further enhanced by a collection of signatures contributed by the public community. These signatures can identify certain characteristics of the analyzed binary such as the execution delay time or the ability to detect virtual environment. All these characteristics are good indicators for the high risk level of an executable but may just be false positives. We consider the binary features of whether the community signatures are triggered or not. In addition, we also consider 3 other features that may be relevant to the behavior characterization: the number of mutex created, the number of processes started and the depth of the process tree. The total number of features in this category is 118.

In summary, we are able to extract 482 features that spans 4 different categories: API calls, registry system, file system and miscellaneous features.

6   Evaluation

6.1   Experiment with LR

We evaluate LR in four different settings: once, multi-once, monthly, and multi-monthly. In the once setting, the samples appeared in the first month of the whole dataset are used to form the train set and the rest of the samples are used to form the test set. The multi-once setting is similar to the once setting except that the samples in the first 6 months are used to form the train set instead. It should be noted that retraining is not involved in the first two settings.

On the other hand, the other two settings do involve retraining, which is a crude mechanism to address the change in behavior of malware over time. Since it is infeasible to carry out retraining upon the arrival of every new sample, we perform retraining on a monthly basis. Due to the characteristic of our dataset, we find that the monthly basis is a good balance to ensure that we have enough samples for the train set and the training time is not too long (the monthly average number of samples is 2.4k). In the monthly setting, we use the samples released in a month to form the test set and the samples released in the immediately preceding month to form the train set. The multi-monthly setting is similar to the monthly setting except that we use the samples in the preceding 6 months to form the train set instead.

For a quick evaluation, we make use of the LR implementation, provided by the TensorFlow library [1] to train and test the LR regressors. TensorFlow is a framework for training large scale neural network, but in our case, we only utilize a single layer network with sigmoid activation, binary cross-entropy loss and two regularizations of L1-norm and L2-norm. 20% of each train set is dedicated for validation and the maximum number of epochs that we use is 100. We stop the training early if the validation does not get improved in 3 consecutive epochs.

6.2   Experiment with FTRL Algorithms

We use the standard procedure to evaluate FTRL-DP and FTRL-Proximal (jointly referred to as FTRL algorithms). Each new sample is tested on the current model giving rise to an error, which is then used to make modification to the current model right after. This evaluation is usually referred to as the mistake-bound model.

Due to their simplicity, FTRL-DP and FTRL-Proximal can be implemented in not more than 40 lines of python code. The implementation makes heavy use of the numpy library, which is mostly written in C++. As TensorFlow also has C++ code under the hood, we believe that the running time comparison between the two cases is sensible. Evaluated on the same computer, it actually turns out that the running time of FTRL algorithms is much lower than that of LR. We use the same amounts of three regularizations for both FTRL-DP and FTRL-Proximal. For FTRL-DP, we report the best possible setting for parameter γ.

The computer used for all the experiments has 16 GB RAM and operates with a 1.2 GHz hexa-core CPU. The running times of all experiments are shown in Table 1. The mean cumulative absolute errors are reported in Fig. 3.

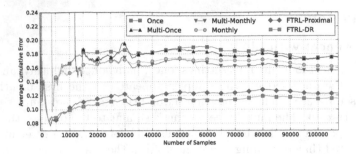

Fig. 3. Mean cumulative absolute errors of FTRL-DP, FTRL-Proximal and different settings of LR.

7   Discussion

7.1   Prediction Accuracy

We use the mean cumulative absolute error (MCAE) to compare the performance between FTRL-DP, FTRL-Proximal and different batch settings of LR, which are reported in Fig. 3. The MCAE is defined in Eq. 8, in which y_t is the actual risk level of an executable and p_t the risk level predicted by the algorithms. In Fig. 3, the horizontal line shows the cumulative number of samples and the vertical line the MCAE. There are four notable observations that we can see from Fig. 3.

$$\frac{1}{n} \sum_{t=1}^{n} |y_t - p_t| \tag{8}$$

Firstly, the more data that we train the LR model on, the better performance we can achieve. This observation is evidenced by the fact that, in most of the time, the error line of the multi-once setting stays below the error line of the once setting, and the error line of the multi-monthly setting stays below the error line of the monthly setting. A possible explanation for this observation is that the further we go back in time to obtain more data to train the model on, the less variance the model becomes, which results in the robustness to noise, and consequently, higher prediction accuracy.

Secondly, the retraining procedure does help to improve prediction accuracy. It is evidenced by the fact that the monthly setting outperforms the once setting, and similarly the multi-monthly setting outperforms the multi-once setting. This observation is a supporting evidence for the phenomenon of evolving malware behavior. As a consequence, the most recent samples would be more relevant to the current samples, and training on most recent samples would result in a more accurate prediction model.

From the first two observations, we can conclude that the further we go back in time to obtain more samples and the more recent the samples are, the better the trained model would perform. This conclusion can be exploited to improve prediction accuracy by going further and further back in time and retraining the model more often. However, this approach would become unpractical at some point when the training time required to frequently update an accurate model via periodic retrainings would become too long to be practical. It turns out that this issue can be elegantly addressed by the FTRL algorithms, which produces much higher prediction accuracy at considerable lower running time.

Thirdly, FTRL algorithms (worse MCAE of 0.123) are shown to outperform the LR algorithm in all settings (best MCAE of 0.156). The error lines corresponding to the performance of FTRL algorithms consistently stays below other error lines. The gain in the prediction accuracy of FTRL algorithms over all settings of LR can be explained by the contribution of all previous samples to its objective function. In different batch settings of LR, only the losses associated with the samples within a certain time window contribute to the respective objective functions.

Finally, the fourth observation is that FTRL-DP (MCAE of 0.116) outperforms FTRL-Proximal (MCAE of 0.123). The gain in performance of FTRL-DP over FTRL-Proximal can be explained by the ability of FTRL-DP to cope with concept drift via the use of a specially designed adaptive mechanism. This mechanism makes use of an exponential decaying function to favor the most recent solutions over older ones. The effective result is that the most recent samples would contribute more to the current solution thereby alleviating the problem of concept drift.

7.2 Running Time

In terms of running time (training time and testing time combined), FTRL-DP and FTRL-Proximal are clearly advantageous over LR. From Table 1, we can see that the running times of FTRL algorithms are much lower than that of LR, especially compared to the settings with retraining involved (monthly and multi-monthly). The reason for this result is that FTRL algorithms only needs to see each sample once to update the current weight vector whereas in the case of LR, it requires multiple passes over each sample to ensure convergence to the optimal solution.

Table 1. Running time of FTRL-DP, FTRL-Proximal and different LR settings.

Experiment	Running time
LR Multi-monthly	44 m 31 s
LR Monthly	14 m 14 s
LR Multi-once	55 s
LR Once	42 s
FTRL-Proximal	28 s
FTRL-DP	26 s

8   Conclusions and Future Work

The evolving nature of malware over time makes the malware detection problem more difficult. According to previous studies, batch learning based methods often perform poorly when encountered zero-days samples. Our research is motivated to fill in this gap by proposing FTRL-DP – a variant of the FTRL-Proximal algorithm – to address this problem. We evaluated two learning paradigms using an extensive dataset generated by more than 100k malware samples executed on Cuckoo sandbox. The experimental results show that FTRL algorithms (worse MCAE of 0.123) outperforms LR in the typical setting of batch learning as well as the settings with retrainings involved (best MCAE of 0.156). The gain in performance of FTRL algorithms over different batch settings of LR can be accounted for by its objective function taking into account the contribution of all previous samples. Furthermore, the improvement of FTRL-DP over FTRL-Proximal can be explained by the usage of an adaptive mechanism that regularizes the weight by favoring recent samples over older ones. In addition, FTRL algorithms are also more advantageous in terms of running time.

It can be noticed that all above methods are black-box solutions, which do not gain domain experts any insights. An interesting development of this work is to enable the direct interaction with a domain expert using a visualization. The domain expert could prioritize or discard weight alterations suggested by the learning algorithm via the interactive exploration of malware behavior. This visual analytics approach would lead to a transparent solution where the domain expert can benefit most of his knowledge in collaboration with black-box automated detection solutions.

A   Proof of Theorem 1

Proof. To remind the optimization objective of FTRL-DP:

$$w_{t+1} = \operatorname*{argmin}_{w} g_{1:t}^{\top} w + \lambda_1 \|w\|_1 + \frac{1}{2}\lambda_2 \|w\|_2^2 + \frac{1}{2}\lambda_p \sum_{s=1}^{t} \gamma^{t-s}\|w - w_s\|_2^2$$

$$w_{t+1} = \operatorname*{argmin}_{w} \left(g_{1:t}^{\top} - \lambda_p \sum_{s=1}^{t} \gamma^{t-s} w_s^{\top}\right) w + \lambda_1 \|w\|_1 + \frac{1}{2}\left(\lambda_2 + \lambda_p \frac{1 - \gamma^t}{1 - \gamma}\right)\|w\|_2^2$$

$$+ \frac{1}{2}\lambda_p \sum_{s=1}^{t} \gamma^{t-s}\|w_s\|_2^2$$

Omitting the constant term $\frac{1}{2}\lambda_p \sum_{s=1}^{t} \gamma^{t-s}\|w_s\|_2^2$, we have:

$$w_{t+1} = \operatorname*{argmin}_{w} z_t^{\top} w + \lambda_1 \|w\|_1 + \frac{1}{2}\left(\lambda_2 + \lambda_p r_t\right)\|w\|_2^2 \tag{9}$$

In Eq. 9, $z_t = g_{1:t}^{\top} - \lambda_p \sum_{s=1}^{t} \gamma^{t-s} w_s^{\top}$ and $r_t = \frac{1 - \gamma^t}{1 - \gamma}$. Each component of w contribute independently to the objective function of 9 hence can be solve separately:

$$w_{t+1,i} = \operatorname*{argmin}_{w_i} z_{t,i} w_i + \lambda_1 \|w_i\|_1 + \frac{1}{2}(\lambda_2 + \lambda_p r_t)\|w_i\|_2^2 \tag{10}$$

Note that w_i in 10 refers to the ith component of w. Let $f(w_i) = z_{t,i} w_i + \lambda_1 \|w_i\|_1 + \frac{1}{2}(\lambda_2 + \lambda_p r_t)\|w_i\|_2^2$. There are two cases:

- If $\|z_{t,i}\|_1 \leq \lambda_1$, we have:

$$f(w_i) \geq -\|z_{t,i} w_i\|_1 + \lambda_1 \|w_i\|_1 + \frac{1}{2}(\lambda_2 + \lambda_p r_t)\|w_i\|_2^2$$

$$f(w_i) \geq -\lambda_1 \|w_i\|_1 + \lambda_1 \|w_i\|_1 + \frac{1}{2}(\lambda_2 + \lambda_p r_t)\|w_i\|_2^2 = \frac{1}{2}(\lambda_2 + \lambda_p r_t)\|w_i\|_2^2 \geq 0$$

$f(w_i)$ achieves the minimum at $w_i = 0$
- If $\|z_{t,i}\|_1 \geq \lambda_1$, $z_{t,i}$ and w_i must have opposite signs at the minimum of $f(w_i)$ as otherwise w_i can always have sign flipped to further reduce $f_i(w_i)$. Therefore, it is equivalent to solving:

$$w_{t+1,i} = \operatorname*{argmin}_{w_i} z_{t,i} w_i - \operatorname{sign}(z_{t,i})\lambda_1 \|w_i\|_1 + \frac{1}{2}(\lambda_2 + \lambda_p r_t)\|w_i\|_2^2$$

which achieves minimum at zero gradient or $w_i = -\frac{z_{t,i} - \operatorname{sign}(z_{t,i})\lambda_1}{\lambda_2 + \lambda_p r_t}$.
This concludes the proof.

References

1. Abadi, M., Agarwal, A., Barham, P., Brevdo, E., Chen, Z., Citro, C., Corrado, G.S., Davis, A., Dean, J., Devin, M., et al.: Tensorflow: large-scale machine learning on heterogeneous distributed systems. arXiv preprint arXiv:1603.04467 (2016)
2. Allix, K., Bissyandé, T.F., Klein, J., Le Traon, Y.: Are your training datasets yet relevant? In: Piessens, F., Caballero, J., Bielova, N. (eds.) ESSoS 2015. LNCS, vol. 8978, pp. 51–67. Springer, Cham (2015). doi:10.1007/978-3-319-15618-7_5
3. Bekerman, D., Shapira, B., Rokach, L., Bar, A.: Unknown malware detection using network traffic classification. In: IEEE Conference on Communications and Network Security (CNS), pp. 134–142 (2015)
4. Feldman, R.: Techniques and applications for sentiment analysis. Commun. ACM 4, 82–89 (2013)
5. Gama, J., Žliobaitė, I., Bifet, A., Pechenizkiy, M., Bouchachia, A.: A survey on concept drift adaptation. ACM Comput. Surv. (CSUR) 46(4), 44 (2014)
6. Guarnieri, C., Schloesser, M., Bremer, J., Tanasi, A.: Cuckoo sandbox-open source automated malware analysis. Black Hat USA (2013)
7. Iliopoulos, D., Adami, C., Szor, P.: Darwin inside the machines: malware evolution and the consequences for computer security. arXiv:1111.2503 [cs, q-bio] (2011)
8. Kantchelian, A., Tschantz, M.C., Afroz, S., Miller, B., Shankar, V., Bachwani, R., Joseph, A.D., Tygar, J.D.: Better malware ground truth: Techniques for weighting anti-virus vendor labels. In: Proceedings of the 8th ACM Workshop on Artificial Intelligence and Security, pp. 45–56. ACM (2015)
9. Korkmaz, Y.: Automated detection and classification of malware used in targeted attacks via machine learning. Ph.D. thesis, Bilkent University (2015)

10. Makandar, A., Patrot, A.: Malware analysis and classification using artificial neural network. In: 2015 International Conference on Trends in Automation, Communications and Computing Technology (I-TACT 2015), vol. 01, pp. 1–6 (2015)

11. McMahan, H.B.: A survey of algorithms and analysis for adaptive online learning. arXiv preprint arXiv:1403.3465 (2014)

12. McMahan, H.B., Holt, G., Sculley, D., Young, M., Ebner, D., Grady, J., Nie, L., Phillips, T., Davydov, E., Golovin, D., et al.: Ad click prediction: a view from the trenches. In: Proceedings of the 19th ACM SIGKDD International Conference on Knowledge Discovery and Data Mining, pp. 1222–1230. ACM (2013)

13. Mirkovic, J., Benzel, T.: Deterlab testbed for cybersecurity research and education. J. Comput. Sci. Coll. **28**(4), 163–163 (2013)

14. Mohaisen, A., Alrawi, O., Mohaisen, M.: AMAL: high-fidelity, behavior-based automated malware analysis and classification. Comput. Secur. **52**, 251–266 (2015)

15. Narayanan, A., Yang, L., Chen, L., Jinliang, L.: Adaptive and scalable android malware detection through online learning. In: 2016 International Joint Conference on Neural Networks (IJCNN), pp. 2484–2491. IEEE (2016)

16. Nari, S., Ghorbani, A.A.: Automated malware classification based on network behavior. In: 2013 International Conference on Computing, Networking and Communications (ICNC), pp. 642–647 (2013)

17. Norouzi, M., Souri, A., Samad Zamini, M.: A data mining classification approach for behavioral malware detection. J. Comput. Netw. Commun. **2016**, 1–9 (2016)

18. Rafique, M.Z., Chen, P., Huygens, C., Joosen, W.: Evolutionary algorithms for classification of malware families through different network behaviors. In: Proceedings of the 2014 Conference on Genetic and Evolutionary Computation, pp. 1167–1174. ACM (2014)

19. Saini, A., Gandotra, E., Bansal, D., Sofat, S.: Classification of PE files using static analysis. In: Proceedings of the 7th International Conference on Security of Information and Networks, p. 429. ACM (2014)

20. Saxe, J., Berlin, K.: Deep neural network based malware detection using two dimensional binary program features. In: 10th International Conference on Malicious and Unwanted Software (MALWARE), pp. 11–20 (2015)

21. Shalev-Shwartz, S.: Online learning and online convex optimization. Found. Trends Mach. Learn. **4**(2), 107–194 (2011)

22. Valenti, R., Sebe, N., Gevers, T., Cohen, I.: Machine learning techniques for face analysis. In: Cord, M., Cunningham, P. (eds.) Machine Learning Techniques for Multimedia, pp. 159–187. Springer, Heidelberg (2008). doi:10.1007/978-3-540-75171-7_7

23. You, I., Yim, K.: Malware obfuscation techniques: a brief survey. In: International Conference on Broadband, Wireless Computing, Communication and Applications (BWCCA), pp. 297–300 (2010)

Real-Time Validation of Retail Gasoline Prices

Mondelle Simeon and Howard J. Hamilton[✉]

Department of Computer Science, University of Regina, Regina, Canada
Howard.Hamilton@uregina.ca

Abstract. We provide a method of validating current gasoline (petrol) prices for a crowd-sourced app that primarily provides current gasoline prices. To validate prices reported by users of the app, we propose an approach that validates each price report in real time as it is entered by a consumer by comparing it to the current prediction of what the price is expected to be at the specified store at the present time. To do so, a forecast model is used to predict, with high accuracy, a price range for each store in real-time so that when a price is entered by a consumer it can be flagged if it falls outside the predicted range. We present the first experimental results concerning predicting the current price in real time at all stores in a city. The results indicate that there is a significant improvement in reducing the forecast error when using our Price Change Rules model over the modified Most Common Action model.

Keywords: Price forecasting · Price prediction · Predictive accuracy · Forecasting gasoline prices · Data mining · Machine learning

1 Introduction

We provide a method of validating current gasoline (petrol) prices for a crowd-sourced app that primarily provides current gasoline prices. For brevity, we refer to any retail location that sells gasoline as a *store*. Prices are reported to a central server by consumers who visit stores. This approach to obtaining data carries the risk that consumers may enter incorrect prices either intentionally or accidentally, resulting in inaccurate data. Based on these incorrect prices, consumers may make faulty decisions about where to purchase their fuel, resulting in a negative experience with the app. Thus it is imperative to limit these occurrences. One approach is to validate each price report in real time as it is entered by a consumer by comparing it to the current prediction of what the price should be at the present time. To do so, a forecast model is used to predict, with high accuracy, a price range for each store in real-time so that when a price is entered by a consumer it can be reported if it falls outside the predicted range.

Gasoline is a unique commodity because it commands the attention of consumers on a weekly or daily basis. A small difference in the price of gasoline at competing stores can determine which of them receives a customer's business. Given the importance of gasoline to consumers' daily lives, many researchers have studied the behavior of crude oil prices, wholesale gasoline prices, and

© Springer International Publishing AG 2017
A. Yamamoto et al. (Eds.): DS 2017, LNAI 10558, pp. 33–47, 2017.
DOI: 10.1007/978-3-319-67786-6_3

retail gasoline prices. However, for the retail gasoline market, most of these studies have mainly focused on identifying factors which generally affect gas prices. This project aims to identify predictive factors that affect retail prices at individual stores and use these factors to predict highly accurate prices on demand for any store in real time. By real time, we mean that the price is validated before being shown to a waiting interactive user. Unless otherwise stated, all gasoline prices mentioned in this article are United States (US) prices quoted per gallon of gasoline and include all applicable taxes.

The remainder of this paper is organized as follows. Section 2 reviews the relevant literature on the retail gasoline market and defines the Most Common Action (MCA) model and algorithm. Section 3 outlines the Price Change Rules (PCR) prediction model. Section 4 presents the results of an evaluation of the MCA and PCR models on several years worth of data from five North American cities. Section 5 presents conclusions and suggestions for future research.

2   Related Research

In this section, we present previous studies that describe the connection between common factors such as crude oil and store location with respect to gasoline prices. Then we describe the Most Common Action model and a modified version of it for real-time forecasting.

2.1   Factors Affecting Gasoline Prices

Several research studies have examined factors that affect retail gasoline prices. The US Federal Trade Commission issued a report that summarized the dynamic factors such as supply, demand, competition, and regulations significantly affecting gasoline prices, as collected from various research sources [4]. This study finds that crude oil price is the most important factor affecting gasoline prices in the United States. Specifically, changes in crude oil prices are responsible for 85% of the changes seen in the US gasoline market. The report also suggests that disruptions to oil supply pipelines across the US have the ability to cause significant price increases.

Another study of gasoline prices determined that the presence of independent, unbranded stores drove prices down [5]. Specifically, the presence of an independent store led to prices that were on average five cents lower than usual. The lower prices were observed at other stores located in a circular neighborhood with a radius of one mile (1.6 km) centered on the independent store. Conversely, the replacement of an unbranded store by a branded store led to prices in the same neighborhood that were on average five cents higher than previously. The study concluded that little difference in prices could be attributed to the demographics of the area or the specific characteristics of the stores.

A separate study focused on the effect of competition density on the retail gasoline market [1]. It examined two theories associated with the presence of price dispersion, namely the monopolistic competition theory and the search-theoretical theory, in order to determine which one more closely described the gasoline market. According to the monopolistic competition theory, when consumers perceive differentiated products across sellers it creates imperfect competition and when assuming diverse demand elasticity this results in price dispersion. In contrast, according to the search-theoretic theory, it suggests price dispersion is generated when consumers do not know the location of a low price [1]. When gas price data were examined, price dispersion decreased as store density increased. Self-serve gasoline prices also decreased as density increased. These findings are more consistent with the monopolistic competition theory, than the search-theoretic theory. However, they appear to contradict previous findings, which found that the search-theoretical theory more accurately reflects the search costs and consumer preferences of the gasoline market.

In a similar study, retail gas prices in three California cities were examined in an effort to discover the effects of competition in the retail gasoline market [2]. The findings show that an increase in price by 2 cents led to a decrease in sales volume that was different depending on the level of competition in the area. In an area with a low density of competition, stores saw a 2.4% reduction in sales volume while in areas with a high density of competition, stores experienced an 8.4% decrease. This study concluded that competition between stores tends to lower prices because consumers have a greater number of alternatives from which to choose. These findings are consistent with two other studies [3,7].

Finally, another study examined the effects of spatial factors on competition and the price of gasoline [8]. This study found that instead that spatial concentration does not matter. In other words, the store location does not, in itself, affect the retail gas price to an appreciable extent. These findings run contrary to microeconomic theory that, suggests that business outlets located proximate to one another should charge lower prices.

2.2   The Most Common Action Model

One pattern observed in another study [9] is that each store can be characterized by a single price change category corresponding to the daily price change made most frequently. This price change category is called the *most common action*. The Most Common Action (MCA) model predicts the most common action as the next price change category [9]. To limit the possible types of price changes to be considered, this model uses z price change categories, where z is a small integer. Table 1 shows the number of price changes made for each of $z = 11$ price change categories for 334 price changes at a store. The most common action observed is price change category 4.

The algorithm to determine the predicted price with the MCA model is shown in Algorithm 1. The predicted price change category for a store on day d is determined by finding the most common action over h earlier days, where h is the *history size*, i.e. a number such as 365 or 730. The method counts the number of days on which the change in price matches each of the price change categories and then chooses the category with the largest count as the most common action.

Table 1. Frequency of changes in price change categories

Price change category (a)	Frequency of price changes (F_a)
0	2
1	7
2	12
3	60
4	121
5	80
6	5
7	4
8	3
9	6
10	34

Algorithm 1. The Most Common Action Algorithm

1: Input: P = price data, $p = |P|$, z = category count, h = history size
2: $A = \{0, ..., z - 1\}$ // possible actions
3: $\forall a \in A : F_a = 0, T_a = 0$
4: $m = 0$, $k = p - 1$
5: **while** $m < h$ and $k > 0$ **do**
6:    **if** both P_k and P_{k-1} are not null **then**
7:       $m + +$
8:       $\delta = (P_k - P_{k-1})$
9:       $a = \text{Category}(\delta)$
10:      $F_a = F_a + 1$
11:      $T_a = T_a + \delta$
12:   **end if**
13:   $k - -$
14: **end while**
15: $a_{max} = \underset{a \in A}{\text{argmax}}\ F_a$
16: Predicted Action $= a_{max}$
17: Predicted Price Change $= T_{a_{max}}/F_{a_{max}}$

Ties are broken arbitrarily. The predicted price change is the average of all the price changes in the predicted price change category that occurred during the h days.

A detailed step-by-step description of the algorithm is as follows. The input is a consecutive series of end-of-day prices $P = \{P_0, \ldots, P_{p-1}\}$, the number of price change categories z, and the number of earlier days h (line 1). If the end-of-day price for a day is not known, a special null value should be provided. The algorithm initializes a frequency counter (F_a) for each price change category a to zero; as well, it initializes the total of all price differences for each price change category to zero (line 3). The main loop (lines 5–14) goes back over the preceding days one by one, from the most recent backwards (lines 4 and 13), attempting to find h days where a price difference (delta) can be calculated. It continues as long as h such days have not been found (i.e., $m < h$) and there are still days to check (i.e., $k > 0$) (line 5). Given consecutive non-null prices, a delta can be computed (line 6). The count of such deltas is incremented (line 7). The delta δ is calculated by subtracting the preceding price from the current one (line 8). The corresponding price change category (or action) a for a delta is determined by the Category function which consults an external table (line 9). As well, the frequency for the action (F_a) is incremented (line 10) and the total of the deltas for the action T_a is updated (line 11). After all deltas in the window have been examined, the action with the highest frequency (a_{max}) is determined (line 15). The argmax function is assumed to break ties arbitrarily and return one of the actions with the maximum count. This action is the predicted action (line 16). The predicted price change is calculated as the average of the deltas in window where the action is the same as the predicted action (line 17). The predicted price can be determined outside the algorithm as the sum of the most recent non-null price and the predicted price change (not shown).

For our experiments, we modified the MCA model so that it predicted the most common action for every price report received, instead of only one for the end of each day.

3   The PCR Real-Time Prediction Model

In this section, we describe the PCR model for predicting prices in real time. We explain the concept of Price Change Rules and then show how to predict prices using these rules. Next, we describe the PCR algorithm, and then explain how to evaluate the PCR model for the task of in forecasting prices in real-time.

3.1   Price Change Rules

The Price Change Rules (PCR) Model is premised on the observation that while various stores make the same price changes on the same day, some specific stores consistently make the price changes first and other stores consistently make similar price changes later. We refer to the first type of store as a *leader* and the second type as a *follower*. The model uses this observation to predict the price change at a store. As with the MCA, this model uses z price change categories to limit the possible types of price changes to be considered. For example, if $z = 7$, the seven price change categories might be as shown in Table 2. We used these price change categories for PCR in our experiments.

In Table 2, the second column shows the range for each price change category shown in the first column. Here, the first three price change categories represent decreases in price, the fourth represents no change in price, and the last three represent increases in price. The range for each of these price change categories includes the end point if "[" or "]" is shown and excludes it if "(" or ")" is shown.

Table 2. Price changes

Price change category	Range
0	[$\leq$–10 cents]
1	[–5 to –10 cents)
2	(0 to –5 cents)
3	[0 cents]
4	(0 to 5 cents)
5	[5 to 10 cents)
6	[$\geq$10 cents]

Two stores S_i and S_j make a *matched change* if both make a price change in the same price change category from exactly the same previous price on the same day. A *price change rule* (or simply a *rule*) is a representation that is created to describe matched changes that are observed between two stores. As a simplification, we assume that if a store makes more than one price change in the same category on the same day, only the first is analyzed. A price change rule has three notable features: frequency, direction, and strength. The *frequency* of the rule is the number of times (days) both stores make a matched change. The *direction* of a rule indicates which store made the change at an earlier time during the day. Although the matched change was made by both stores on the same day, the specific times of the price reports are used to determine which store was first. Ties are resolved arbitrarily.

As previously mentioned, the store that makes a matched change first is called a *leader*, and the other store a *follower*. Every rule can be described from the point of view of the leader or the follower. Therefore we define two types of price change rules named *leading* and *following* rules. A *leading rule* has the form "S_i leads S_j (N_1/N)" and indicates that stores S_i and S_j made matching changes on N separate days, and on N_1 days, S_i made the change before S_j. The value N_1/N as a percentage represents the *strength* of the rule. The corresponding *following rule* has the form "S_i follows S_j ((N_1)/N)" and indicates that store S_i and S_j made matching changes on N days, and on N_1 days, S_i made the change after S_j. Any leading rule can be rewritten as a following rule and vice versa. For simplicity, we will use only following rules for the remainder of this paper. Table 3 shows the forms of two possible following rules between two stores S_i and S_j that describe the same situation (assuming there are no ties in the update times).

Given a rule of the form "S_i follows S_j (N_1/N) $\times$ 100%", for each of the x times that S_i changes its price after S_j, then the time difference between the price reports received for S_i and S_j is recorded. Thus a determination can be made, for instance, of the average amount of time that S_i changes its price after S_j. Additional measures such as the median time or interquartile range of the time can also be determined. This information can be used to decide if a future

Table 3. Price change rule

Rule	Store	Direction	Store	Frequency	Strength (%)
1	S_i	follows	S_j	N_1	$N_1/N \times 100\%$
2	S_j	follows	S_i	$N - N_1$	$(N - N_1)/N \times 100\%$

price report of S_i falls in an expected time period after a price report of S_j. It can also be used to decide if a price report for S_j is too old to be considered.

The Leading Set for a store S_k is a relatively permanent set of rules describing cases where store S_k often leads another store. When a new price is reported for store S_k, a rule is added to the Active Following Set for all stores listed in the Leading Set. The Active Following Set is the highly variable set of all following rules that are active for store S_k. When a price must be predicted for store S_k, the set of rules in its Active Following Set are consulted, as explained in the next subsection.

3.2 Prediction Using Price Change Rules

Price change rules are used to determine a predicted price for a store at a given time. Two measures are also used to make this determination, namely the price equality frequency and the inverse power distance. For two stores, S_i and S_j, the *price equality frequency* measures the number of days both stores had the same end-of-day price. The *spherical distance* (d) between the locations of S_i and S_j is calculated using the Haversine formula as follows:

$$d = 2 \times r \times \sin^{-1}\sqrt{\sin^2(\frac{\phi_i - \phi_j}{2}) + \cos(\phi_i)\cos(\phi_j)\sin^2(\frac{\lambda_i - \lambda_j}{2})} \quad (1)$$

where ϕ_i and ϕ_j are the latitudes of store S_i and store S_j, λ_i and λ_j are their longitudes, and r is the radius of the earth (6372.8 km).

The *inverse power distance* (I) between stores is calculated such that smaller distances get a higher weight. For power k, we use the following:

$$I = \frac{1}{(d+1)^k} \quad (2)$$

where d is the spherical distance between a pair of stores. Denominator $(d + 1)$ is used to ensure that the formulation works even when the distance is 0. After preliminary testing with a variety of values of k from 1 to 10, we obtained the best results with $k = 4$, which we used in our experiments.

The *potency* of a following rule is modeled by an exponential function such that higher frequencies have an exponentially higher weight. We calculate the potency as follows:

$$C = b^{(N_1/N)} \quad (3)$$

where b is a base for the exponent, N_1 is the frequency of the following rule for a price change category, and N is the maximum number of times a store is

Table 4. Available price change rules for store S_k at time t

Rule	Store ID	Time received	Actual price change (A)	Inverse distance (I)	Price change rule frequency (C)
1	S_{k_1}	t_2	x_2	y_2	z_2
2	S_{k_2}	t_3	x_3	y_3	z_3
...	...		...	...	...
m	S_{k_m}	t_m	x_m	y_m	z_m

followed in that price change category. After preliminary testing with a variety of values of b from 1 to 20, we obtained the best results with $b = 15$, which we used in our experiments.

Table 4 shows a set of m rules for store S_k. For instance, rule 1 shows that Store S_{k_1} made the actual price change x_2 at time t_2.

To determine the contribution to the predicted price change at store S_k that is due to a specific following rule from store S_i at time t_i, we consider three factors: (1) the actual price change A made at store S_i at time t_i, (2) the inverse distance I between stores S_k and S_i, and (3) the potency C of the price change rule. First, the inverse distance I and potency C are multiplied together to give a weight E. Similarly, the product D of the actual price change A, the inverse distance I, and the potency C of each rule is calculated. To determine the predicted price, we consider all m rules that are available for store S_k at the current time (in the Active Following Set and satisfied the time filter). The predicted price is obtained by dividing the sum of the product D over all m rules by the sum of the weights E over all m rules, which is calculated as follows:

$$prediction = \frac{\sum_{i=1}^{m} D}{\sum_{i=1}^{m} E} \qquad (4)$$

The details of the calculation are shown in Table 5.

Table 5. Predicted price change for store S_k at time t

Rule	Store ID	Time received	Actual price change (A)	Inverse distance (I)	Price change rule potency (F)	Product (D)	Weight (E)
1	S_{k_1}	t_2	x_2	y_2	z_2	$x_2 \times y_2 \times z_2$	$y_2 \times z_2$
2	S_{k_2}	t_3	x_3	y_3	z_3	$x_3 \times y_3 \times z_3$	$y_3 \times z_3$
...	...		...	...	...	...	...
m	S_{k_m}	t_m	x_m	y_m	z_m	$x_m \times y_m \times z_m$	$y_m \times z_m$
Sums						$\sum_{i=1}^{m} D$	$\sum_{i=1}^{m} E$
Predicted price change						$\frac{\sum_{i=1}^{m} D}{\sum_{i=1}^{m} E}$	

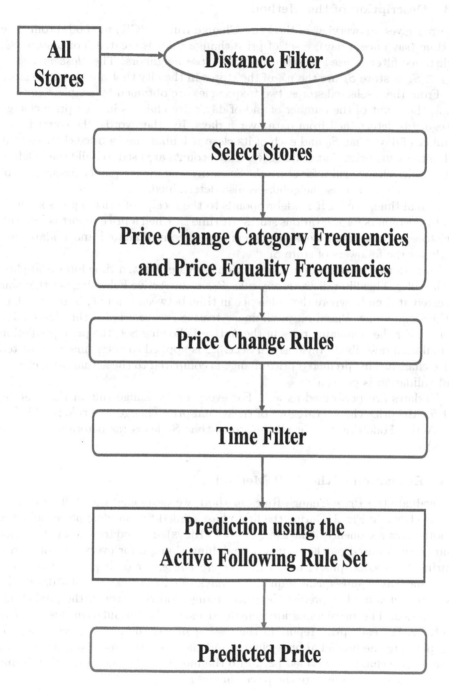

Fig. 1. Overview of the PCR method

3.3   Description of the Method

Figure 1 gives an overview of the Price Change Rules (PCR) method. From time to time (say once a day), a set of price change rules is created. For a store S_k, a distance filter is used to select only its close neighbors. The *close neighbors* $S_{k,p} \subset S_k$ of store S_k are the $p\%$ of the stores in the city that are closest to store S_k. From these selected stores, two frequencies are obtained by simple counting. First, the count of the number of end-of-day price changes in each price change category is determined from data over n days. In other words, the count is the number of days that S_k and each of its close neighbors make price changes that fall in the same price change category. These counts are used to build the Leading Set of price change rules for store S_k. Secondly, the price equality frequencies of S_k with each of its close neighbors is also determined.

In real time, the PCR model responds to the receipt of price reports submitted by customers for the various stores. At time t_i, when a price report is received for store S_k, two main steps are taken: the price is validated and updates are made for the followers of store S_k.

To perform the validation, a prediction is made. First, a time filter is applied to the rules. This filter discards any rule, R_i, in the Active Following Set that was received at time t_i, where the difference in time between t_i and t_j is greater than a threshold value. Based on preliminary testing, we selected a threshold value of 48 h. For the remaining rules in the Active Following Set, the price prediction formulation described previously in Sect. 3.2 is applied to determine a predicted price change. This predicted price change is compared to the actual price change and validation is performed.

Updates are performed as well. For every price change rule in the Leading Set for the price change category of the actual price change, the rule is added to the Active Following Set for all the stores that S_k leads (as determined by the rules).

3.4   Evaluation of the PCR Method

To evaluate the Price Change Rules method, we developed the Build and Test model shown in Fig. 2. During the first phase, called the *building phase*, all price reports from a sequence of days (e.g., $w = 730$ days) are used to generate the price change rules based on the counts of matching changes for every pair of stores. During the second phase, called the *testing phase*, for each price report from an immediately subsequent sequence of days (e.g., the next $n = 430$ days), the PCR model is used to predict the price change and the error in the predictions is recorded. The predictions are generated using the formulation described in Sect. 3.2. For each price report in the testing phase, the actual price change is compared to the predicted price change and the forecast error determined. Each actual price change is also fed back into the model at the end of the day during the testing phase to update the price change rules.

The *forecast error* (e) is the difference between the actual and the forecast (or predicted) value in a time series. Let y_i denote the ith observation and $\hat{y}_i$ denote a forecast of y_i. The forecast error is simply $e_i = y_i - \hat{y}_i$.

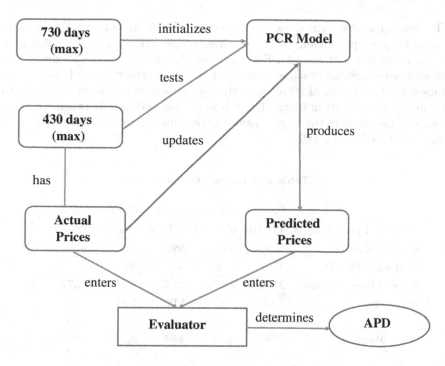

Fig. 2. Build and Test model

The *mean absolute error* (MAE) is used to measure the forecast error over n predictions. It is defined as follows:

$$MAE = \frac{1}{n} \sum_{i=1}^{n} |e_i|.$$

4   Experimental Evaluation

In this section, an evaluation of the Price Change Rules model is presented. We first describe the data used in the experiments and then we present the experimental results. We also compared the performance to that of the Most Common Action (MCA) method, as described in Sect. 2. This method was already in use by the makers of the app.

Data were available for five North American cities; for brevity, they are named City1 to City5. The data for each city were treated as a separate data set when evaluating the PCR and MCA models. In all cases, we considered only "regular" gasoline (the most commonly purchased gasoline). For City1 and City2, retail prices are available between 1 January 2011 and 8 March 2013 while for City3, City4, and City5, prices are available between 1 January 2010 and 8 March 2013. While many stores have at least one price report per day, over the given time period, there are also many days where some stores have no price reports.

Table 6 summarizes the data. The prices for City5 are converted from price per litre to price per gallon. There is considerable variation in price not only between cities but within cities. For example, the range between cheapest and most expensive regular gasoline is at least 57.2%. Further, there is also great variation in the number of stores in each city, with a minimum of 212 in City4 and a maximum of 1091 in City5. There is much less variation between the cities in terms of the share of the largest brand with a minimum of 17% in City4 to a maximum of 24.7% in City5.

Table 6. Summary statistics

		City1	City2	City3	City4	City5
Number of price reports		443,310	679,610	674,480	663,941	1,663,489
Number of stores		316	849	380	212	1091
Number of brands		45	114	53	30	10
Share of largest brand		21.2%	24.6%	18.4%	17%	24.7%
Price	Min	2.64	1.12	1.09	1.84	3.78
	Max	4.15	4.19	6.42	4.3	6.05
	Mean	3.50	3.38	3.84	3.52	4.61

Tables 7 to 8 show the evaluation results for the PCR model on the five datasets from City1 to City5. Table 7 shows the number of price reports predicted in each of the cities, and the number and percentage of these predictions that matched the most frequent rule in the rule set, some other rule in the rule set, or no rule in the rule set at that time. The most frequent rule predicts the price change category with the highest frequency in the list of price change rules. For instance, in City1, 291,237 price reports were predicted, 155,502 matched the most frequent rule, 120,060 matched some other rule, and 15,675 did not match any rule present at the time, and these values represent 53.39%, 41.22%, and 5.38% of the total price reports, respectively.

These results show that the highest percentage of predictions that matched the most frequent rule was 85.64% in City5, while the lowest was in City1, with 53.39%. In all five cities, less than 7.5% of the predicted prices did not match any of the rules, with City5 having the lowest percentage at 3.06% and City2, having the highest at 7.23%. We also assessed the accuracy of the predicted prices. Given that the most frequent rule suggested a prediction in category i, we predicted the median price change of all price changes made in that category. Thus, although the price category is $[-5$ to -10 cents$]$, the median might be -6 cents.

Table 8 shows the results for the Build and Test evaluation method for the PCR and Most Common Action (MCA) models. We calculated the MAE as described in Sect. 3.4. The third column gives the percentage of price reports

Table 7. Number of predictions

City	# of Prices	# of predictions			% of predictions		
	Total	Matches rule			Matches rule		
		Most frequent	Other	None	Most frequent	Other	None
City1	291,237	155,502	120,060	15,675	53.39	41.22	5.38
City2	413,092	309,100	74,120	29,872	74.83	17.94	7.23
City3	384,066	272,485	92,829	18,752	70.95	24.17	4.88
City4	343,646	191,867	138,259	13,520	55.83	40.23	3.93
City5	856,928	733,887	96,812	26,229	85.64	11.30	3.06

Table 8. Summary of results

City	The PCR method		The MCA model	
	MAE(¢)	% < 5 cents	MAE (¢)	% < 5 cents
City1	2.092	85.80	3.971	75.87
City2	0.856	93.63	1.262	90.67
City3	1.134	90.29	1.703	85.81
City4	1.374	91.46	3.234	81.22
City5	0.393	99.52	0.919	98.69

for which the predicted error was less than 5 cents. For PCR, the highest forecast error and lowest percentage within 5 cents was in City1 with 2.092 cents and 85.80%, respectively, while the lowest forecast error and highest percentage within 5 cents was in City5 with 0.393 cents and 99.52%, respectively.

The mean absolute error (MAE) for City1 with the PCR model is 2.092 cents but 3.971 cents for the MCA model. This difference shows that the PCR model improves the error by 1.879 cents over the MCA model. Improvements of 0.406, 0.569, 1.860, and 0.526 cents are also observed for City2, City3, City4, and City5, respectively. All of these differences are statistically significant.

The PCR model predicts price changes at a store based entirely on price changes made at other stores. Detailed examination of cases where the model does poorly shows that if matching changes occur frequently, which they commonly do, the predictions are highly accurate, but in cases where they do not, the predictions can be highly inaccurate. The main weakness of the PCR approach may be that it only considers competitor actions in predicting the price.

5 Conclusion

This paper described a forecast model for forecasting real-time retail gas prices. The Price Change Rules Model assumes that knowledge of a store's competitors price changes on the same day can be employed to predict the price change for that store. The model creates price change rules for each pair of stores based on past price changes in the same category. Each real-time price report at a given store triggers the addition of a price change rule to the rule set for every other store that follows it. The predicted price is determined as a weighted average of all the price change rules in the rule set. The evaluation results showed that the model significantly reduced the forecast error for all five cities, compared to the Most Common Action Model modified for real-time forecasting.

There are three main areas for continued research. The first area is the modeling of the relationship between wholesale costs and competitive considerations. As mentioned, the PCR model forecasts price changes at a store that are based entirely on price changes made at other stores. Previous research on retail price analysis has found that the changes in price are also significantly affected by wholesale costs and competitive behavior. Wholesale gasoline is a commodity that is traded on the open market. Thus, its price can change by the minute, which may influence the cost framework for a retailer. Retailers purchase gasoline on different schedules based on volume of sales and storage capacity [6]. Considering the volatility of wholesale prices, the cost of each delivery can vary significantly even in a short time period. Now while wholesale costs may be a significant factor, retail prices on a day-to-day basis are also heavily influenced by competition between retailers. How much of the changes in wholesale prices are passed on to the consumer and how soon? Does competition outweigh wholesale costs? The answers to these questions will assist in modeling the relationship between wholesale costs and competitive considerations.

The second area is incorporating additional information about the stores. For example, the brand of the store could be considered as well as the availability of other services, such as a convenience store or carwash.

Finally, the third area is the modeling of longer-term trends in retail prices. Predicting gas prices beyond one day ahead becomes increasingly difficult because of the volatility of both crude oil and whole sale prices as well as competition considerations. However, it may be possible to identify trends in the changes in price that would indicate prices are increasing, or decreasing, or remaining the same. Determining this information at the store level, would provide additional information to customers that can assist them in making decisions about when and where to fuel their vehicles.

References

1. Barron, J.M., Taylor, B.A., Umbeck, J.R.: Number of sellers, average prices, and price dispersion. Int. J. Ind. Organ. **22**(8–9), 1041–1066 (2004)
2. Barron, J.M., Umbeck, J.R., Waddell, G.R.: Consumer and competitor reactions: evidence from a field experiment. Int. J. Ind. Organ. **26**(2), 517–531 (2008)
3. Deck, C.A., Wilson, B.J.: Experimental gasoline markets. J. Econ. Behav. Organ. **67**(1), 134–149 (2008)
4. FTC: Gasoline price changes: the dynamic of supply, demand, and competition, June 2005

5. Hastings, J.: Vertical relationships and competition in retail gasoline markets: empirical evidence from contract changes in Southern California: reply. Am. Econ. Rev. **100**(3), 1277–79 (2010)
6. NACS: What influences gasoline prices (2013)
7. Pinkse, J., Slade, Margaret Emily, Brett, Craig: Spatial price competition: a semiparametric approach. Econometrica **70**(3), 1111–1153 (2002)
8. Schultheis, A., Johnson, D.K.N., Lybecker, K.M., Nadar, D.: Should i buy here, or keep driving? The effect of geographic market density on retail gas prices. Colorado College Working Paper 2011–2015, December 2011
9. Simeon, M.: Are these reported prices correct? Forecasting retail gas prices (2017, in Preparation)

Regression

General Meta-Model Framework
for Surrogate-Based Numerical Optimization

Žiga Lukšič[1], Jovan Tanevski[2(✉)], Sašo Džeroski[2], and Ljupčo Todorovski[1,2]

[1] University of Ljubljana, Ljubljana, Slovenia
ziga.luksic@live.com, ljupco.todorovski@fu.uni-lj.si
[2] Jožef Stefan Institute, Ljubljana, Slovenia
{jovan.tanevski,saso.dzeroski}@ijs.si

Abstract. We present a novel, general framework for surrogate-based numerical optimization. We introduce the concept of a modular meta model that can be easily coupled with any optimization method. It incorporates a dynamically constructed surrogate that efficiently approximates the objective function. We consider two surrogate management strategies for deciding when to evaluate the surrogate and when to evaluate the true objective. We address the task of estimating parameters of non-linear models of dynamical biological systems from observations. We show that the meta model significantly improves the efficiency of optimization, achieving up to 50% reduction of the time needed for optimization and substituting up to 63% of the total number of evaluations of the objective function. The improvement is a result of the use of an adaptive management strategy learned from the history of objective evaluations.

1 Introduction

Numerical optimization is a task of finding the values of numerical parameters that minimize or maximize a real-valued objective function. It is an omnipresent task in various domains of engineering and science. The methods for numerical optimization rely on numerous evaluations of the objective function, which presents a problem when the evaluation is non-trivial. This occurs when the evaluation involves either an expensive real-world experiment or a computationally complex procedure. In the first case, the number of possible evaluations is limited due to the cost of the experiment. In the second case, although the number of possible evaluations is in principle unlimited, the time need to perform a computationally complex evaluation practically limits its repeated execution. The focus of our interest is on tasks that are problematic due to the latter limitation.

In this paper we address the task of estimating parameters of ordinary differential equations [12], which is often approached as a numerical optimization problem. The objective function involves computationally expensive methods for simulation of differential equations. The task of parameter estimation is central to the task of mathematical modeling, which is in turn an essential part of the discovery of knowledge about the complex behavior and function of biological systems [16]. Another example of such task is hyper-parameter tuning [11], i.e.,

© Springer International Publishing AG 2017
A. Yamamoto et al. (Eds.): DS 2017, LNAI 10558, pp. 51–66, 2017.
DOI: 10.1007/978-3-319-67786-6_4

selecting a parameter setting for a machine learning algorithm that leads to its optimal predictive performance on a given data set. In this case, the objective function involves computationally expensive evaluation of the predictive performance of the machine learning algorithm on a data set.

Surrogate-based approaches to numerical optimization address exactly the case where the evaluation of the objective function is non-trivial. A surrogate function is a close approximation of the objective function that is computationally more efficient to evaluate [14]. The optimization task can be streamlined by evaluating such surrogates instead of the true objective function. We can cluster the surrogate-based approaches in two groups based on the surrogate management strategy, i.e., the way they decide when to use the surrogate and when to use the true objective function. During optimization, *wrapper approaches* evaluate only the surrogate, but validate the identified optimal points using the true objective function [3,21]. The control over the use of surrogates is thus very limited. In contrast, *embedded approaches* modify the optimization algorithm by incorporating strategies that decide between evaluating the surrogate or the true objective function within the optimization method [13]. While the embedded approaches can deploy arbitrary strategy for surrogate management, they require modifications of the core optimization method.

In this paper, we present an alternative framework for surrogate-based optimization that allows for the use of arbitrary surrogate management strategies without modifying the optimization method. As a consequence, a surrogate management strategy can be combined with any core optimization method without additional efforts for re-implementing and/or modifying the optimization method. The framework employs a meta model that incorporates a dynamically constructed surrogate, the procedure for its construction and a decision function that implements a surrogate management strategy.

We aim at showing that the proposed framework can solve complex numerical optimization problems with non-trivial objective functions while significantly reducing the number of true objective evaluations. We test the utility of the proposed framework on three task of estimating the parameters of models of dynamical biological systems represented by ordinary differential equations. We couple the meta model with Differential Evolution [22], as the core algorithm for numerical optimization, and Random Forest regression as method for learning surrogates. According to [8], Differential Evolution *"due to its competitive performance stands out as a good choice for the core optimizer"* in surrogate-based approaches. Random Forests [5] are known as a strong, robust and versatile method reported to work well in a variety of contexts, domains and data sets.

Section 2 introduces the tasks of numerical optimization and estimating parameters of differential equations and overviews the related work on surrogate-based optimization. We then introduce our general framework for surrogate-based numerical optimization in Sect. 3. We next report on the empirical evaluation of the proposed framework in Sect. 4. Finally, Sect. 5 concludes the paper with a summary and an outline of directions for further research.

2   Background and Related Work

Before overviewing the surrogate-based optimization methods and placing our contribution in its context, we introduce two central tasks of interest: numerical optimization and estimating parameters of differential equations.

2.1   Numerical Optimization

We consider the task of single-objective optimization of unconstrained, continuous, nonlinear and deterministic problems. Numerical optimization is the task of finding the solution $x^* \in \mathbb{R}^k$ in a given k-dimensional continuous space of solutions X that leads to the extremum of an objective function $F : x \in X \to \mathbb{R}$. The objective function can be either minimized or maximized: in the former case, the result of optimization is $x^* = \mathrm{argmin}_{x \in X} F(x)$. If the analytic solution for the minimum of F is intractable, numerical methods are applied. These can be clustered into two groups of local and global optimization methods.

Local optimization methods [19] are commonly used due to their efficiency. Such methods rely on the derivative of the objective function with respect to the problem parameters. The derivative is estimated by sampling or by direct calculation. They quickly converge to the point with the minimal value of the objective function in the neighborhood of the initial point. Given a hard non-linear problem, the local solution might not represent the global optimum. Therefore, local methods are frequently restarted from multiple initial points to increase the probability of finding the global solution, sacrificing the efficiency of the process.

Global optimization methods are concerned with finding the global optimum of an objective function [20]. They can be deterministic or stochastic. While the deterministic methods guarantee finding the global optimum, they do not guarantee that it will be found in a finite amount of time. Stochastic methods efficiently find the optimal regions of an objective function, but they do not guarantee the global optimality of the solution. A good property of global stochastic methods, in particular metaheuristics [25], is their ability to consider black-box objective functions. As a result, they have gained popularity and have been successfully applied to problems from various domains [4].

2.2   Estimating Parameters of Ordinary Differential Equations

The mathematical modeling of biological systems is an essential part of the discovery of knowledge about the complex behavior and function of biological systems [16]. The mathematical formalism that has been widely accepted as most adequate for representing the interactions within a dynamical biological system is the formalism of ordinary differential equations (ODEs) [18]. The estimation of the parameters of a model of a dynamical system from observations, also known as system identification, is central to the task of mathematical modeling [12].

A model of a dynamical system is described by a set of coupled ODEs $\dot{\mathcal{V}} = G(\mathcal{V}, x)$, where $v_i \in \mathcal{V}$ denote state variables, $\dot{v}_i = dv_i/dt$ their time derivatives, the functions $g_i \in G$ describe the structure of the model and x denotes the real-valued constant parameters of the model. Given an initial condition $\mathcal{V}_0$ (values of $\mathcal{V}$ at time t_0), the model can be integrated to obtain trajectories of values V_T representing the simulated behavior of the dynamical system at time points T. Analytic solution for a set of non-linear ODEs is rarely an option, so computationally expensive numerical approximation methods for ODE integration are typically applied.

The problem of estimating the parameters of ODEs from observations can be formulated as a numerical optimization problem. Given the observations O_T of variables $\mathcal{V}$ at time points T, the objective function is the likelihood of x to lead to simulated behaviour V_T, i.e., $F(x) = -\mathcal{L}(O_T|V_T)$, where $\mathcal{L}$ is a likelihood function. In practice, due to the complexity of the models, the likelihood-based function is approximated by a least-squares function $F(x) = \sum_{v \in \mathcal{V}} \|O_T^{(v)} - V_T^{(v)}\|^2$, where $O_T^{(v)}$ and $V_T^{(v)}$ denote the observed and simulated values of variable v at time points T. Recall however, that V_T is obtained using computationally intensive ODE integration method, which can be severely limiting.

Regarding the choice of a parameter estimation method for problems coming from the domain of systems biology, global stochastic and hybrid methods based on metaheuristics are considered as most promising in the literature [2,7]. Out of the many different metaheuristic methods, Evolutionary Strategies and Differential Evolution have been identified as the most successful [24,26].

2.3 Surrogate-Based Numerical Optimization

Surrogate-based optimization approaches are used to solve numerical optimization problems when the number of available evaluations of the objective function is limited. This limited availability is often related to the limited resources for performing the evaluation. The limited resources might involve physical equipment when the evaluation of the objective function involves performing experiments (in engineering domains) or computational time when evaluating computationally complex objectives (in computational domains). Surrogate-based approaches replace the true objective function F with a surrogate P, i.e., a predictive model that approximates the true objective function. The numerical optimization method then interchangeably employs F and P to obtain the evaluation of the objective function given a series of candidate solutions $x \in X$. Thus, in addition to F and P the surrogate-based optimization employs a decision function D that decides when to use F and when P. It also involves decision about when the approximation model is learned and updated from a training set based on a sample of the available evaluations of the true objective function.

In the literature on surrogate-based optimization, the decision function D is referred to as a surrogate management strategy [13]. Figure 1 depicts the clustering of the state-of-the-art of surrogate-based methods into two groups of wrapper (B) and embedded (C) approaches. To better understand the figure, consider first the simple situation of a numerical optimization algorithms that do not use surrogates (A). In such an environment, the optimization method interacts only with the true objective function F by requesting numerous

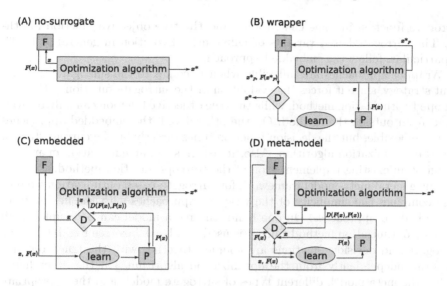

Fig. 1. Numerical optimization without surrogates (A), two state-of-the-art surrogate-based approaches to numerical optimization, wrapper (B) and embedded (C) and the framework based on meta models, proposed in this paper (D). In the four illustrations, F denotes the objective function, S the surrogate, and D the decision function that corresponds to the surrogate management strategy. The arrows denote the flow of values between the different components of the optimization approach.

evaluations of candidate solutions x. At the end, the method reports the optimal solution x^* that minimizes the value of the objective function.

Wrapper approaches place the surrogate management strategy outside the optimization method. Following this approach, the wrapper first initializes the surrogate P using a sample of candidate solutions x and their respective objective evaluations $F(x)$. In consecutive iterations the wrapper runs the optimization method using the surrogate P, obtaining a solution x_P^*, which is evaluated using the true objective function. The solution x_P^* and its evaluation $F(x_P^*)$ are then added to the surrogate training set and the surrogate is updated before running the next iteration. Examples of wrapper-approach methods are the methods for constrained numerical optimization COBRA [21] and SOCOBRA [3]. Both are based on the earlier work on efficient global optimization (EGO) methods [14] that also follow the wrapper approach.

Embedded approaches rely on encoding the management strategy within the optimization method. Following this approach, the decision whether to use the surrogate or the true objective function is based on the various artifacts of the algorithm [13]. In particular, population-based evolutionary optimization methods use surrogates to evaluate the offspring candidates for the next generation of individuals. On the other hand, the selection of the top candidates to be actually included in the next generation, is based on the evaluation of the true objective function. A simpler, generation-based management strategy evaluates the surrogate function in some generations, and the true objective function in others. The surrogate-based variants of Differential Evolution in general [8,17,23] in particular, follow the embedded approach.

Wrapper approaches are inflexible when it comes to the surrogate management strategy, since it forces the evaluation of the surrogate function within the wrapped optimization method, while the true objective function can only be evaluated from outside the method. On the other hand, the embedded approaches are more flexible, but the decision function relies directly on the current state of the core optimization algorithm. Also, it requires re-implementation or modification of an existing implementation of the core optimization method.

The general meta-model framework for surrogate-based optimization we propose combines the simplicity of the wrapper approaches with the flexibility of the embedded approaches. On one hand, the meta model can be coupled with any core optimization method since it is used as a black box (see Fig. 1(D)). The surrogate and decision functions are coupled together with the true objective function independently from the optimization algorithm. On the other hand, within the meta model, different types of surrogate models and the appropriate procedure for their construction can be seamlessly integrated. Subsequently, an arbitrary complex surrogate management strategy can be applied to the dynamically constructed surrogate model and the true objective function, generating a single meta-model evaluation. The decision can be different for each request for evaluation from the optimization method as it is based on the history of meta-model evaluations.

3   Meta Model for Surrogate-Based Optimization

We first introduce the *meta-model* framework for surrogate-based optimization. We next introduce two meta models that use two different surrogate management strategies. The first one is a simple, *"uninformed"* meta model that uses only the length of the evaluation history to decide whether to evaluate the surrogate function or the true objective function. The second one is a more complex, adaptive management strategy, called a *relevator*. The decision function for the relevator uses a predictive model trained using the history of evaluations of the true objective function.

3.1   Meta-Model Framework

The function $MetaModel \colon \mathbb{R}^k \times (\mathbb{R}^{k+2})^* \to \mathbb{R}$ is defined by the three functions (F, S, D), corresponding to the components of the *meta model*:

- objective function $F : \mathbb{R}^k \to \mathbb{R}$,
- surrogate function $S : \mathbb{R}^k \times (\mathbb{R}^{k+2})^* \to \mathbb{R}$,
- decision function $D : \mathbb{R}^k \times (\mathbb{R}^{k+2})^* \to \{0, 1\}$.

In our meta-model framework the function *MetaModel* is defined as:

$$MetaModel(x, h) = \begin{cases} F(x); & D(x, h) = 1 \\ S(x, h); & D(x, h) = 0 \end{cases} \tag{1}$$

Note the difference between the values of D in Fig. 1 and its role in the meta model in Eq. 1. In the latter case, the value of D determines whether the meta model returns the value of the true objective function F or the value of the surrogate P. In Fig. 1, the inputs and outputs of D correspond to the flow of values between the components.

The surrogate function S takes care of learning and updating the surrogate predictive model $P : \mathbb{R}^k \rightarrow \mathbb{R}$ from the training set sampled from the *history of evaluations* h of the true objective function. In particular, S takes care of collecting the history of meta-model evaluations, i.e., the finite sequence of past evaluations $h \in (\mathbb{R}^{k+2})^*$. Each past evaluation is recorded as $(x_1, \cdots, x_k, MetaModel(x), \delta)$, where $x = (x_1, \cdots, x_k)$ is a point of evaluation, while $\delta = 1$, if the objective function was used for evaluation, and $\delta = 0$, otherwise. The history of evaluations is updated after each evaluation of the meta model. The surrogate training set is the sample of the history of evaluations with $\delta = 1$.

There are three important properties to be considered when constructing a good surrogate function: the type of the prediction model, the size of the training set used for its construction and the frequency of model updates. We aim at selecting a surrogate that closely approximates the true objective function and can be evaluated efficiently. Moreover, the efficiency of the surrogate function depends upon the trade-off between the frequency of surrogate learning and the size of the training set. Having a high update frequency is desirable since the surrogate then always takes into account the most recent history of evaluations. On the other hand, frequent surrogate updates are unproductive unless the learning time is fairly low compared to the evaluation time of the true objective function. To this end, we introduce a user-defined parameter that determines the number of true object evaluations between the consecutive surrogate updates.

When it comes to the size of the training set, the issue of *filtration* of the history of evaluations arises. For example, when using the meta model in conjunction with a population-based method, the population slowly converges towards the minimum of the true objective function. After a number of evaluations, we can focus to the recent evaluations that correspond to the lower values of the objective. Therefore, older history can be safely removed from the training set. In our implementation, the training set includes a user-defined number of the recent points from the history of evaluations as well as a user-defined number of points with the lowest values of the true objective.

3.2   Uninformed Meta Model

The simplest surrogate management strategy (decision function) for a meta model is the one based only on the current index of the evaluation. For instance, meta model can use the surrogate for every third evaluation. Such decision functions do not use any kind of information about the point being evaluated other than the length of the evaluation history. Thus this management strategy is considered to be "*uninformed*".

A decision function D is uninformed, if it can be represented as a composite of the length function $L : (\mathbb{R}^{k+2})^* \to \mathbb{N}$ and a function $\tilde{D} \colon \mathbb{R} \to \{0, 1\}$ such that the meta model (F, S, D) is defined as:

$$MetaModel(x, h) = \begin{cases} F(x); & \tilde{D}(L(h)) = 1 \\ S(x, h); & \tilde{D}(L(h)) = 0 \end{cases} \tag{2}$$

A meta model with an uninformed decision function is an uninformed meta model.

3.3   Relevator Meta Model

An alternative approach is to identify points which are of high "relevance" for the optimization algorithm and decide whether to use the true objective function or the surrogate function to evaluate it. The evaluation of the most relevant points should be performed using the true objective function in order to properly estimate the current state of the optimization. We want to avoid misleading the algorithm into false optima that may appear as artifacts of the evaluation of the surrogate function.

The strategy for making a decision based on the relevance of a point brings up two issues. How is the relevance of a point formally defined and how can the relevance of a point be estimated before evaluating it. During the task of optimization, the points with values that are closest to the lowest seen value are considered as most relevant. In our approach, we calculate the relevance of a point as follows. Let $f = (f_1 \cdots, f_m)$ represent the vector of values of previously evaluated points in the history of evaluations. We define the relevance of the point $x \in \mathbb{R}^k$ relative to these values f as

$$relevance(x, f) = (1 + (F(x) - \min_i f_i)/(\text{avg}_i f_i - \min_i f_i))^{-1} \tag{3}$$

As long as $F(x) > \min_i f_i$ the relevance is bound to the interval $[0, 1]$ where the value of 0 corresponds to a point of low relevance and 1 to a point of high relevance.

The relevator meta model employs machine learning models for predicting the point relevance. From the history of evaluations and the same training set as the one used to learn the surrogate, we learn another model that predicts the point relevance. We refer to this model as the relevator. In addition to the relevator, the decision function also includes a decision threshold that distinguishes the points with high relevance, which should be evaluated using the true objective, from points with low relevance, which should be evaluated with the surrogate. To allow for the definition of a dynamical threshold, we define the threshold Θ as a function of the history of evaluations.

Thus, the decision function of a *relevator* meta model is the indicator function $\mathbf{1}[R(x, h) > \Theta(h)]$, where $R : \mathbb{R}^k \times (\mathbb{R}^{k+2})^* \to [0, 1]$ is the relevance of point x given the history of evaluations h and $\Theta : (\mathbb{R}^{k+2})^* \to [0, 1]$ is a dynamical relevance threshold function. The relevator meta model is then defined as:

$$MetaModel(x, h) = \begin{cases} F(x); & R(x, h) > \Theta(h) \\ S(x, h); & R(x, h) \leq \Theta(h) \end{cases} \qquad (4)$$

We implement the dynamical relevance threshold using an iterative updating procedure with the goal to control for and locally bound the rate of surrogate evaluations. By considering the user-defined number of most recent evaluations, we can either raise or lower the threshold after every meta model evaluation in order to increase or decrease the rate of surrogate evaluations to achieve (locally) the user-defined substitution rate.

4   Empirical Evaluation of the Meta-Model Variants

We empirically compare the performance of the two meta-model variants against the Differential Evolution method without using surrogates on three parameter estimation problems from the domain of systems biology[1]. After introducing the problems, we present the experimental setup and results.

4.1   Parameter Estimation Problems

For the empirical evaluation of the proposed framework we have selected three dynamical biological systems with varying degrees of complexity shown in Fig. 2. The three systems have been well studied in terms of their dynamical properties and identifiability [6,10].

The first system is a synthetic oscillatory network of three protein-coding genes interacting in an inhibitory loop, known as the Repressilator, modeled by Elowitz and Leibler [9]. The system is represented by a set of six ODEs with four constant parameters that are subject to estimation. Each gene (rectangle in Fig. 2 (A)) is modeled by two observable variable properties: the amount of mRNA transcribed by the gene and the amount of protein translated from the mRNA. Each of the three proteins inhibits the transcription of a target mRNA. The inhibition is modeled by a Hill type kinetics, the translation of mRNA to protein and the degradation of both mRNA and protein are modeled by linear kinetics. The transcription is assumed to have an additional constant component due to "leakiness" of the promoter.

The second system is a metabolic pathway representing a biological NAND gate modeled by Arkin and Ross [1]. The model is represented by a set of five ODEs with 15 constant parameters that are subject to estimation. The ODEs correspond to the five observed variables S3-S7 represented by rectangles in

---

[1] The implementation of the framework, the two meta-model variants, the models and data are available at http://source.ijs.si/zluksic/metamodel/.

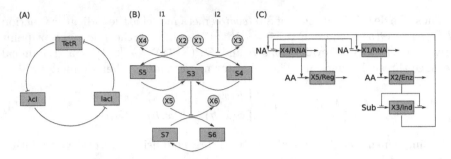

Fig. 2. Diagrams of the three models of dynamical biological systems used for the empirical evaluation: (A) A synthetic oscillatory network - repressilator; (B) Metabolic NAND gate; and (C) S-system model of a genetic network. The rectangles represent observed and modeled variables. The arcs ending with an arrow ($\rightarrow$) represent interactions with positive regulation while the arcs ending with a bar ($\dashv$) represent interactions with negative regulation.

Fig. 2 (B). The metabolites X1-X6 (circles) are assumed to be in steady-state (i.e. $\dot{X}_i = 0$). The system has two inputs I1 and I2, modeled as step functions. The dynamics of the interactions, represented by the arcs in the figure, are modeled by Michaelis-Menten kinetics with non-competitive inhibition.

The third system is a genetic network modeled by Kikuchi *et al.* [15]. The system is represented as a five variable S-system model with 23 constant parameters. S-system model is a set of ODEs in which the interactions in the system are approximated by a multivariate power-law functions of the form $\dot{X}_i = \sum_j s_{ij} \cdot k_j \prod_k X_k^{o_{jk}}$, where s_{ij} are stoichiometric coefficients, k_j are reaction rates and o_{jk} are kinetic orders. In the system represented in Fig. 2 (C) the observed variables are represented by rectangles. The stoichiometric coefficients for the reactions are -1 for reactants and $+1$ for products, the reaction rates and the kinetic orders are subject to estimation. The amounts of nucleic acid (NA), amino acid (AA) and substrate are assumed to be constant.

4.2 Experimental Setup

The uninformed meta model was set to use the true objective function every third meta-model evaluation (66% substitution rate). For the relevator meta model the surrogate function and the relevator are trained using the Weka implementation of Random Forest [5] with default parameters (100 trees with $\text{int}(\log_2(\#\text{parameters}) + 1)$ parameters per split) using 8 threads to reduce build time. The local substitution rate of the relevator was kept between 60% and 70%. We used a step size of 0.001 to adjust the threshold, with the starting value of the threshold set to 0.7. Based on the dimension of the problems, the training set for the repressilator was filtrated so it contains the 2000 most recent evaluations of the true objective function with additional 500 points with the lowest seen values. For the other two problems the training set was increased by factor of 2. Both prediction models were rebuilt after every 500 (repressilator) or 1000 (metabolic, s-system) evaluations of the meta model.

The meta models were coupled with the Differential Evolution optimization method [22] with fixed parameter settings. Based on the dimensionality of the problems, the population size was set to 100 (repressilator) or 200 (metabolic, s-system). The crossover probability was set to 0.8 and the differential weight was set to 0.9. For all experiments the same random seed (42) was used. The models were simulated using the classical explicit fourth order Runge-Kutta integrator with a step size of 10^{-2}. The observation data for the repressilator was obtained by simulating the model in the time interval $[0, 30]$. Samples for all variables were taken at each integer time point. The objective function used was the sum of the root of squared errors — $F(x) = \sum_{t \in T} \|O_t - V_t\|$, where O_t and V_t are the vectors of observed and simulated values of all variables $\mathcal{V}$ at time t.

The observation data for the metabolic pathway model was obtained from Gennemark and Wedelin [10]. It consists of 12 sets of observations obtained by simulating the model using 12 different pairs of input step functions $(I1, I2)$ in the time interval $[0, 150]$ sampled uniformly at 7 time points. The objective function used was the negative log-likelihood calculated as $F(x) = \mathcal{L}(x) = -\frac{1}{2} \sum_{v \in \mathcal{V}} \sum_{t \in T} (O_t^{(v)} - V_t^{(v)})^2 / \sigma_t^{(v)}$, where $O_t^{(v)}$ and $V_t^{(v)}$ are the observed and simulated values of variable v at time point t and $\sigma_t^{(v)} = 10^{-1} \cdot O_t^{(v)}$. To obtain the objective function the estimated likelihood was summed across all datasets.

The observation data for the s-system model was also obtained from Gennemark and Wedelin [10]. It consists of 10 sets of observations obtained by simulating the model using 10 different sets of initial conditions for all variables in the time interval $[0, 0.5]$. Each dataset contains 11 data points for each variable sampled uniformly from the simulations. As in the experiment with the metabolic pathway, to compare the observations to the simulated values we use the negative log-likelihood. The standard deviation of observations at each time point was set to $\sigma_t^{(v)} = 10^{-2}$. Due to the wide ranges of values that can occur during the evaluation of the highly nonlinear system, we transform the likelihood function to obtain the objective function $F(x) = \log(1 + \mathcal{L}(x))$. The transformation preserves the order and maps 0 to 0. As in the previous experiment, the objective function was summed across all datasets.

4.3   Results

Figure 3 depicts the convergence curves for the three parameter estimation problems (rows) obtained using no surrogate, the uninformative, and the relevator meta model (columns). The convergence rate and the obtained optima indicate a superior performance of the relevator meta model.

Even for a relatively simple repressilator problem, it can be observed that using a meta model significantly improves the rate of convergence relative to the number of evaluations of the true objective function. While slight improvements are obtained with the uninformed meta model, the relevator meta model achieves a nearly perfect fit of the observations in less than 30,000 true

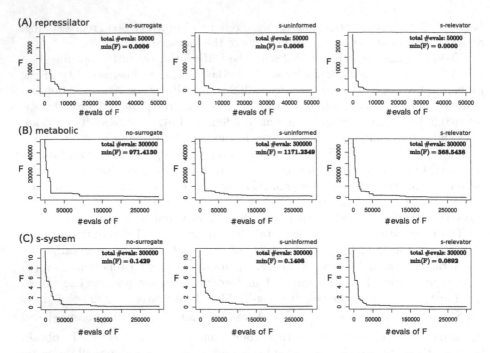

Fig. 3. Dependence of the quality of the best solution found so far on the number of evaluations of the true objective function. Convergence curves for the three parameter estimation problems: (A) Repressilator; (B) Metabolic pathway; and (C) S-System model of a genetic network. The three curves correspond to the no-surrogate method (left), the uninformed meta model (middle), and the relevator meta model (right).

objective evaluations. For the other two problems, the uninformed meta model slows down the convergence of DE without surrogates. In contrast, the relevator meta model outperforms the other two methods by factor of over 2.5 on the metabolic problem and by factor of 1.6 on the s-system problem. In all the experiments that we have conducted, the global rate of substitution was close to the local one.

Figure 4 confirms the superiority of the relevator meta model to the two alternative methods both in rate of convergence and the achieved optimal values. In the case of the repressilator, the relevator reaches the objective value of 10^{-4} after less than 30,000 true function evaluations, whereas the uninformed meta model and DE without surrogates are not able to reach that value in 50,000 evaluations. Similarly, for the other two problems, the number of true objective evaluations that the relevator needs to reach an objective threshold is at least twice lower then the number of evaluations needed by the other two methods. The uninformed model has never reached the lowest objective thresholds, while the DE without surrogates has reached it only for the metabolic problem.

In Table 1, we compare the three methods by observing the time and the number of evaluations needed to achieve the minimal objective value obtained using

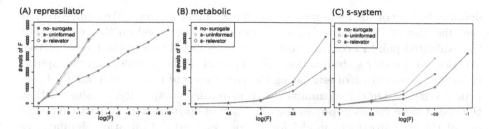

Fig. 4. Transposed convergence curves for the three parameter estimation problems ((A) Repressilator, (B) Metabolic pathway and (C) S-System model of genetic network) show the number of true objective evaluations needed to reach a certain objective value threshold. Points are missing from the end of some of the curves if that method did not reach the threshold in the allocated total number of evaluations.

Table 1. Time (minutes), number of evaluations of the true objective function (F) and of the meta model (MM) needed to achieve the minimal value of the objective obtained using DE without surrogates.

		no-surrogate	s-uninformed	s-relevator
Repressilator	Time	2.66	3.83	1.83
	#evals(F)	50,000	50,179	22,351
	#evals(MM)	50,000	148,533	59,929
Metabolic	Time	283.33	>283.33	135.5
	#evals(F)	300,000	>300,000	137,620
	#evals(MM)	300,000	>1,000,000	363,702
S-system	Time	145.66	95.17	97.00
	#evals(F)	300,000	178,984	181,464
	#evals(MM)	300,000	517,936	498,335

DE without surrogates. Again, the uninformed meta model does not improve the optimization performance for two reasons: it requires more time and more true objective evaluations. For the s-system, it reduces both time and number of evaluations, just like the relevator meta model, which outperforms no-surrogate optimization for all three problems. The results show the time reduction of 30% on the repressilator problem, 50% on metabolic and 33% on s-system. The rates of substitution of the true objective with the surrogate evaluations are 62.7%, 62.2% and 63.6%, respectively.

5   Conclusion

We presented a novel, general meta-model framework for surrogate-based numerical optimization. The framework is modular, easily configurable and independent from the core optimization method. We focused on the basic, defining feature of the meta model, the management strategy. We demonstrated the efficiency of the strategy that supports decisions based on the relevance of the evaluated point in contrast to uninformed substitution of the true objective function with a surrogate function. We approached the prevalent and computationally expensive task of estimating the parameters of models of dynamical biological systems. On three examples with increasing complexity, we showed that the use of meta model improves the efficiency of optimization. In particular, the use of the relevator meta model for surrogate-based optimization significantly and efficiently improves the convergence rate and the final result of the optimization when considering a limited number of evaluations of the true objective function.

Other than that, the components of the meta model and the core optimization method can be easily adapted to a specific problem, such that the efficiency of optimization is maximized. The adaptation introduces problems that can be approach both from the aspect of numerical optimization and machine learning. Such optimization of the framework is a direction for further work.

Particularly, in order to empirically evaluate the generality of the framework, it can be instantiated using different core optimization methods. The evaluation of the improvement in efficiency can then be established with regards to different strategies or to the parameters of the presented general strategies. Other issues concern more specifically the construction of the surrogate model and the decision function. Such is the issue of learning models that can predict the values of the true objective function and the relevance of prediction at the same time. The decision function may use information from the learned surrogate to derive the relevance of prediction. For example by analysis of the variance of the prediction of ensemble components or by considering the learning of other types models that contain information about the certainty of the prediction (Bayesian models, Gaussian process models, etc.). The problem of simultaneous prediction of the value of the objective function and the relevance of the evaluated point can be alternatively posed as a multi-target problem and approached by suitable learning methods. In the direction of multi-target learning, the framework can also be generalized towards the optimization of multi-objective problems. Furthermore, the task of optimization has a temporal dimension. The evolution of the population generates streams of information with increasing relevance. Such information can be efficiently exploited by iterative and online learning methods.

Acknowledgments. The authors acknowledge the financial support of the Slovenian Research Agency (research core funding No. P2-0103, No. P5-0093 and project No. N2-0056 Machine Learning for Systems Sciences) and the Slovenian Ministry of Education, Science and Sport (agreement No. C3330-17-529021).

References

1. Arkin, A., Ross, J.: Statistical construction of chemical reaction mechanisms from measured time-series. J. Phys. Chem. **99**(3), 970–979 (1995)
2. Ashyraliyev, M., Fomekong-Nanfack, Y., Kaandorp, J.A., Blom, J.G.: Systems biology: parameter estimation for biochemical models. FEBS J. **276**(4), 886–902 (2009)
3. Bagheria, S., Konena, W., Emmerich, M., Bäck, T.: Solving the G-problems in less than 500 iterations: improved efficient constrained optimization by surrogate modeling and adaptive parameter control. arXiv https://arxiv.org/abs/1512.09251 (2015)
4. Boussaïd, I., Lepagnot, J., Siarry, P.: A survey on optimization metaheuristics. Inf. Sci. **237**, 82–117 (2013)
5. Breiman, L.: Random forests. Mach. Learn. **45**(1), 5–32 (2001)
6. Buse, O., Pérez, R., Kuznetsov, A.: Dynamical properties of the repressilator model. Phys. Rev. E **81**, 066206 (2010)
7. Chou, I.C., Voit, E.O.: Recent developments in parameter estimation and structure identification of biochemical and genomic systems. Math. Biosci. **219**(2), 57–83 (2009)
8. Das, S., Mullick, S.S., Suganthan, P.N.: Recent advances in differential evolution - an updated survey. Swarm Evol. Comput. **27**, 1–30 (2016)
9. Elowitz, M., Leibler, S.: A synthetic oscillatory network of transcriptional regulators. Nature **403**, 335–338 (2000)
10. Gennemark, P., Wedelin, D.: Efficient algorithms for ordinary differential equation model identification of biological systems. IET Syst. Biol. **1**(2), 120–129 (2007)
11. Hutter, F., Hoos, H.H., Leyton-Brown, K.: Sequential model-based optimization for general algorithm configuration. In: Coello, C.A.C. (ed.) LION 2011. LNCS, vol. 6683, pp. 507–523. Springer, Heidelberg (2011). doi:10.1007/978-3-642-25566-3_40
12. Jaqaman, K., Danuser, G.: Linking data to models: data regression. Nat. Rev. Mol. Cell Biol. **7**(11), 813–819 (2006)
13. Jin, Y.: Surrogate-assisted evolutionary computation: recent advances and future challenges. Swarm Evol. Comput. **1**(2), 61–70 (2011)
14. Jones, D.R.: A taxonomy of global optimization methods based on response surfaces. J. Glob. Optim. **21**, 345–383 (2001)
15. Kikuchi, S., Tominaga, D., Arita, M., Takahashi, K., Tomita, M.: Dynamic modeling of genetic networks using genetic algorithm and S-system. Bioinformatics **19**(5), 643–650 (2003)
16. Kirk, P., Silk, D., Stumpf, M.P.H.: Reverse engineering under uncertainty. Uncertainty in Biology: A Computational Modeling Approach. SMTEB, vol. 17, pp. 15–32. Springer, Cham (2016)
17. Mallipeddi, R., Lee, M.: An evolving surrogate model-based differential evolution algorithm. Appl. Soft Comput. **34**, 770–787 (2015)
18. Murray, J.D.: Mathematical Biology, 2nd edn. Springer, Heidelberg (1993)
19. Nocedal, J., Wright, S.J.: Numerical Optimization, 2nd edn. Springer, Heidelberg (2006)
20. Pintér, J.: Global Optimization in Action: Continuous and Lipschitz Optimization: Algorithms, Implementations and Applications. Springer, Heidelberg (1995)
21. Regis, R.G.: Constrained optimization by radial basis function interpolation for high-dimensional expensive black-box problems with infeasible initial points. Eng. Optim. **46**(2), 218–243 (2013)

22. Storn, R., Price, K.: Differential evolution - a simple and efficient heuristic for global optimization over continuous spaces. J. Glob. Optim. **11**(4), 341–359 (1997)
23. Su, G.: Gaussian process assisted differential evolution algorithm for computationally expensive optimization problems. In: Proceedings of the IEEE Pacific-Asia Workshop on Computational Intelligence and Industrial Application, pp. 272–276 (2008)
24. Sun, J., Garibaldi, J., Hodgman, C.: Parameter estimation using metaheuristics in systems biology: a comprehensive review. IEEE/ACM Trans. Comput. Biol. Bioinform. **9**(1), 185–202 (2012)
25. Talbi, E.: Metaheuristics: From Design to Implementation. Wiley, Hoboken (2009)
26. Tashkova, K., Korošec, P., Šilc, J., Todorovski, L., Džeroski, S.: Parameter estimation with bio-inspired meta-heuristic optimization: modeling the dynamics of endocytosis. BMC Syst. Biol. **5**(1), 1–26 (2011)

Evaluation of Different Heuristics for Accommodating Asymmetric Loss Functions in Regression

Andrei Tolstikov[✉], Frederik Janssen, and Johannes Fürnkranz

Knowledge Engineering Group, TU Darmstadt,
Hochschulstrasse 10, 64289 Darmstadt, Germany
andreit@ke.tu-darmstadt.de

Abstract. Most machine learning methods used for regression explicitly or implicitly assume a symmetric loss function. However, recently an increasing number of problem domains require loss functions that are asymmetric in the sense that the costs for over- or under-predicting the target value may differ. This paper discusses theoretical foundations of handling asymmetric loss functions, and describes and evaluates simple methods which might be used to offset the effects of asymmetric losses. While these methods are applicable to any problem where an asymmetric loss is used, our work derives its motivation from the area of predictive maintenance, which is often characterized by a small number of training samples (in case of failure prediction) or monetary cost-based, mostly non-convex, loss functions.

1 Introduction

Recently an increasing number of regression problems require that different emphasis is placed on over- and under-estimation. For example, consider one of the most important parameters for predictive maintenance, the *remaining useful lifetime* (RUL) of a given component. Given a reliable RUL estimation, specific maintenance or repair actions can be planned as to minimize the overall cost of using a particular piece of equipment. Usually, regression methods try to predict the target value as accurately as possible, and do not distinguish between over- and under-estimation errors. However, for the case of RUL, since the cost of replacement of a component after a failure is usually higher than before a failure, we would prefer to estimate the remaining lifetime pessimistically. Informally, we can say that we want to predict the target value as closely as possible without over-estimation.

One approach to solving such problems with asymmetric loss functions is to try to find closed-form solutions that minimize the given loss. We will briefly discuss this approach in Sect. 3. However, such solutions typically have to make assumptions about the model class. Moreover, for some popular model classes, such as regression trees, closed-form solutions cannot be derived. Therefore, we attempt a different, more general approach in this work: We will explore the

© Springer International Publishing AG 2017
A. Yamamoto et al. (Eds.): DS 2017, LNAI 10558, pp. 67–81, 2017.
DOI: 10.1007/978-3-319-67786-6_5

suitability of generic machine learning methods for regression under asymmetric loss. The aim is to find simple and generally applicable methods that perform better than basic regression methods for symmetric loss functions. We tested several simple heuristics, which aim at offsetting the effects of asymmetric loss.

In this paper, we focus the discussion on *predictive maintenance* (PM), although the problem has also been investigated in other areas, such as computational finance. The specifics of predictive maintenance in this case are that the loss function may be explicitly defined based on the monetary cost involved when a certain failure happens or a maintenance procedure is performed. On the other hand, the difficulty in predictive maintenance often lies in the lack of data describing the failure, since failures might be quite rare and costly. Yet, the total amount of data can be quite high, because data describing normal operation without failures are usually abundant.

We will start with a formal definition of the problem (Sect. 2) and a recapitulation of previous work in this area (Sect. 3). The core contribution of the paper is a comparison of various methods which address this problem empirically, via a static or dynamic shift of the target value. These are discussed in Sect. 4, and experimentally compared in Sect. 5. From the obtained results, we draw our conclusions in Sect. 6.

2   Problem Formulation

The regression task is that given a set of training examples $(\mathbf{x}_i, y_i)$ with $\mathbf{x}_i \in \mathbb{R}^n$ and $y \in \mathbb{R}$, find a function $g(\mathbf{x}_i)$, which minimizes an expected loss $C(g(\mathbf{x}), y)$ for new pairs $(\mathbf{x}, y)$. The input variables $\mathbf{x} = (x_1, x_2, \ldots, x_n)$ are called regressors, the target value y is also known as the regressand. The function $C(y', y)$ specifies the penalty assigned to a sample used in the model training, when the prediction is not perfect. The learning stage finds for a given type of regression function $y' = g(\beta, \mathbf{x})$ such a set of parameters β, that the overall cost of mis-prediction for all samples i is minimized, i.e.,

$$\beta = \arg\min \sum_i C(y' = g(\beta, \mathbf{x}_i), y_i) \qquad (1)$$

Additional constraints may be included to ensure that no over-fitting occurs.

This formulation fits into a wider context of cost-sensitive learning, where various costs pertaining the data and predicted results are considered. However, in machine learning, the problem has typically been considered in the context of classification, where different costs are associated with different ways of mistaking one class for the other [4]. Instead, we are dealing with regression problems and continuously changing loss functions.

In most cases, the loss function is such that $C(y', y) = C(y' - y)$, that is the loss function depends only on the error value. Moreover, the penalty is zero only for perfect prediction ($C(y' = y, y) = 0$) and every error is penalized, i.e., $C(y' \neq y, y) > 0$. In terms of RUL prediction, if we want to reduce the number of over-estimations, we should assign a higher loss to the training instances which

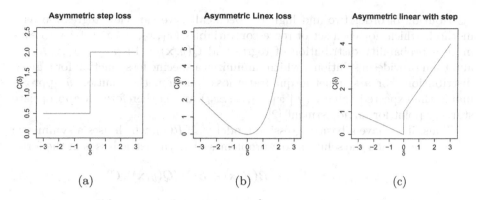

Fig. 1. Examples of asymmetric loss functions. *Step loss* (a) is used when we want to completely avoid over-predicting, *linex loss* (b) when only the costs for large positive and negative errors are significantly different, and *asymmetric linear loss with step* (c) offers the opportunity to model actual monetary cost.

over-estimate the target value, whereas an under-estimating by the same error should result in a smaller penalty. Thus, in this case, the loss function becomes asymmetric, i.e., $C(\delta) > C(-\delta) \; \forall \delta > 0$.

Examples of asymmetric loss function include *step loss*, which describes a situation where we would like to equally penalize all cases of over- or under-estimation (Fig. 1 (a)), or the *linex function*, which consists of a linear form in the negative range, an exponential function in the positive reange, connected by a symmetric quadratic function for small positive or negative values (Fig. 1 (b)). For a predictive maintenance domain, it would be of particular interest to use a loss function that reflects the actual gain or cost of equipment operation or failure. In many cases, such functions can be quite easy to define, e.g., via *piecewise linear functions*. However, this also means that these function most likely will not be smooth or convex, which restricts the use of gradient-based numerical methods (cf. Fig. 1 (c)).

3 Related Work

The first analytical work exploring the effect of asymmetric loss functions on regression was by Granger [6]. The author showed that in some simple cases of asymmetric cost functions and a Gaussian distribution of the estimation error, one can obtain an optimal solution by first solving the problem with a conventional method using symmetric quadratic loss, and then shifting its predictions by a certain value, which depends on the loss function and standard deviation of error.

Christoffersen and Diepold [2] extended this work and showed that for every loss function which depends only on the estimation error $C(y' - y)$, the optimal estimation under asymmetric loss also can be obtained by adding to a mean a certain value depending on the loss function and conditional moments of error distribution of order two and higher. Specifically, we can view regression as a method, which, for each set of regressor variables $(x_1, x_2, \ldots, x_n)$, tries to estimate a probability distribution of regressand $Q(y, \mathbf{x}) = \Pr(y \mid x_1, x_2, \ldots, x_n)$, and then provides a solution $\hat{y}$ which minimizes a specific loss function for a given distribution. For a symmetric quadratic loss function, the solution $\hat{y}$ approximates the expected value $E(y \mid x_1, x_2, \ldots, x_n)$ and is therefore an appropriate starting point for an adjustment [2].

Thus, if we have a basic regression model $y' = R(x)$, which uses a symmetric loss function, then the solution to a problem with asymmetric loss is in the form:

$$\hat{y} = g(\mathbf{x}) = R(\mathbf{x}) - B(R, \mathbf{x}) + S(\lambda_k(Q(y, \mathbf{x})), C) \tag{2}$$

where $B(R, \mathbf{x})$ is a bias of regression R, and the adjustment shift $S(.)$ depends on moments of variance λ_k with respect to y of probability distribution $Q(y, \mathbf{x})$ and the loss function $C(.)$. The importance of this result is that it shows that a solution of form (2) can give an optimal solution. However, it is still not a trivial problem, as we still need to find the probability distribution $Q(y, \mathbf{x}) = \Pr(y \mid x_1, x_2, \ldots, x_n)$. Attempts to address this problem include, for example [11], which extends [2] by proposing a method for estimating a conditional probability distribution for a given combination of regressors using the bootstrap [9]. However, it assumes that there is a sufficient number of data points for each combination of regressors $(x_1, x_2, \ldots, x_n)$, which may not be the case for many real-life problems.

One important implication of these theoretical results with respect to applications in predictive maintenance is that in order to have an optimal adjustment, we need to estimate the error probability distribution conditional on the regressors. Probability distribution estimation requires a large volume of data for each point, and thus might be infeasible for predictive maintenance applications. The reason for this is that datasets for fault prediction consist of points collected after actual failures, which implies that the cost of getting each point can be quite high. Thus, we have to find methods that allow to correct the basic model using only a small number of available training points.

The scope of our paper is close to [8], which compares multiple heuristics for the case of asymmetric loss functions. There is a difference in focus, however. In [8] the author targets systems with possibly varying loss functions, and suggests to create a model of variance first, and later derive corrections from local or global variances for specific loss functions. This approach is more flexible but may increase error due to assumption of Gaussian distribution of residuals. Some of the methods we are testing are based on the KNC method proposed in [8].

Another related work is [12], where authors propose to compute polynomial adjustments based only on a predicted value. Taking into account the application domain of the paper (positive values such as housing prices) and restating the underlying assumption about the data (variability of house prices does depend on the house price), it is reasonable to assume that this approach might work for similar datasets. However, we do not include this method in our comparison, since it requires a convex loss function.

In the following, we explore several simpler approaches which try to estimate a good prediction shift empirically from the given dataset.

4 Empirical Approaches Based on Prediction Shift

Machine learning offers a wealth of methods which can be used for regression. Each of them offers unique capabilities, which may be helpful for addressing a specific estimation problem. Most of them explicitly or implicitly assume symmetric cost functions. We would like to preserve this breadth of capabilities, but add to it the ability to handle the case of asymmetric loss. One approach could be to adapt individual techniques so that they can compute a direct solution to the global optimization given by (1). However, in many cases this may be impractical, since both assumed model and loss function might not be convex and differentiable, which would make finding the global minimum extremely difficult.

Thus, we would like to adopt a two-step approach. At the first step a known and proven machine learning method is used to estimate the required value, providing a regression function $R(\mathbf{x})$. At the second step, a shift function $S(\mathbf{x})$ is used to correct the original regression model. The computation of this function is based on the recognition results and achieved errors from the first model. Thus the complete solution to Eq. 1 is in the form:

$$g(\mathbf{x}) = R(\mathbf{x}) + S(\mathbf{x}) \tag{3}$$

This is a reasonable approach, since this is a simplified version of (2), which estimates the regression method bias and optimal shift terms of (2) together.

Formally, we would like to solve a problem of minimizing either the average or combined loss, given a solution of a basic black-box machine learning method $y' = R(\mathbf{x})$ and its prediction errors on the training samples $(\delta_i \equiv y'_i - y_i, \mathbf{x}_i)$. From this, we need to obtain a shift function $S(\mathbf{x})$ from a class of functions $S(\gamma, \mathbf{x})$, which minimizes the asymmetric loss given the errors of the basic model. Formally, we want to find the parametrization γ such that

$$\gamma = \arg\min \left(\sum_i C(\delta_i + S(\gamma, \mathbf{x})) \right). \tag{4}$$

There are several straight-forward approaches for finding a suitable shift for compensating an asymmetric loss. We later compare variants of the following basic methods, which are described in more detail in the following sections and summarized in Table 1:

- *Constant shift:* Given the results of a basic machine learning method, find a constant (the same for every instance) correction shift, which reduces the expected loss. This constant shift is then applied to the estimation given by the basic model.

– *Pointwise shift:* Having the results of a model with a (presumably) symmetric
loss, transform the training data by adding a specific shift to each training
point, and then re-train the model. Two cases can be considered: First, the
shift may correct a significant prediction failure, or, second, the shift for each
point may reflect the difference between the loss accrued with an asymmetric
function compared to the loss accrued in the symmetric case.
– *Learned model-based shift:* Having a first-cut result of how well the regression
approximates the data, learn a meta model which, for a given basic model,
attempts to find the value or at least the sign of the error for each instance,
and then compensate according to the loss function.
– *Assumed error model-based shift:* This is based on the analytical results mentioned in the Sect. 3. Since the optimal shift for a given loss function depends
on the moments of error distribution, we can obtain a model of moments
under certain assumptions. In this paper, we assume that at each point of the
state space the error distribution is Gaussian, and that the standard deviation
for each point linearly depends on the regressor values.
– *KNC Methods:* Direct (i.e. using direct distribution instead of variance) variations of the methods introduced in [8], which attempt to estimate the conditional probability distribution based on nearest-neighbour errors.

Being most obvious, the constant shift method serves as a baseline heuristic
for the comparison of the other methods. Another possible baseline is the direct
computation of optimum solution $g(\beta, \mathbf{x})$ of Eq. 1, using for example, a gradient
descent method. However, the direct optimum is difficult to find for complex
forms of regression and loss functions, and the comparison is only applicable
within the same type of regression functions.

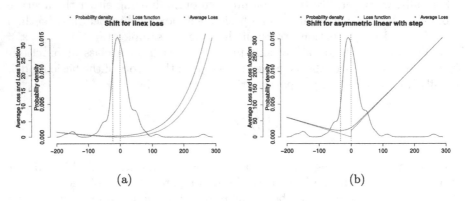

(a)   (b)

Fig. 2. Examples of computing constant shift for different loss functions. In green
color is the probability density of errors produced by a given primary machine learning
method. The probability distribution is shifted with respect to the loss function, and for
each shift the average loss is computed. Since this is a simple uni-dimensional function,
the optimum shift can be easily found even for inconvenient loss functions. The dotted
lines show the zero shift (red) and the shift achieving minimum average cost (blue).
(Color figure online)

4.1   Constant Shift

Assume that we have a loss function $C(\delta)$ and for a given regression method and training dataset we have an approximation function $y' = R(\mathbf{x})$. The total loss for a dataset can then be computed by summing up the loss for all instances as $\sum_i C(y'_i - y_i)$.

Figure 2 shows two examples of computing a dependency of the average loss on the shift value using the CPU dataset from the UCI collection [10]. Here, linear regression was used as a primary model and the error density displayed is for the training set.

4.2   Pointwise Shift

The approach in the previous section applies the same constant shift to all predictions. In this section, we discuss two approaches for making the magnitude of the shift depend on the instance that is shifted.

Shift of Over-Predicted Instances of Training Set. This method, referred to later as *shift to zero*, is ignorant of the loss function. Instead, it aims at reducing the number of over- (or under-) predicted instances.

Suppose the basic regression model $y' = R(\mathbf{x})$ was trained using the set of pairs $(\mathbf{x}_i, y_i)$. The subsequent model $\hat{y} = \hat{R}(\mathbf{x})$ is trained using the same algorithm as R, but using set of pairs $(\mathbf{x}_i, \hat{y}_i)$, where, for the case of avoiding over-prediction

$$\hat{y}_i = \begin{cases} y_i & \text{if } R(\mathbf{x}_i) \geq y_i \\ R(\mathbf{x}_i) & \text{if } R(\mathbf{x}_i) < y_i. \end{cases}$$

Non-linear Shift of Individual Instances of Training Set. The idea of this approach is to externally emulate the effects asymmetric loss function without considering the details of the chosen regression method. We assume that we know both the loss function optimized by the method, which we denote as $M(\delta)$, as well as the required asymmetric loss function $C(\delta)$. Typically, we can just assume that $M(\delta)$ is symmetric and quadratic. Another assumption is that both loss functions have the same loss value at the zero point. Furthermore, we denote the parts of loss functions for positive and negative δ as $M^+(\delta)$, $C^+(\delta)$ and $M^-(\delta)$, $C^-(\delta)$, respectively.

First, we use the training set to create a regression model and then obtain the error values $\Delta y_i = y'_i - y_i$ for each of the training instances. The total accrued loss for the training set under the method loss function is $L_M = \sum_i M(\Delta y_i)$, and assumed to be optimal for the method and the dataset. Then we modify the label values y of the training set so that the total cost stays the same under the required loss function

$$L_R = \sum_i M(\Delta \hat{y}_i) = L_M$$

The labels are modified by substituting each y_i with $\hat{y}_i$ so that

$$\hat{y}_i = y_i + \begin{cases} \delta : C^-(\delta) = M^-(\Delta y_i) & \text{if } \Delta y_i < 0 \\ \delta : C^+(\delta) = M^+(\Delta y_i) & \text{if } \Delta y_i > 0 \end{cases}$$

Then, we obtain a new regression model from this modified training set, and use this newly learned model for prediction.

4.3 Learned Model-Based Shift

This is another generic machine learning approach, where we want to learn how much offset is needed for every point to compensate the difference between whatever loss function used by a basic regression method and the required loss function $C(\delta)$.

Here, we train the basic model on a training set of pairs $(\mathbf{x}_i, y_i)$, then use an adjustment set of different pairs $(\mathbf{x}_j, y_j)$ for obtaining the errors $\delta_j = y'_j - y_j$ and the corresponding losses $l_j = C_{asc}(\delta_j)$. We use an ascending version of the loss function $C(\mathbf{x})$ defined as

$$C_{asc}(\delta) = \begin{cases} C(\delta) & \text{if } \delta \geq 0 \\ -C(\delta) & \text{if } \delta < 0 \end{cases}$$

Then we use pairs $(l_j, \mathbf{x}_j)$ to train a regression function $L(\mathbf{x})$ based on these observed losses. For the final regression, we can use the loss to off-set the prediction either in all cases (method L2), or only when we predict that an over- or under-prediction occurs (method L1).

4.4 Assumed Error Model-Based Shift

As mentioned before, this approach makes use of the analytical results described in Sect. 3. The general idea is as follows: since the optimal shift at a point depends on two functions, the moments of error probability distribution and the loss function, we can separate the computation of the shift from the modelling of the error distribution. However, there must be an assumed model of probability distribution, which connects these two stages. Formally, we assume point error distribution can be represented by a function E

$$\Pr(\delta \mid \mathbf{x}) = E(\delta, P(l, \mathbf{x}))$$

where δ is the error value, $\mathbf{x}$ is a vector of regressors, P is a function of parameters p of a probability distribution, and l are parameters describing the dependency of p on $\mathbf{x}$. We need to assume both the type of the distribution E and the function type of P, and then fit a maximum likelihood model so that

$$l = \arg\min \left(-\sum_i \log(E(\delta_i, P(l, \mathbf{x}_i))) \right)$$

Then, having fixed the parameters l and the function $p = P(l, \mathbf{x})$, we can substitute the general $E(\delta, P(l, \mathbf{x}))$ with a specific $\hat{E}(\delta, \mathbf{x})$, and then find for specific loss and error functions $C(\delta)$ and $\hat{E}(\delta, \mathbf{x})$, a dependency of the local shift on the distribution parameters $S(\mathbf{x})$, which minimizes expected loss

$$S(\mathbf{x}) = \arg\min \int_{-\infty}^{\infty} C(\delta)\hat{E}(\delta - S, \mathbf{x})\, d\delta \tag{5}$$

In the simplest scenario, which we explore here, E is a normal distribution, and P models standard deviation as linear function $\sigma = \mathbf{a} \cdot \mathbf{x} + b$. Then the combination error function will be either (method AM)

$$E(\delta, \mathbf{x}) = \frac{1}{(\mathbf{a} \cdot \mathbf{x} + b)\sqrt{2\pi}} e^{-\frac{(\delta - \mu)^2}{2(\mathbf{a} \cdot \mathbf{x} + b)^2}} \tag{6}$$

or, if we enforce an assumption of unbiased basic regression (method AZ)

$$E(\delta, \mathbf{x}) = \frac{1}{(\mathbf{a} \cdot \mathbf{x} + b)\sqrt{2\pi}} e^{-\frac{\delta^2}{2(\mathbf{a} \cdot \mathbf{x} + b)^2}} \tag{7}$$

The shift value from Eq. 5 becomes a function of standard deviation $S(\mathbf{x}) = \hat{S}(\mathbf{a} \cdot \mathbf{x} + b) = \hat{S}(\sigma)$, obtained from a single dimension optimization. Figure 3 shows examples of loss functions and dependency $\hat{S}(\sigma)$ of optimal shift on σ under assumption of Gaussian noise.

The benefit of this approach is that error data modelling is performed independently of the loss function, and, once the error model is found, any loss function can be used. Moreover, since finding an optimal shift for a given parameter of the error probability distribution involves only a single dimension optimization, it can be easily done for various types of loss functions, including non-convex, discontinuous or non-differentiable functions.

Correction for Non-Gaussian Noise. The assumption of Gaussian noise may mean that the shifts computed for specific values of the standard deviation would be either too big or too small for a particular error distribution. We can try to offset the effect of this assumption by correcting the shift using the information obtained while computing the constant shift model from the Sect. 4.1. Constant shift computation does not assume a specific model for the error distribution and is optimal for a given distribution. That is, for the complete prediction set, with error standard deviation σ_{total}, the computed optimal shift for Gaussian noise should be $\hat{S}(\sigma_{total})$, whereas the constant shift method provides value S_{CS}. We therefore can adjust the shift computation by a factor of $S_{CS}/\hat{S}(\sigma_{total})$ (methods AMC and AZC)

$$S(\mathbf{x}) = \frac{S_{CS}}{\hat{S}(\sigma_{total})}\hat{S}(\mathbf{a} \cdot \mathbf{x} + b)$$

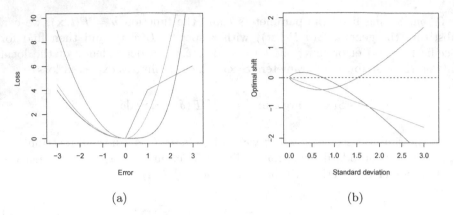

(a)   (b)

Fig. 3. Examples of asymmetric loss functions and corresponding computed optimal shift $\hat{S}(\sigma)$ under the assumption of Gaussian noise. Red is an example with quadratic loss on the left side and piece-wise linear on the right. It is an example of a *difficult* loss function, as it is both non-differentiable and non-convex. Blue has quadratic on the left and x^4 on the right. Green is the asymmetric quadratic. Notice, that for some loss functions the direction of the shift may change. (Color figure online)

4.5   k-Nearest Neighbors Based Methods

KNC [8] uses estimation errors $\delta_1, \ldots, \delta_k$ of k nearest neighbours of each sample i of the training or adjustment set as a sample of the conditional distribution. In [8] this sample is used to compute the variance. We suggest to use it to directly estimate the required shift as (methods Ka and Kt)

$$S_i = \arg\min \left(\sum_j C(\delta_j - S_i) \right) \tag{8}$$

Since k is small and this is only single dimensional optimization, it can be done for a wide class of loss functions.

Other versions tested here include:

- *Univariate* direct KNC - version of uKNC from [8], which uses only predicted value to find nearest neighbours (methods uKa and uKt).
- Using only features which are useful for primary regression method (methods Kfa and Kft).
- The last variation is a combination an attempt to combine a constant shift (which is essentially a simplest linear model) method with KNC. It uses linear regression on values S_i (methods Kla and Klt).

5   Experimental Comparison of Methods

We implemented the heuristics described above, and tested them on thirteen, mostly small-sized (as common in predictive maintenance applications) datasets from the UCI [10] and WEKA [7] repositories.[1] Each loss function was adapted to each dataset so that a significant cost difference was observed between over- and under-predicted instances. Specifically, the cost of over-prediction was on average 10 times higher for training instances of linear regression. Since some of the heuristics require an additional dataset for parameter adjustment, each dataset was randomly divided into three roughly equal parts, one of them used for model training, one for adjustment and one for testing. In case the adjustment set was not needed, it was used as additional training data. All possible assignments of dataset parts were used and results averaged with the geometrical mean of improvement ratios.

5.1   Regression Methods Used

As base learners, we used the following basic regression methods from the Weka library [7]:

- **Linear regression:** Loss function – quadratic.
- **M5 prime** [5]: Loss function – quadratic (model tree of linear regressions).
- **k-Nearest Neighbor:** No explicit loss function. However, since mean value of nearest neighbors are used, it is essentially equivalent to quadratic loss function.

5.2   Loss Functions Used

We used the following types of asymmetric loss functions:

- Asymmetric quadratic function in the form, for $a \in (0, +\infty)$

$$C(\delta, a) = \begin{cases} \frac{1}{a}\delta^2 & \text{if } \delta < 0 \\ a\delta^2 & \text{if } \delta \geq 0 \end{cases}$$

- Asymmetric polynomial in the form, for $a \in (0, +\infty)$

$$C(\delta, a) = \begin{cases} \delta^2 & \text{if } \delta < 0 \\ \delta^{2+a} & \text{if } \delta \geq 0 \end{cases}$$

- Asymmetric linear in the form, for $a \in (0, +\infty)$

$$C(\delta, a) = \begin{cases} \frac{\delta}{a} & \text{if } \delta < 0 \\ a\delta & \text{if } \delta \geq 0 \end{cases}$$

- Asymmetric linex in the form, for $a \neq 0$

$$C(\delta, a) = e^{a\delta} - a\delta - 1$$

- Asymmetric linear function with a step (linstep)

$$C(\delta, a, b, s) = \begin{cases} -a\delta & \text{if } \delta < 0 \\ b\delta + s & \text{if } \delta \geq 0 \end{cases}$$

For $a > 0$, the linex function behaves as a linear function for large negative δ, as exponential for large positive δ, and as $(a\delta)^2/2$ for $\delta \to 0$. For negative a, linear and exponential parts are exchanged. This loss function is appropriate for the cases when for small errors there is no cost difference, but there is a significant difference between large positive and negative errors.

---

[1] The datasets used are *auto93.arff, autoMpg.arff, autoPrice.arff, cloud.arff, cpu.arff, echoMonths.arff, elevators.arff, housing.arff, meta.arff, pyrim.arff, strike.arff, triazines.arff,* and *veteran.arff*.

The asymmetric linear function with a step is especially relevant for real-life predictive maintenance scenarios as it reflects actual monetary costs. Specifically, different slopes on both sides reflect the difference between the lost utility from yet functional part, if the said part is replaced before the failure, versus the cost of loss of use of the whole piece of equipment, when it cannot be used due to a failure. The step reflects the cost of repair after the failure versus the cost of replacement of yet functional part. For many optimization methods this function is rather inconvenient to work with, as it is discontinuous and not differentiable at certain points, and not convex.

5.3   Results

The results of experiments are summarized in Fig. 4, which shows the average performance over all loss functions and Fig. 5, which shows the results for the individual loss functions. The type of diagram used for these figures is proposed by Demšar [3], and drawn by R *scmamp* package [1]. It is useful for performance comparison of multiple methods on multiple datasets. It shows the average rank

Table 1. Abbreviations of method names used in diagrams.

Abbr	Method	Described in
CSA	Constant shift computed using adjustment set	Sect. 4.1
CST	Constant shift computed using training set	Sect. 4.1
PZA	Point-wise shift computed using adjustment set	Sect. 4.2
PZT	Point-wise shift computed using training set	Sect. 4.2
PZa	Nonlinear point-wise shift using adjustment set	Sect. 4.2
PZt	Nonlinear point-wise shift using training set	Sect. 4.2
L1	Learned shift with one-sided correction	Sect. 4.3
L2	Learned shift with two-sided correction	Sect. 4.3
AM	Assumed model	Sect. 4.4
AMC	Assumed model, with correction	Sect. 4.4
AZ	Assumed model, zero mean	Sect. 4.4
AZC	Assumed model, zero mean, with correction	Sect. 4.4
Ka	Direct KNC computed using adjustment set	Sect. 4.5
Kt	Direct KNC computed using training set	Sect. 4.5
uKa	Univariate direct KNC computed using adjustment set	Sect. 4.5
uKt	Univariate direct KNC computed using training set	Sect. 4.5
Kfa	Direct KNC with feature selection, using adjustment set	Sect. 4.5
Kft	Direct KNC with feature selection, using training set	Sect. 4.5
Kla	Linear model of direct KNC, using adjustment set	Sect. 4.5
Klt	Linear model of direct KNC, using training set	Sect. 4.5

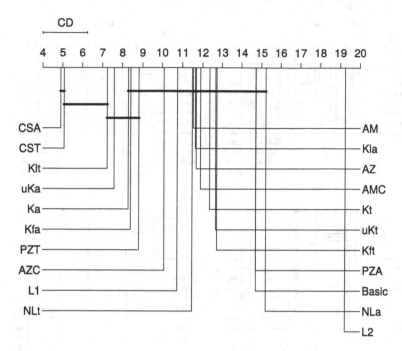

Fig. 4. Demšar [3] diagram, showing the critical distance of a Friedman test for the average ranking of methods. The scale on top is the average rank of a method. Better methods are on the left. The rank differences between methods connected with a thick line are not significant. The abbreviations for methods are explained in Table 1.

of each method (in our case each adjustment heuristic) and uses this as the basis for an estimate of the significance of the observed performance difference. A lower average rank (left side of the diagram) means better performing method. The thick lines connecting some of the results which are within the *critical difference* (CD at the left top corner) bounds. Thus, for methods that are connected in this way, the rank differences are not significant (significance level $\alpha = 0.05$).

For space efficiency, the diagrams show only abbreviated names of the methods; tthe abbreviations are explained in Table 1.

Somewhat surprisingly, the constant shift method proved to be the most consistent method of adjustment. Apparently, the case of similar moments of variance across the entire space is quite common. The KNC-based methods provided also good results, which can be expected, taking into account that they essentially try to estimate local error distributions. Non-linear point correction and meta-model adjustment of expected over-prediction performed better in some cases, but often also provided extremely poor solutions, and therefore had a poor average improvement results. It is, however, possible that for some known scenarios such methods might be beneficial. For example, the overall performance of the non-linear shift for asymmetric quadratic shift was the best.

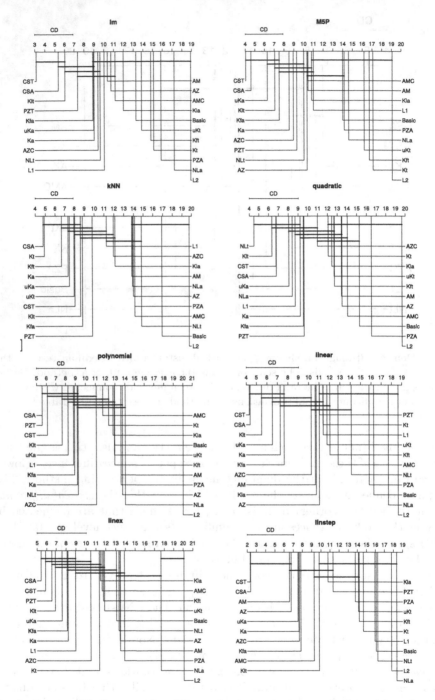

Fig. 5. Demšar [3] diagrams for different basic regression methods and loss function types. M5P, lm, and kNN are basic regression methods, linear, quadratic, polynomial, linex, and linstep are asymmetric loss functions described in Sect. 5.2

6   Conclusion

In this paper, we made a first step towards analyzing the problem of predicting with asymmetric continuous loss functions. This prediction problem commonly occurs in predictive maintenance applications, where the task is to estimate the remaining useful lifetime of a system component. The prediction should approximate the real lifetime as closely as possible, but should never overpredict the real value because that would mean that the component breaks before it is repaired or replaced.

We reviewed theoretical results that show that under some common assumptions, the problem can be viewed as finding an optimal shift value for the predictions of a model that has been trained with a conventional regression learner. In the following, we investigated and empirically compared a few simple heuristics for shift estimation, namely the use of a constant shift, a pointwise shift, learned and assumed shift models.

Acknowledgements. This work was supported by the German Federal Ministry of Education and Research (BMBF) under the "An Optic's Life" project (no. 16KIS0025). Thanks to the reviewers of this and a previous version of the paper, in particular for the pointers to prior work.

References

1. Calvo, B., Santafe, G.: scmamp: Statistical comparison of multiple algorithms in multiple problems. R J. **8**(1), 248–256 (2016)
2. Christoffersen, P.F., Diebold, F.X.: Optimal prediction under asymmetric loss. Econom. Theor. **13**, 808–817 (1997)
3. Demšar, J.: Statistical comparisons of classifiers over multiple data sets. J. Mach. Learn. Res. **7**, 1–30 (2006)
4. Elkan, C.: The foundations of cost-sensitive learning. In: Proceedings of the 17th International Joint Conference on Artificial Intelligence (IJCAI 2001), vol. 2. pp. 973–978. Morgan Kaufmann Publishers Inc, San Francisco, CA, USA (2001)
5. Frank, E., Wang, Y., Inglis, S., Holmes, G., Witten, I.H.: Using model trees for classification. Mach. Learn. **32**(1), 63–76 (1998)
6. Granger, C.W.J.: Prediction with a generalized cost of error function. Oper. Res. Q. **20**(2), 199–207 (1969)
7. Hall, M., Frank, E., Holmes, G., Pfahringer, B., Reutemann, P., Witten, I.H.: The Weka data mining software: An update. SIGKDD Explor. **11**(1), 10–18 (2009)
8. Hernández-Orallo, J.: Probabilistic reframing for cost-sensitive regression. ACM Trans. Knowl. Discov. Data **8**(4), 17.1–17.55 (2014)
9. Léger, C., Romano, J.P.: Bootstrap choice of tuning parameters. Annal. Inst. Stat. Math. **42**(4), 709–735 (1990)
10. Lichman, M.: UCI machine learning repository (2013). http://archive.ics.uci.edu/ml
11. McCullough, B.: Optimal prediction with a general loss function. J. Comb. Inf. Syst. Sci. **25**(14), 207–221 (2000)
12. Zhao, H., Sinha, A.P., Bansal, G.: An extended tuning method for cost-sensitive regression and forecasting. Decis. Support Syst. **51**(3), 372–383 (2011)

Differentially Private Empirical Risk Minimization with Input Perturbation

Kazuto Fukuchi[1(✉)], Quang Khai Tran[2], and Jun Sakuma[1,3]

[1] Department of Computer Science,
Graduated School of System and Information Science,
University of Tsukuba, 1-1-1 Tennodai, Tsukuba, Ibaraki 305-8577, Japan
kazuto@mdl.cs.tsukuba.ac.jp , jun@cs.tsukuba.ac.jp
[2] Intelligent Systems Laboratory, Secom Co., Ltd.,
10-16, Shimorenjaku 8-chome, Mitaka City, Tokyo 181-8528, Japan
ku-chan@secom.co.jp
[3] JST CREST, Ks Gobancho 6F, 7, Gobancho, Chiyoda-ku, Tokyo 102-0076, Japan

Abstract. We propose a novel framework for the differentially private ERM, input perturbation. Existing differentially private ERM implicitly assumed that the data contributors submit their private data to a database expecting that the database invokes a differentially private mechanism for publication of the learned model. In input perturbation, each data contributor independently randomizes her/his data by itself and submits the perturbed data to the database. We show that the input perturbation framework theoretically guarantees that the model learned with the randomized data eventually satisfies differential privacy with the prescribed privacy parameters. At the same time, input perturbation guarantees that local differential privacy is guaranteed to the server. We also show that the excess risk bound of the model learned with input perturbation is $O(1/n)$ under a certain condition, where n is the sample size. This is the same as the excess risk bound of the state-of-the-art.

Keywords: Differential privacy · Empirical risk minimization · Local privacy · Linear regression · Logistic regression

1 Introduction

In recent years, differential privacy has become widely recognized as a theoretical definition for output privacy [5]. Let us suppose a *database* collects private information from data contributors. *Analysts* can submit queries to learn knowledge from the database. Query-answering algorithms that satisfy differential privacy return responses such that the distribution of outputs does not change significantly and is independent of whether the database contains particular private information submitted by any single data contributor. Based on this idea, a great

Q.K. Tran—This work was done when he was a master's student in the Department of Computer Science, Graduated School of SIE, University of Tsukuba.

A. Yamamoto et al. (Eds.): DS 2017, LNAI 10558, pp. 82–90, 2017.
DOI: 10.1007/978-3-319-67786-6_6

deal of effort has been devoted to guaranteeing differential privacy for various problems. For example, there are algorithms for privacy-preserving classification [8], regression [12], etc.

Differentially private empirical risk minimization (ERM), or more generally, differentially private convex optimization, has attracted a great deal of research interest in machine learning, for example, [1,2,8,11]. These works basically follow the standard setting of differentially private mechanisms; the database collects examples and builds a model with the collected examples so that the released model satisfies differential privacy.

Recently, the data collection process is also recognized as an important step in privacy preservation. With this motivation, a *local privacy* was introduced as a privacy notion in the data collection process [3,9,13]. However, the existing methods of differentially private ERM are specifically derived for satisfying differential privacy of the released model, and thus there is no guarantee for the local privacy.

In this work, we aim to preserve the local privacy of the data and the differential privacy of the released model simultaneously in the setting of releasing the model constructed by ERM. The goal of this paper is to derive a differentially private mechanism with an utility guarantee, at the same time, the mechanism satisfies the local privacy in the data collection process.

Table 1. Comparison of differentially private ERM. All methods assume that ℓ_2 norm of the parameters is bounded by η, the loss function is ζ-Lipschitz continuous. n and d denote the number of examples and the dimension of the parameter, respectively.

Method	Perturbation	Privacy	Utility	Additional requirements
Objective [2,11]	obj. func.	(ϵ, δ)-DP for model	$O\left(\frac{\eta\zeta\sqrt{d\log(1/\delta)}}{\epsilon n}\right)$	λ-smooth
Gradient Descent [1]	grad.	(ϵ, δ)-DP for model	$O\left(\frac{\eta\zeta\sqrt{d\log^2(n/\delta)}}{\epsilon n}\right)$	
Input (proposal)	example	(α,δ)-DP for model $(\beta\sqrt{\epsilon},\delta)$-DLP for data s.t. $O(\sqrt{n\alpha}) = \beta$	$O\left(\frac{\eta\zeta\sqrt{d\log(1/\delta)}}{\epsilon\alpha n}\right)$	λ-smooth quadratic loss

Related Work. Chaudhuri et al. [2] formulated the problem of differentially private empirical risk minimization (ERM) and presented two different approaches: output perturbation and objective perturbation. Kifer et al. [11] improved the utility of objective perturbation by adding an extra ℓ_2 regularizer into the objective function. Moreover, they introduced a variant of objective perturbation that employs Gaussian distribution for the random linear term, which improves dimensional dependency from $O(d)$ to $O(\sqrt{d})$ whereas the satisfying privacy is relaxed from $(\epsilon, 0)$-differential privacy to (ϵ, δ)-differential privacy (Table 1, line 1). Objective perturbation is work well for smooth losses, whereas Bassily et al. [1] proved that it is suboptimal for non-smooth losses. They developed the optimal algorithm of (ϵ, δ)-differentially private ERM, named *differentially* *private gradient descent.* It conducts the stochastic gradient decent where the gradient is perturbed by adding a Gaussian noise. They showed that the expected empirical excess risk of the differentially private gradient descent is optimal up to multiplicative factor of $\log n$ and $\log(1/\delta)$ even for non-smooth losses (Table 1, line 2). They also provides the optimal mechanisms that satisfy $(\epsilon, 0)$-differential privacy for strong and non-strong convex losses. Jain et al. [8] showed that for the specific applications, the dimensional dependency of the excess risk can be improved from polynomic to constant or logarithmic. These studies assume that the database collects raw data from the data contributors, and so no attention has been paid to the data collection phase.

Recently, a new privacy notion referred to as *local privacy* [3,9,13] has been presented. In these studies, data are drawn from a distribution by each contributor independently and communicated to the data collector via a noisy channel; local privacy is a privacy notion that ensures that data cannot be accurately estimated from individual privatized data. [3] has introduced a private convex optimization mechanism that satisfies the local privacy. Their method has guarantee of differential privacy for the model, whereas its privacy level is same as the differential local privacy.

Our Contribution. In this study, we propose a novel framework for the differentially private ERM, input perturbation (Table 1, line 3). In contrast to the existing methods, input perturbation allows data contributors to take part in the process of privacy preservation of model learning. The mechanism of input perturbation is quite simple: each data contributor independently randomizes her/his data with a Gaussian distribution, in which the noise variance is determined by a function of privacy parameters (ϵ, δ), sample size n, and some constants related to the loss function.

In this paper, we prove that models learned with randomized examples following our input perturbation scheme are guaranteed to satisfy $(\alpha\epsilon, \delta)$-differential privacy under some conditions, especially, (ϵ, δ)-differential privacy if $\alpha = 1$ (Table 1, line 3, column 3). The guarantee of differential privacy is proved using the fact that the difference between the objective function of input perturbation and that of objective perturbation is probabilistically bounded. To achieve this approximation with randomization by independent data contributors, input perturbation requires that the loss function be quadratic with respect to the model parameter, w (Table 1, line 3, column 5).

From the perspective of data contributors, data collection with input perturbation satisfies the local privacy with the privacy parameter $(\beta\epsilon, \delta)$ where $\beta = O(\sqrt{n}\alpha)$ (Table 1, line 3, column 3). In the input perturbation framework, not only differential privacy of the learned models, but also privacy protection of data against the database is attained. From this perspective, we theoretically investigate the influence of input perturbation on the excess risk.

We compared the utility analysis of input perturbation with those of the output and objective perturbation methods in terms of the expectation of the excess empirical risk. We show that the excess risk of the model learned with input perturbation is $O(1/\alpha n)$ (Table 1, line 3, column 4). If $\alpha = 1$, the utility and the privacy guarantee of the model are equivalent to that of objective perturbation.

All proofs defer to the full version of this paper due to space limitation.

2 Problem Definition and Preliminary

Let $\mathcal{Z} = \mathcal{X} \times \mathcal{Y}$ be the domain of examples. The objective of supervised prediction is to learn a parameter $\boldsymbol{w}$ on a closed convex domain $\mathcal{W} \subseteq \mathbb{R}^d$ from a collection of given examples $D = \{(\boldsymbol{x}_i, y_i)\}_{i=1}^n$, where $\boldsymbol{w}$ parametrizes a predictor that outputs $y \in \mathcal{Y}$ from $x \in \mathcal{X}$. Let $\ell : \mathcal{W} \times \mathcal{Z} \to \mathbb{R}$ be a loss function. Learning algorithms following the empirical risk minimization principle choose the model that minimizes the empirical risk:

$$J(\boldsymbol{w}; D) = \frac{1}{n} \sum_{i=1}^n \ell(\boldsymbol{w}, (\boldsymbol{x}_i, y_i)) + \frac{1}{n}\Omega(\boldsymbol{w}), \qquad (1)$$

where $\Omega(\boldsymbol{w})$ is a convex regularizer. We suppose that the following assumptions hold throughout this paper: (1) $\mathcal{W}$ is bounded, i.e., there is η s.t. $\|\boldsymbol{w}\|_2 \le \eta$ for all $\boldsymbol{w} \in \mathcal{W}$, (2) ℓ is doubly continuously differentiable w.r.t. $\boldsymbol{w}$, (3) ℓ is ζ-Lipschitz, i.e., $\|\nabla\ell(\boldsymbol{w}, (x, y))\|_2 \le \zeta$ for any $\boldsymbol{w} \in \mathcal{W}$ and $(x, y) \in \mathcal{Z}$, and (4) ℓ is λ-smooth, i.e., $\|\nabla^2\ell(\boldsymbol{w}, (x, y))\|_2 \le \lambda$ for any $\boldsymbol{w} \in \mathcal{W}$ and $(x, y) \in \mathcal{Z}$ where $\|\cdot\|$ is the ℓ_2 matrix norm.

Three stakeholders appear in the problem we consider: *data contributors*, *database*, and *model user*. Each data contributor owns a single example $(\boldsymbol{x}_i, y_i)$. The goal is that the model user obtains the model $\boldsymbol{w}$ learned by ERM, at the same time, privacy of the data contributors is ensured against the database and the model user. Let us consider the following process of data collection and model learning.

1. All the stakeholders reach an agreement on the privacy parameters (ϵ, δ) before data collection
2. Each data contributor independently perturbs its own example and sends it to the database
3. The database conducts model learning at the request of the model user with the collected perturbed examples and publishes the model

Note that once a data contributor sends her perturbed example to the database, she can no longer interact with the database. This setting is suitable for real use, for example, if the data contributors sends their own data to the database via their smartphones, the database is difficult to always interact with the data contributors due to instability of internet connection. In this process, the privacy concerns arise at two occasions; when the data contributors release their own data to the database (data privacy), and when the database publishes the learned model to the model user (model privacy).

Model privacy. The model privacy is preserved by guaranteeing the (ϵ, δ)-differential privacy. It is a privacy definition of a randomization mechanism $\mathcal{M}$

which is a stochastic mapping from a set of examples D to an output on an arbitrary domain $\mathcal{O}$. Given two databases D and D', we say D and D' are neighbor databases, or $D \sim D'$, if two databases differ in at most one element. Then, differential privacy is defined as follows:

Definition 1 $((\epsilon, \delta)$**-differential privacy** [4]$)$. A randomization mechanism $\mathcal{M}$ is (ϵ, δ)-differential privacy, if, for all pairs (D, D') s.t. $D \sim D'$ and for any subset of ranges $S \subseteq \mathcal{O}$,

$$\Pr[\mathcal{M}(D) \in S] \leq \exp(\epsilon)\Pr[\mathcal{M}(D') \in S] + \delta. \tag{2}$$

Data privacy. For the definition of the data privacy, we introduce the differential local privacy [3,9,13]. Because of the data collection and model learning process, the non-interactive case of the local privacy should be considered, where in this case, individuals release his/her private data without seeing the other individuals' private data. Under the non-interactive setting, the differential local privacy is defined as follows.

Definition 2 $((\epsilon, \delta)$**-differential local privacy** [7,10,13]$)$. A randomization mechanism $\mathcal{M}$ is (ϵ, δ)-differentially locally private, if, for all pairs (z, z') s.t. $z \neq z'$ and for any subset of ranges $S \subseteq \mathcal{O}$,

$$\Pr[\mathcal{M}(z) \in S] \leq \exp(\epsilon)\Pr[\mathcal{M}(z') \in S] + \delta. \tag{3}$$

Utility. To assess utility, we use the *empirical excess risk*. Let $\hat{w} = \arg\min_{w \in \mathcal{W}} J(w; D)$. Given a randomization mechanism $\mathcal{M}$ that (randomly) outputs w over $\mathcal{W}$, the empirical excess risk of $\mathcal{M}$ is defined as $J(\mathcal{M}(D); D) - J(\hat{w}; D)$.

3   Input Perturbation

In this section, we introduce a novel framework for differentially private ERM. The objective of the input perturbation framework is three-fold:

- (data privacy) The released data from the data contributors to the database satisfies $(O(\sqrt{n}\epsilon), \delta)$-differentially locally private,
- (model privacy) The model resulted from the process eventually meets (ϵ, δ)-differentially private,
- (utility) The expectation of the excess empirical risk of the resulting models is $O(1/n)$, which is equivalent to that obtained with non-privacy-preserving model learning.

By adjusting ϵ, the input perturbation satisfies $(\alpha\epsilon, \delta)$-differential privacy and $(\beta\epsilon, \delta)$-differential local privacy with the $O(1/\alpha n)$ excess empirical risk where $\beta = O(\sqrt{n}\alpha)$.

3.1 Loss Function for Input Perturbation

The strategy of input perturbation is to minimize a function that is close to the objective function of the objective perturbation method. The requirements on the loss and objective function thus basically follow the objective perturbation method with the Gaussian noise [11]. Input perturbation allows any (possibly non-differential) convex regularizer as supported by objective perturbation. However, for simplicity, we consider the non-regularized case where $\Omega(\boldsymbol{w}) = 0$.

In addition to the requirements from the objective perturbation, input perturbation requires a restriction; the loss function is quadratic in $\boldsymbol{w}$. Let $\boldsymbol{q}(\boldsymbol{x}_i, y_i)$ and $\boldsymbol{p}(\boldsymbol{x}_i, y_i)$ be d dimensional vectors and $s(\boldsymbol{x}_i, y_i)$ be a scalar. Then, our quadratic loss function has a form:

$$\ell(\boldsymbol{w}, (\boldsymbol{x}, y)) = \frac{1}{2}\boldsymbol{w}^T \boldsymbol{q}(\boldsymbol{x}, y)\boldsymbol{q}(\boldsymbol{x}, y)^T \boldsymbol{w} - \boldsymbol{p}(\boldsymbol{x}, y)^T \boldsymbol{w} + s(\boldsymbol{x}, y).$$

3.2 Input Perturbation Method

In this subsection, we introduce the input perturbation method. Algorithm 1 describes the detail of input perturbation; Algorithm 2 describes model learning with examples randomized with input perturbation. In Algorithm 1, each data contributor transforms owing example $(\boldsymbol{x}_i, y_i)$ into $(\boldsymbol{q}_i, \boldsymbol{p}_i)$, where $\boldsymbol{q}_i = \boldsymbol{q}(\boldsymbol{x}_i, y_i), \boldsymbol{p}_i = p(\boldsymbol{x}_i, y_i)$. Then, she adds perturbation to $(\boldsymbol{q}_i, \boldsymbol{p}_i)$ in Step 3. We denote the example after perturbation by $(\tilde{\boldsymbol{q}}_i, \tilde{\boldsymbol{p}}_i)$, which is submitted to the database independently by each data contributors.

In Algorithm 2, the database collects the perturbed examples $\tilde{D} = \{\tilde{\boldsymbol{q}}_i, \tilde{\boldsymbol{p}}_i\}_{i=1}^n$ from the n data contributors. Then, the database learns a model with these randomized examples by minimizing

$$J^{in}(\boldsymbol{w}; \tilde{D}) = \frac{1}{n}\sum_{i=1}^n \left(\frac{1}{2}\boldsymbol{w}^T \tilde{\boldsymbol{q}}_i \tilde{\boldsymbol{q}}_i^T \boldsymbol{w} - \tilde{\boldsymbol{p}}_i^T \boldsymbol{w} + s_i \right) + \frac{\Delta_{in}}{2n}\|\boldsymbol{w}\|_2^2. \tag{4}$$

In the following subsections, we show the privacy guarantee of the input perturbation in the sense of the differential local privacy and the differential privacy. The utility analysis of models obtained following the input perturbation framework is also shown.

3.3 Privacy of Input Perturbation

In this subsection, we analyze the privacy of the input perturbation in the sense of the data privacy and the model privacy.

Data privacy of input perturbation. In Algorithm 1, each data contributor of the input perturbation adds a Gaussian noise into the released data. Adding a Gaussian noise into the released data satisfies (ϵ, δ)-differential local privacy as well as the Gaussian mechanism [6]. As a result, we get the following corollary that shows the level of the differential local privacy of Algorithm 1.

Algorithm 1. Input Perturbation

Public Input: $\epsilon, \delta, d, n, \eta, \zeta$ and λ
Input of data contributor i: $\boldsymbol{x}_i, y_i$
Output of data contributor i: $\tilde{\boldsymbol{q}}_i, \tilde{\boldsymbol{p}}_i$

1: $\gamma, \delta' \leftarrow \frac{\delta}{2}, a = \sqrt{\frac{\log(2/\gamma)}{n}}$, $\sigma_b^2 \leftarrow \frac{\zeta^2(8\log 2/\delta'+4\epsilon)}{\epsilon^2}$, $\sigma_u^2 > \left(\frac{\sqrt{2da}\lambda+\sqrt{2da^2\lambda^2+\frac{2\lambda}{\epsilon}(1-2a)}}{(1-2a)}\right)^2$

2: Sampling of noise vectors: $\boldsymbol{r}_i \sim \mathcal{N}(0, \frac{\sigma_b^2}{n}\mathbb{I})$, $\boldsymbol{u}_i \sim \mathcal{N}(0, \frac{\sigma_u^2}{n}\mathbb{I})$
3: $\tilde{\boldsymbol{q}}_i \leftarrow \boldsymbol{q}_i + \boldsymbol{u}_i, \tilde{\boldsymbol{p}}_i \leftarrow \boldsymbol{p}_i - \boldsymbol{r}_i$ where $\boldsymbol{q}_i = q(\boldsymbol{x}_i, y_i)$ and $\boldsymbol{p}_i = p(\boldsymbol{x}_i, y_i)$
4: Submission: Send $\tilde{\boldsymbol{q}}_i, \tilde{\boldsymbol{p}}_i$ to the database

---

Algorithm 2. Model Learning on Input Perturbation

Require: $\epsilon, \delta, d, n, \eta, \zeta$ and λ
1: All stakeholders agree with (ϵ, δ) and share parameters d, n, η, ζ and λ.
2: The database collects $(\tilde{\boldsymbol{q}}_i, \tilde{\boldsymbol{p}}_i)$ from the data contributors with Algorithm 1.
3: The database learns $\boldsymbol{w}^{in} = \arg\min_{\boldsymbol{w} \in \mathcal{W}} J^{in}(\boldsymbol{w}; \tilde{D})$ with $\Delta_{in} = \Delta - \frac{2\lambda}{\epsilon}$.
4: Return $\boldsymbol{w}^{in}$.

Corollary 1. *Suppose that q and p in Algorithm 1 are in the bounded domain with the size parameter B. Then, Algorithm 1 satisfies $(2c\sqrt{n}(\lambda/\sigma_u + \zeta/\sigma_b), 2\delta)$-differential local privacy, where $c > \sqrt{2\ln(1.25/\delta)}$.*

Since we have $\lambda/\sigma_u + \zeta/\sigma_b \to (\sqrt{\frac{\lambda}{2}} + \sqrt{\frac{\epsilon}{8\log(2/\delta')+4\epsilon}})\sqrt{\epsilon}$ as $n \to \infty$, Algorithm 1 is $(O(\sqrt{n\epsilon}), \delta)$-differentially locally private.

Model privacy of input perturbation. The following theorem states the guarantee of differential privacy of models that the database learns from examples randomized by the input perturbation scheme.

Theorem 1. *Let $\tilde{D}$ be examples perturbed by Algorithm 1 with privacy parameters ϵ and δ. Then, if $\Delta > \frac{2\lambda}{\epsilon}$ and $\gamma = \frac{\delta}{2}$, the output of Algorithm 2 satisfies (ϵ, δ)-differential privacy.*

3.4 Utility Analysis

The following theorem shows the excess empirical error bound of the model learned by input perturbation:

Lemma 1. *Let $\boldsymbol{w}^{in}$ be the output of Algorithm 2. If $\Delta > \frac{2\lambda}{\epsilon}$ and examples are randomized by Algorithm 1, w.p. at least $1 - \gamma - \beta$ the bound of the excess empirical risk is*

$$J(\boldsymbol{w}^{in}; D) - J(\hat{\boldsymbol{w}}; D) \leq \frac{4d\zeta^2(8\log\frac{4}{\delta} + 4\epsilon)\log\frac{1}{\beta}}{n\epsilon^2\Delta} + \frac{\Delta}{2n}\|\hat{\boldsymbol{w}}\|_2^2 + \frac{\sigma_u^2 - \frac{2\lambda}{\epsilon}}{2n}\|\hat{\boldsymbol{w}}\|_2^2$$

$$+ \frac{\sigma_u^2\sqrt{\log\frac{4}{\gamma}} + \sigma_u^2\frac{\log\frac{4}{\gamma}}{\sqrt{n}} + \sigma_u\lambda\sqrt{2d\log\frac{2}{\gamma}}}{n\sqrt{n}}\|\hat{\boldsymbol{w}}\|_2^2$$

In the right side of the bound, the first two terms of $O(1/n)$ are the same as the excess empirical risk of objective perturbation [11]. The third term of $O(1/n)$ and the last term of $O(1/n^{3/2})$ are introduced by input perturbation. The same holds with expectation of the excess risk, as stated in the following theorem.

Theorem 2. *If the assumptions from Lemma 1 hold, and $n \geq 16 \log \frac{8}{\delta}$, the expectation of the excess empirical risk $E\left[J(\boldsymbol{w}^{in}; D) - J(\hat{\boldsymbol{w}}; D)\right] = O\left(\frac{\zeta \|\hat{\boldsymbol{w}}\|_2 \sqrt{d \log(1/\delta)}}{\epsilon n}\right)$ by setting $\Delta = \Theta\left(\frac{\sqrt{\zeta^2 d \log(1/\delta)}}{\epsilon \|\hat{\boldsymbol{w}}\|_2}\right)$ and σ_u as the lowest value specified in Algorithm 1.*

4   Conclusion

In this study, we propose a novel framework for differentially private ERM, input perturbation. In contrast to objective perturbation, input perturbation allows data contributors to take part in the process of privacy preservation of model learning. From the privacy analysis of the data releasing of the data contributors, the data collection process in the input perturbation satisfies $(O(\sqrt{n}\epsilon, \delta)$-differential local privacy. Thus, from the perspective of data contributors, data collection with input perturbation can be preferable.

Acknowledgments. This work is supported by JST CREST program and is partly supported by JSPS KAKENHI No. 16H02864.

References

1. Bassily, R., Smith, A., Thakurta, A.: Private empirical risk minimization: Efficient algorithms and tight error bounds. In: Proceedings - Annual IEEE Symposium on Foundations of Computer Science, FOCS, pp. 464–473. IEEE, October 2014
2. Chaudhuri, K., Monteleoni, C., Sarwate, A.D.: Differentially private empirical risk minimization. J. Mach. Learn. Res. **12**, 1069–1109 (2011)
3. Duchi, J.C., Jordan, M.I., Wainwright, M.J.: Local privacy and statistical minimax rates. In: 2013 IEEE 54th Annual Symposium on Foundations of Computer Science (FOCS), pp. 429–438. IEEE (2013)
4. Dwork, C., Kenthapadi, K., McSherry, F., Mironov, I., Naor, M.: Our data, ourselves: privacy via distributed noise generation. In: Vaudenay, S. (ed.) EUROCRYPT 2006. LNCS, vol. 4004, pp. 486–503. Springer, Heidelberg (2006). doi:10.1007/11761679_29
5. Dwork, C., McSherry, F., Nissim, K., Smith, A.: Calibrating noise to sensitivity in private data analysis. In: Halevi, S., Rabin, T. (eds.) TCC 2006. LNCS, vol. 3876, pp. 265–284. Springer, Heidelberg (2006). doi:10.1007/11681878_14
6. Dwork, C., Roth, A., et al.: The algorithmic foundations of differential privacy. Found. Trends Theor. Comput. Sci. **9**(3–4), 211–407 (2014)
7. Evfimievski, A., Gehrke, J., Srikant, R.: Limiting privacy breaches in privacy preserving data mining. In: Proceedings of the twenty-second ACM SIGMOD-SIGACT-SIGART symposium on Principles of database systems, pp. 211–222. ACM (2003)

8. Jain, P., Thakurta, A.G.: (Near) dimension independent risk bounds for differentially private learning. In: Proceedings of The 31st International Conference on Machine Learning, pp. 476–484 (2014)
9. Kairouz, P., Oh, S., Viswanath, P.: Extremal mechanisms for local differential privacy. In: Advances in Neural Information Processing Systems, pp. 2879–2887 (2014)
10. Kasiviswanathan, S.P., Lee, H.K., Nissim, K., Raskhodnikova, S., Smith, A.: What can we learn privately? SIAM J. Comput. **40**(3), 793–826 (2011)
11. Kifer, D., Smith, A., Thakurta, A.: Private convex empirical risk minimization and high-dimensional regression. J. Mach. Learn. Res. **1**, 41 (2012)
12. Lei, J.: Differentially private m-estimators. In: Advances in Neural Information Processing Systems, pp. 361–369 (2011)
13. Wainwright, M.J., Jordan, M.I., Duchi, J.C.: Privacy aware learning. In: Advances in Neural Information Processing Systems, pp. 1430–1438 (2012)

Label Classification

On a New Competence Measure Applied to the Dynamic Selection of Classifiers Ensemble

Marek Kurzynski[✉] and Pawel Trajdos

Department of Systems and Computer Networks,
Wroclaw University of Science and Technology,
Wyb. Wyspianskiego 27, 50-370 Wroclaw, Poland
marek.kurzynski@pwr.edu.pl

Abstract. In this paper a new method for calculating the classifier competence in the dynamic mode is developed. In the method, first decision profile of the classified object is calculated using K nearest objects from the validation set. Next, the decision profile is compared with the support vector produced by the classifier. The competence measure reflects the outcome of this comparison and rates the classifier with respect to the similarity of its support vector and decision profile of the test object in a continuous manner. Three different procedures for calculating decision profile and three different measures for comparing decision profile and support vector are proposed, which leads to nine methods of competence calculation. Two multiclassifier systems (MC) with homogeneous and heterogeneous pool of base classifiers and with dynamic ensemble selection scheme (DES) were constructed using the methods developed. The performance of constructed MC systems was compared against seven state-of-the-art MC systems using 15 benchmark data sets taken from the UCI Machine Learning Repository. The experimental investigations clearly show the effectiveness of the combined multiclassifier system in dynamic fashion with the use of the proposed measures of competence regardless of the ensemble type used.

Keywords: Multiclassifier system · Dynamic ensemble selection · Measure of competence

1 Introduction

In the last two decades, multiclassifier (MC) systems which combine responses of set of classifiers have been intensively developed. The reason is that different classifiers offer complementary information about the object to be classified and therefore MC system can achieve better classification accuracy than any single classifier in the ensemble.

MC system has three general phases [2]: (1) generation in which the training set is used to generate a pool of classifiers; (2) selection in which a single classifier (or an ensemble of classifiers) is selected to perform the classification; (3)

© Springer International Publishing AG 2017
A. Yamamoto et al. (Eds.): DS 2017, LNAI 10558, pp. 93–107, 2017.
DOI: 10.1007/978-3-319-67786-6_7

combination (or integration) in which the final decision is made based on the predictions of the classifiers. It must be noted that selection and integration phases may be facultative, since for the classifier combination two main approaches used are classifier fusion and classifier selection [13]. In the first method, all classifiers in the ensemble contribute to the decision of the MC system, e.g. through sum or majority voting [11]. In the second approach, a single classifier is selected from the ensemble and its decision is treated as the decision of the MC system. The selection of classifiers can be either static or dynamic. In the static selection scheme, classifier is selected for all test objects, whereas dynamic classifier selection (DCS) approach explores the use of different classifiers for different test objects [6].

Recently, dynamic ensemble selection (DES) methods have been developed which first dynamically select an ensemble of classifiers from the entire set (pool) and then combine the selected classifiers by majority voting [3,4,12,18]. In this way a DES based system takes advantage of both selection and fusion approaches. In most methods, the base classifiers are selected from the pool on the basis of their individual accuracy measure called competence in a local region of the feature space. These methods differ in algorithms for determining classifier competence and ways of defining the local regions.

In [23] two methods were proposed where the local accuracy (competence) of classifier is calculated as a simple percentage of correct classified samples from the validation set. In the first method called OLA (overall local accuracy), local accuracy is calculated in the region containing K-nearest validation objects of a test object. Whereas in the LCA (local class accuracy) method, classifier competence is determined considering only these validation objects from the K-nearest neighbors set which belong to the same class into which an unknown object is assigned. In [20–22] two methods using probabilistic model were developed. The idea of the first method is based on relating the response of the classifier with the response obtained by random guessing. The measure of competence reflects this relation and rates the classifier with respect to random guessing in a continuous manner. In this way, it is possible to evaluate a group of classifiers against a common reference point. Competent (incompetent) classifiers gain with such approach meaningful interpretation, i.e. they are more (less) accurate than the random classifier. In the second method, first a randomized reference classifier (RRC) is constructed which, on average, acts like the classifier evaluated. Next the competence of the classifier evaluated is calculated as the probability of correct classification of the respective RRC. Two interesting methods called *A priori* and *A posteriori* selection scheme was presented in [9]. In the *A priori* method, a classifier is selected based on its accuracy within the local region, without considering the class assigned to the unknown pattern. Similarly, in the *A posteriori* method, local accuracies are estimated using the class posterior probabilities and the distances of the samples in the defined local region. In [17] an interesting ranking-based approach to determine competence measure was proposed. In the method the ranking of base classifiers is done by estimating parameters related to the correctness of the classifiers in the pool. An interesting method called

MCB (Multiple Classifier Behavior) was proposed in [10]. In this method the competence is defined as the classification accuracy calculated for a subset of a validation set which is generated as follows. First, the MCB is calculated for a test object and its K-nearest validation objects as a vector whose elements are class labels assigned by all classifiers in the ensemble. Next, similarity between the MCB's are calculated using the averaged Hamming distance. Finally, the objects in the validation set that are the most similar to the test object are used to generate the subset. The original KNORA-Eliminate (KE) method belonging to the category of oracle-based methods was proposed in [12]. The oracles are represented by the K-nearest neighbors of the unknown pattern in the validation set and the KE method selects only those classifiers which are able to recognize the entire K-neighborhood of the test pattern.

In this paper a new method for calculating the classifier competence in the feature space is presented. In the proposed method, first the so-called decision profile of the classified object is determined using K-nearest validation objects. The decision profile provides the chance that the recognized object belongs to the specified class. In the probabilistic model the natural concept of decision profile is based on *a posteriori* probabilities of classes at the point x. Next, the decision profile is compared with the response produced by the classifier (support vector or values of discriminant functions) [7] and the competence is calculated according to the similarity rule: the closer the response to the profile is, the more competent the classifier is [14,15]. Three different procedures for calculating a decision profile and three different measures for comparing the decision profile and the support vector are proposed in this study.

In a nutshell, originality of the proposed approach consists in a different use of the validation set. In the state-of-the-art-methods described above, the validation set is directly used for calculating local accuracy of a classifier (i.e. its local competence) via ranking-based, accuracy-based, probabilistic-based, behavior-based and oracle-based measures. However, in the proposed method, validation set is used for evaluating the classification profile of the test point and competence of the classifier is determined by similarity of its response to this evaluation.

The paper is divided into four sections and organized as follows. In Sect. 2 the measures of classifier competence are presented and two multiclassifier systems using proposed measures of competence in a dynamic fashion are developed. The performance of proposed MCS's were compared with seven multiple classifier systems using 15 datasets taken from the UCI Machine Learning Repository. The results of computer experiments are described in Sects. 3, and 4 concludes the paper.

2   Multiclassifier System

2.1   Preliminaries

In the multiclassifier (MC) system we assume that a set of trained classifiers $\Psi = \{\psi_1, \psi_2, \ldots, \psi_L\}$ called base classifiers is given.

A classifier ψ_l is a function $\psi_l : \mathcal{X} \to \mathcal{M}$ from a metric feature space $\mathcal{X} \subseteq \mathbb{R}^{dim}$ to a set of class labels $\mathcal{M} = \{1, 2, \ldots, M\}$. Classification is made according to the maximum rule

$$\psi_l(x) = i \Leftrightarrow d_{li}(x) = \max_{j \in \mathcal{M}} d_{lj}(x), \tag{1}$$

where $[d_{l1}(x), d_{l2}(x), \ldots, d_{lM}(x)]$ is a vector of class supports (classifying function) produced by ψ_l. Without loss of generality we assume that $d_{lj}(x) \geq 0$ and $\sum_j d_{lj}(x) = 1$.

In this paper, we propose MC systems which use a dynamic ensemble selection scheme and trainable combining methods based on a competence measure $c(\psi_l|x)$ of each base classifier ($l = 1, 2, \ldots, L$) evaluating the competence of classifier ψ_l at a point $x \in \mathcal{X}$. Competence measure is normalized, i.e. $0 \leq c(\psi_l|x) \leq 1$. $c(\psi_l|x) = 0(1)$ denotes the most incompetent (competent) classifier ψ_l.

For the training methods of combining the base classifiers, it is assumed that a validation set

$$\mathcal{V} = \{(x_1, j_1), (x_2, j_2), \ldots, (x_N, j_N)\}; \quad x_k \in \mathcal{X}, \ j_k \in \mathcal{M} \tag{2}$$

containing pairs of feature vectors and their corresponding class labels is available.

2.2 Measure of Competence

K-neighborhood. Let first introduce the concept of K-neighborhood of object $x \in \mathcal{X}$ which is defined as the set of K nearest neighbors of the point x from validation set $\mathcal{V}$, viz.

$$\mathcal{S}_K(x) = \{x_{n_1}, x_{n_2}, \ldots x_{n_K} \in \mathcal{V} :$$

$$\max_{k=1,2,\ldots,K} ||x_{n_k} - x|| \leq \min_{x_l \notin \mathcal{S}_K(x)} ||x_l - x||\}, \tag{3}$$

where $|| \cdot ||$ denotes the distance in the feature space $\mathcal{X}$. The neighborhood size K is a parameter of the method – its value can be selected experimentally.

Decision Profile. Decision profile of object $x \in \mathcal{X}$

$$\delta(x) = [\delta_1(x), \delta_2(x), \ldots, \delta_M(x)], \ \delta_j(x) \geq 0, \ \sum_j \delta_j(x) = 1 \tag{4}$$

denotes the vector of normalized values where the jth value $\delta_j(x)$ is interpreted as a measure of chance that object x belongs to the jth class ($j \in \mathcal{M}$). In the probabilistic model the natural value of $\delta_j(x)$ is *a posteriori* probability of jth class at the point x.

We propose the following methods for calculating decision profile at the point x using its K-neighborhood.

The Fraction-based Method (FM)

In this approach, $\delta_j(x)$ is calculated as the fraction of objects from the jth class in the set $\mathcal{S}_K(x)$. Let $M_j^{(K)}(x)$ be the number of validation objects from $\mathcal{S}_K(x)$ belonging to the j-th class. Then

$$\delta_j(x) = \frac{M_j^{(K)}(x)}{K}, \quad j \in \mathcal{M}. \tag{5}$$

The Ranking-based Method (RM)

In the RM method $\delta_j(x)$ is equal to the normalized sum of ranks of objects belonging to the jth class in the set $\mathcal{S}_K(x)$. Let $r(x_k)$ be the rank of validation object $x_k \in \mathcal{S}_K(x)$. The nearest neighbour has the rank equal to K, the rank of the furthest neighbor is equal to 1. Then

$$r_j(x) = \sum_{x_k \in \mathcal{S}_K(x):j_k=j} r(x_k) \tag{6}$$

is the sum of ranks of validation objects from the K neighborhood of x belonging to the jth class. And next

$$\delta_j(x) = \frac{r_j(x)}{\sum_{k=1}^{K} k}. \tag{7}$$

The Potential Function Method (PM)

Let $H(x, x_k)$ be a non-negative potential function [16] decreasing with the increasing distance between x and x_k. In this study, a Gaussian potential function with the Euclidean distance is used:

$$H(x, x_k) = \exp(-||x - x_k||^2). \tag{8}$$

Then, we can calculate $\delta_j(x)$ as a normalized sum of potential functions (8) for objects belonging to the jth class from the set $\mathcal{S}_K(x)$, namely:

$$\delta_j(x) = \frac{\sum_{x_k \in \mathcal{S}_K(x):j_k=j} \exp(-||x - x_k||^2)}{\sum_{j \in \mathcal{M}} \sum_{x_k \in \mathcal{S}_K(x):j_k=j} \exp(-||x - x_k||^2)}. \tag{9}$$

Distance Between Decision Profile and Vector of Supports. In order to evaluate ψ_l at x and determine its competence $c(\psi_l|x)$, we must compare decision profile $\delta(x)$ and vector of supports $d_l(x)$ and calculate distance $dist[\delta(x), d_l(x)]$. Competence measure is a normalized function of this distance decreasing with the increasing distance between $\delta(x)$ and $d_l(x)$. In particular, $c(\psi_l|x)$ is equal to 1 (0) if distance is equal to 0 (is the greatest one).

We propose three different methods for calculating distance $dist[\delta(x), d_l(x)]$ and the resulting measures of competence.

Euclidean Distance (ED)

We adopt the Euclidean distance

$$dist[(\delta(x), d_l(x)] = ||\delta(x) - d_l(x)|| \tag{10}$$

and hence we get

$$c(\psi_l|x) = \frac{\sqrt{2} - ||\delta(x) - d_l(x)||}{\sqrt{2}}. \tag{11}$$

Max-Max Distance (MD)

Let j be the class number for which classifier ψ_l produced the greatest support value at the point x (i.e. $d_{lj}(x) = \max_{k \in \mathcal{M}}(d_{lk}(x))$). Similarly, let i be the class number with the greatest value in the decision profile $\delta(x)$ at x. Then, the max–max distance is defined as:

$$dist[(\delta(x), d_l(x)] = |d_{lj}(x) - \delta_j(x)| + |d_{li}(x) - \delta_i(x)|. \tag{12}$$

Hence we have the following formula for competence measure:

$$c(\psi_l|x) = \frac{2 - |d_{lj}(x) - \delta_j(x)| + |d_{li}(x) - \delta_i(x)|}{2}. \tag{13}$$

Hamming Distance (HD)

Let $h(\psi_l(x)) = [j_1', j_2', \ldots, j_M']$ and $h(x) = [j_1'', j_2'', \ldots, j_M'']$ be the vectors of class numbers ordered according to the decreasing values of supports produced by ψ_l at x and decision profile of x, respectively. Distance between $d_l(x)$ and $\delta_j(x)$ is defined as the Hamming distance between vectors $h(\psi_l(x))$ and $h(x)$, namely

$$dist[(\delta(x), d_l(x)] = D_H[h(\psi_l(x)), h(x)]. \tag{14}$$

Hence we get the following form of competence measure

$$c(\psi_l|x) = \frac{M - D_H[h(\psi_l(x)), h(x)]}{M}. \tag{15}$$

Example. Consider a classification problem with three classes ($M = 3$). Figure 1 presents 6-neighborhood of an object x in the two-dimensional feature space. Additional unit grid will help to determine distances between objects. Suppose that classifier ψ produced supports $d_1(x) = 0.3$, $d_2(x) = 0.6$ and $d_3(x) = 0.1$. Our purpose is to determine the competence $c(\psi|x)$ of the classifier ψ at the point x using presented methods.

From Fig. 1 we simply get the Euclidean distances between x and validation objects:

$$||x - x_1|| = 2, \ ||x - x_2|| = 5, \ ||x - x_3|| = 2.83,$$
$$||x - x_4|| = 2.23, \ ||x - x_5|| = 3, \ ||x - x_6|| = 3,6.$$

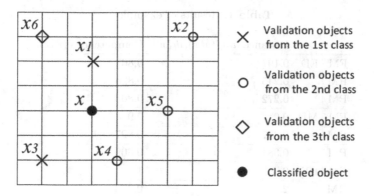

Fig. 1. Illustration of Example: 6-neighborhood of an object x.

First, we calculate decision profiles for the proposed methods.
FM method:

$$M_1^{(5)} = 2, M_2^{(5)} = 3, M_3^{(5)} = 1 \text{ and hence } \delta_1(x) = 1/3, \delta_2(x) = 1/2, \delta_3(x) = 1/6.$$
(16)

RM method:

$$r_1(x) = 10, \ r_2(x) = 9, \ r_3(x) = 2$$

and hence

$$\delta_1(x) = 10/21, \ \delta_2(x) = 9/21, \ \delta_3(x) = 2/21.$$
(17)

PM method:

$$H(x, x_1) + H(x, x_3) = 0.135 + 0.059 = 0.194, \ H(x, x_6) = 0.027$$

$$H(x, x_2) + H(x, x_3) + H(x, x_5) = 0.006 + 0.108 + 0.049 = 0.163$$

and hence

$$\delta_1(x) = \frac{0.194}{0.384} = 0.505, \ \delta_2(x) = \frac{0.163}{0.384} = 0.424, \ \delta_3(x) = \frac{0.027}{0.384} = 0.071, \quad (18)$$

Now, using formulas (10)–(15) and calculated decision profiles (16), (17) and (18), we can calculate competence $c(\psi|x)$ of classifier ψ at the point x. Results for all combination of calculating decision profile methods and concept of distance between decision profile and vector of supports are presented in Table 1.

2.3   DES Systems

The proposed measure of competence can be applied in any multiclassifier system in selection/fusion algorithm provided that the feature space $\mathcal{X}$ is a metric space. In this subsection we describe two multiclassifier systems based on the DES strategy.

Table 1. Results of example.

		Distance $dist[(\delta(x), d_l(x)]$	Competence $c(\psi|x)$
FM	ED	0.197	0.86
RM		0.245	0.826
PM		0.272	0.808
FM	MD	0.2	0.9
RM		0.347	0.827
PM		0.381	0.809
FM	HD	0	1
RM		2	0.333
PM		2	0.333

Multiclassifier System with Fusion at the Decision Level (MC1). In this system, first a subset $\Psi^*(x)$ of base classifiers with the competences greater than the random guess is selected for a given x:

$$\Psi^*(x) = \{\psi_{l1}, \psi_{l2}, \ldots, \psi_{lT}\}, \quad \text{where} \quad c(\psi_{lt}|x) > 1/M. \tag{19}$$

The selected classifiers are combined using the weighted majority voting rule where the weights are equal to the competences. This fusion method leads to the following class supports ($j = 1, 2, \ldots M$):

$$d_j^{(MC1)}(x) = \sum_{t=1}^{T} c(\psi_{lt}|x) \lfloor \psi_{lt}(x) = j \rfloor, \tag{20}$$

where $\lfloor \cdot \rfloor$ denotes the Iverson bracket.

The MC1 system $\psi^{(MC1)}$ classifies x using the maximum rule:

$$\psi^{(MC1)}(x) = i \Leftrightarrow d_i^{(MC1)}(x) = \max_{j \in \mathcal{M}} d_j^{(MC1)}(x). \tag{21}$$

Multiclassifier System with Fusion at the Continuous-Value Level (MC2). The MC2 system is identical to the MC1 system except that selected classifiers (19) are combined at the continuous-value level ($j = 1, 2, \ldots M$):

$$d_j^{(MC2)}(x) = \sum_{t=1}^{T} c(\psi_{lt}|x) d_{lt,j}(x). \tag{22}$$

Final decision – as previously – is made according to the maximum rule:

$$\psi^{(MC2)}(x) = i \Leftrightarrow d_i^{(MC2)}(x) = \max_{j \in \mathcal{M}} d_j^{(MC2)}(x). \tag{23}$$

The MC2 system with competence measures developed will be applied in the experimental investigations.

3 Experiments

3.1 Experimental Setup

The performance of the developed MC systems was evaluated in experiments using 15 benchmark data sets. In the first experiment, the MC2 system was evaluated using different methods for calculating decision profile (FM, RM and PM) and different distances between decision profile and support vector (ED, MD and HD). The methods that showed the best performance were identified. In the second experiment, the methods identified were compared with other competence–based MC systems. The experiments were conducted in MATLAB using PRTools 4.1 [8]. In both experiments, the value of $K = 5 \times M$ (M denotes the number of classes) was used as the neighborhood size.

The 15 benchmark data sets were taken from the UCI Machine Learning Repository [1]. We selected the same data sets which were used in experimental investigations presented in [21]. A brief description of the data sets used is given in Table 2.

Table 2. The data sets used in the experiments.

Data set	#Objects	#Features	#Classes
Blood transfusion	748	4	2
Breast cancer Wisconsin	699	9	2
Clouds	5000	2	2
Dermatology	366	34	6
EColi	336	7	8
Glass	214	9	6
Ionosphere	351	34	2
OptDigits	3823	64	10
Page blocks	5473	10	5
Pima Indians	768	8	2
Segmentation	2310	19	7
Spam	4601	57	2
Vowel	990	10	11
Wine	178	13	3
Yeast	1484	8	10

For each data set, feature vectors were normalized to zero mean and unit standard deviation. Two-fold cross-validation was used to extract training and test sets from each data set. For the calculation of the competences, a two-fold stacked generalization method was used [24]. In the method, the training set is split into two sets A and B of roughly equal sizes. Set A is first used for the training of the classifiers in the ensemble while set B is used for the calculation of the competences. Then, set B is used for the training while the competences are calculated using set A. Finally, the competences calculated for both sets are stacked together and the classifiers in the ensemble are trained using the union of sets A and B (i.e. the original training set). In this way, the competences of the classifiers are calculated for all the feature vectors in the original training set, but the data used for the calculation is unseen during the classifier training.

The experiments were conducted using two ensemble types: homogeneous and heterogeneous. The homogeneous ensemble consisted of 50 feed-forward back-propagation neural network classifiers with one hidden layer and the maximum number of learning epochs set to 80. Each neural network classifier was trained using randomly selected 70% of the objects from the training data set. The heterogeneous ensemble consisted of the following 11 base classifiers [7]:

- (1) linear classifier based on normal distribution with the same covariance matrix for each class;
- (2) quadratic classifier based on normal distribution with different covariance matrix for each class;
- (3) nearest mean classifier;
- (4–6) k-nearest neighbors classifiers with k = 1, 5, 10;
- (7, 8) Parzen density based classifier with the Gaussian kernel and the optimal smoothing parameter h_{opt} (and the smoothing parameter $h_{opt}/2$);
- (9) pruned decision tree classifier with the Gini splitting criterion;
- (10–11) feed-forward backpropagation neural network classifier containing one hidden layer with 10 neurons (two hidden layers with 5 neurons each) and the maximum number of learning epochs set to 80;

The performance of the systems constructed was compared with the following seven MC systems:

1. **Overall local accuracy method (OLA1)** [23]. In this method the competence at a test point x is calculated as the percentage of the correct recognition of the K-nearest validation samples of x;
2. **Local class accurracy method (LCA)** [23]. In this method the competence is estimated for each base classifier as the percentage of correct classifications within the local region (the K neighborhood), but considering only examples from the class as classifier gives for the unknown pattern;
3. **Overall local accuracy method (OLA2)** [19]. In this method the competence is calculated as in OLA1 approach but validation objects from the K-neighborhood are additionally weighted by their Euclidean distances to the unknown object x;
4. **Multiple classifier behavior method (MCB)** [10]. In this method the competence is calculated using a similarity function to measure the degree of similarity of the output profiles of all base classifiers;
5. **Oracle KNORRA-eliminate method (ORE)** [12]. In this method all classifiers are selected that correctly classify all samples in the local region (the K neighborhood). If no classifiers are selected, the local region is reduced until at least one classifier is selected;

6. **Randomized reference classifier method (RRC)** [21]. In this method the competence of base classifier is calculated as the probability of correct classification of randomized reference classifier (RRC) which - on average - acts as a modeled base classifier;

7. **Random guessing based method (RGM)** [22]. In this method the competence is calculated in relation to the random guessing method – the classifier is considered as competent (incompetent) if it is more (less) accurate than the random classifier.

3.2 Results and Discussion

Classification accuracies were averaged over 5 repetitions of two-fold cross-validation. Statistical differences in rank between the systems were obtained using the Friedman test with Iman and Davenport correction combined with the post hoc Holm's stepdown procedure [5]. The average ranks of the systems and a critical rank difference calculated using the Bonferroni-Dunn test are visualised. The level of $p < 0.05$ is considered as statistically significant.

The average ranks obtained from the first experiment for the nine methods proposed and for the homogeneous and heterogeneous ensembles are presented in Figs. 2A and B, respectively. The use of the potential function method for calculating decision profile and max-max distance between decision profile and support vector (PM-MD) resulted in the best average rank regardless of the ensemble type used. The average rank of PM-MD method is significantly better than average ranks for FM-MD, RM-MD, PM-HD, FM-HD and RM-HD methods. Methods with the Hamming distance achieved the worst average ranks regardless of the method for calculating decision profile and the ensemble type used. Thus, for the second experiment the PM-MD method was selected.

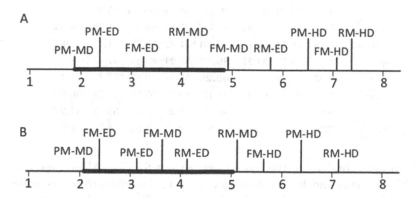

Fig. 2. Average ranks of the MC2 systems for different methods of calculating decision profile and different distances between decision profile and support vector for homogeneous (A) and heterogeneous (B) ensemble of base classifiers. The interval (thick line) is the critical rank difference (2.991) calculated using the Bonferroni-Dunn test ($p < 0.05$).

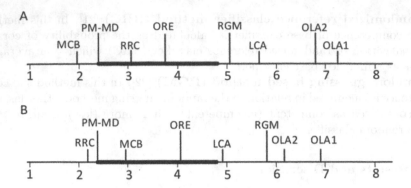

Fig. 3. Average ranks of MC systems compared for homogeneous (A) and heterogeneous (B) ensemble of base classifiers. The interval (thick line) is the critical rank difference (2.394) calculated using the Bonferroni-Dunn test ($p < 0.05$).

Table 3. Classification accuracies (in percent) and average ranks of the PM-MD system and the eight MCSs for the homogeneous ensemble. The best result for each data set is in bold.

Dataset	OLA1	LCA	OLA2	MCB	ORE	RRC	RGM	PM-MD
Blood	76.18	75.29	76.44	**77.15**	76.28	76.44	76.32	77.02
Breast	94.21	93.88	94.92	97.25	95.29	**97.88**	95.94	97.12
Clouds	62.92	63.15	64.22	**65.80**	64.00	63.64	63.82	65.27
Dermat	68.88	70.44	69.35	**75.28**	72.61	73.91	71.45	74.82
EColi	67.62	69.48	70.35	76.44	71.43	**78.01**	76.22	77.25
Glass	50.72	54.22	52.83	**69.95**	57.86	67.22	60.28	67.38
Iono	83.95	84.25	84.15	**88.57**	85.47	87.62	86.15	86.47
OptDig	81.28	86.55	87.12	88.32	86.48	88.21	87.35	**89.42**
Page	94.92	95.12	95.23	96.21	95.80	**96.35**	95.05	95.92
Pima	65.23	65.55	64.92	67.48	65.49	66.30	65.78	**68.52**
Segment	87.88	86.54	88.75	**96.23**	91.32	95.72	91.44	94.96
Spam	81.84	83.57	84.21	**89.12**	85.29	88.85	87.17	88.24
Vowel	49.92	53.22	50.5	**60.25**	55.86	59.45	57.73	**61.15**
Wine	91.28	93.15	91.22	94.47	89.52	95.84	92.78	**96.03**
Yeast	49.85	53.22	50.81	56.28	52.28	56.36	55.42	**57.27**
Av.Rank	7.26	5.68	6.83	1.98	3.79	3.06	5.05	2.35

The results obtained in the second experiment for the PM-MD method and seven MC systems and for the homogeneous and heterogeneous ensembles are presented in Tables 3 and 4 and in Figs. 3A and B, respectively. The system constructed achieved the second best average ranks for both types of classifier ensemble. The average rank of PM-MD method is significantly better than average ranks for LCA, RGM, OLA1 and OLA2 methods regardless of the ensemble type used.

Table 4. Classification accuracies (in percent) and average ranks of the PM-MD system and the eight MCSs for the heterogeneous ensemble. The best result for each data set is in bold.

Data base	OLA1	LCA	OLA2	MCB	ORE	RRC	RGM	PM-MD
Blood	75.70	77.23	75.93	77.43	76.48	**78.26**	77.58	78.12
Breast	95.12	95.84	95.48	95.17	**96.42**	96.28	94.87	95.54
Clouds	74.50	75.13	74.21	79.27	77.12	79.07	75.12	**80.02**
Dermat	93.18	93.28	93.42	**96.31**	94.82	96.27	93.88	95.92
EColi	82.13	84.55	82.86	84.18	86.12	**86.24**	83.15	84.88
Glass	64.18	65.28	64.05	67.40	67.15	67.35	64.88	**67.48**
Iono	82.98	83.17	82.75	86.12	85.94	**86.95**	83.15	86.92
OptDig	87.92	91.15	88.24	95.31	95.15	97.43	90.65	**97.48**
Page	89.24	92.38	90.82	95.84	95.21	**96.24**	91.13	96.18
Pima	67.21	68.89	67.15	69.12	68.73	**69.45**	67.45	69.32
Segment	84.02	87.55	86.58	96.41	95.11	95.32	89.55	**97.12**
Spam	88.21	89.45	88.85	**92.17**	90.32	91.91	90.05	91.72
Vowel	82.24	85.92	84.72	88.32	83.51	**90.18**	85.77	89.71
Wine	95.42	96.41	96.15	97.05	96.84	97.17	96.32	**98.03**
Yeast	55.66	57.05	55.51	56.94	56.83	**57.79**	57.12	57.11
Av.Rank	6.96	4.98	6.24	2.98	4.16	2.30	5.87	2.48

4   Conclusion

Nowadays, many researches have been focused on MC systems and consequently, many new solutions have been dedicated to each of the two main approaches: classifiers fusion and classifiers selection. In the proposed solutions the fundamental role plays the assessment of competence of base classifiers which is crucial in the DES scheme and in the combining of base classifiers. In the paper a new method for calculating the competence of a classifier in the feature space was presented. In the proposed method, first the K-neighborhood is used to determine the so-called decision profile of a test object. The decision profile is an evaluation of the chance that the recognized object belongs to particular classes. Next, the decision profile is compared with the response produced by the classifier and the competence is calculated according to the similarity rule. The MC systems with DES scheme using the proposed competence measure were developed and experimentally evaluated using 15 benchmark datasets. Experimental results showed that the idea of calculating the competence of a classifier by comparing its response with the decision profile of the classified object is a correct method and leads to the accurate and efficient multiclassifier systems.

Acknowledgment. This work was supported by the statutory funds of the Department of Systems and Computer Networks, Wroclaw University of Technology.

References

1. Lichman, M.: UCI Machine Learning Repository. University of California, School of Information and Computer Science, Irvine (2013). http://archive.ics.uci.edu/ml
2. Britto, A., Sabourin, R., de Oliveira, L.: Dynamic selection of classifiers - a comprehensive review. Pattern Recogn. **47**(11), 3665–3680 (2014)
3. Cavalin, P., Sabourin, R., Suen, C.: Dynamic selection approaches for multiple classifier systems. Neural Comput. Appl. **22**(3–4), 673–688 (2013)
4. Cruz, R., Sabourin, R., et al.: META-DES: a dynamic ensemble selection framework using meta-learning. Pattern Recogn. **48**, 1925–1935 (2015)
5. Demsar, J.: Statistical comparison of classifiers over multiple data sets. J. Mach. Learn. Res. **7**, 1–30 (2006)
6. Didaci, L., Giacinto, G., Roli, F., Marcialis, G.: A study of the performance of dynamic classifier selection based on local accuracy estimation. Pattern Recogn. **38**, 2188–2191 (2005)
7. Duda, R., Hart, P., Stork, D.: Pattern Classification. Wiley-Interscience, New York (2001)
8. Duin, R., Juszczak, P., et al.: PR-Tools 4.1, A Matlab Toolbox for Pattern Recognition. Delft University of Technology (2007). http://prtools.org
9. Giacinto, G., Roli, F.: Methods for dynamic classifier selection. In: Proceedings of the 10th International Conference on Image Analysis and Processing, pp. 659–664 (1999)
10. Giacinto, G., Roli, F.: Dynamic classifier selection based on multiple classifier behaviour. Pattern Recogn. **34**, 1879–1881 (2001)
11. Kittler, J., Hatef, M., Duin, R., Matas, J.: On combining classifier. IEEE Trans. Pattern Anal. Mach. Intell. **PAMI–20**, 226–239 (1998)
12. Ko, A., Sabourin, R., Britto, A.: From dynamic classifier selection to dynamic ensemble selection. Pattern Recogn. **41**(5), 1718–1731 (2008)
13. Kuncheva, L.: Combining Pattern Classifiers: Methods and Algorithms. Wiley-Interscience, Hoboken (2004)
14. Kurzynski, M.: On a new competence measure applied to the combining multiclassifier system. Int. J. Sig. Process. Syst. **4**(3), 185–191 (2016)
15. Kurzynski, M., Krysmann, M., Trajdos, P., Wolczowski, A.: Multiclassifier system with hybrid learning applied to control of bioprosthetic hand. Comput. Biol. Med. **69**, 286–297 (2016)
16. Meisel, W.: Potential functions in mathematical pattern recognition. IEEE Trans. Comput. **C–18**, 911–918 (1969)
17. Sabourin, M., Mitiche, A., Thomas, D., Nagy, G.: Classifier combination for hand-printed digit recognition. In: Proceedings of the 2nd International Conference on Document Analysis and Recognition, pp. 163–166 (1993)
18. dos Santos, E., Sabourin, R., Maupin, P.: A dynamic over produce-and-choose strategy for the selection of classifier ensembles. Pattern Recogn. **41**(10), 2993–3009 (2008)
19. Smits, P.: Multiple classifier systems for supervised remote sensing image classification based on dynamics classifier selection. IEEE Trans. Geosci. Remote Sens. **40**, 801–813 (2002)
20. Woloszynski, T., Kurzynski, M.: On a new measure of classifier competence applied to the design of multiclassifier systems. In: Foggia, P., Sansone, C., Vento, M. (eds.) ICIAP 2009. LNCS, vol. 5716, pp. 995–1004. Springer, Heidelberg (2009). doi:10.1007/978-3-642-04146-4_106

21. Woloszynski, T., Kurzynski, M.: A probabilistic model of classifier competence for dynamic ensemble selection. Pattern Recogn. **44**, 2656–2668 (2011)
22. Woloszynski, T., Kurzynski, M., et al.: A measure of competence based on random classification for dynamic ensemble selection. Inf. Fusion **13**, 207–213 (2012)
23. Woods, K., Kegelmeyer, W., Bowyer, K.: Combination of multiple classifiers using local accuracy estimates. IEEE Trans. Pattern Anal. Mach. Intell. **PAMI–19**, 405–410 (1997)
24. Wolpert, D.: Stacked generalization. Neural Netw. **5**, 214–259 (1992)

Multi-label Classification Using Random Label Subset Selections

Martin Breskvar[1,2(✉)], Dragi Kocev[1,2], and Sašo Džeroski[1,2]

[1] Department of Knowledge Technologies, Jožef Stefan Institute, Ljubljana, Slovenia
[2] Jožef Stefan International Postgraduate School, Ljubljana, Slovenia
Martin.Breskvar@ijs.si

Abstract. In this work, we address the task of multi-label classification (MLC). There are two main groups of methods addressing the task of MLC: problem transformation and algorithm adaptation. Methods from the former group transform the dataset to simpler local problems and then use off-the-shelf methods to solve them. Methods from the latter group change and adapt existing methods to directly address this task and provide a global solution. There is no consensus on when to apply a given method (local or global) to a given dataset. In this work, we design a method that builds on the strengths of both groups of methods. We propose an ensemble method that constructs global predictive models on randomly selected subsets of labels. More specifically, we extend the random forests of predictive clustering trees (PCTs) to consider random output subspaces. We evaluate the proposed ensemble extension on 13 benchmark datasets. The results give parameter recommendations for the proposed method and show that the method yields models with competitive performance as compared to three competing methods.

Keywords: Multi-label classification · Structured outputs · Output space decomposition · Predictive clustering trees · Ensemble methods

1 Introduction

Supervised learning is a very actively researched area of machine learning. Its goal is to learn models able to provide predictions for previously unseen examples of data. Single-target prediction scenarios are very common and applicable in many domains. However, not all solutions to problems can be *fitted* into one predicted variable. It is very possible that a more complex representation of the data is needed. This is a challenge because it requires methods to predict more than one variable of interest. In that sense, we move towards structured output prediction (SOP) tasks. Examples of SOP tasks are MT regression (MTR), multi-label classification (MLC), time series prediction etc.

This work focuses on solving the MLC task where a given example can be annotated with one or more labels. For instance, a gene could have more than one function, an image can contain different objects, a document can belong to several categories, a disease can manifest with multiple symptoms, etc. This

© Springer International Publishing AG 2017
A. Yamamoto et al. (Eds.): DS 2017, LNAI 10558, pp. 108–115, 2017.
DOI: 10.1007/978-3-319-67786-6_8

particular area of research attracts the attention of the community due to the increasing number of possible applications in various domains (multimedia, biology, medicine, semantic web, legislation,...). Traditional MLC approaches consider individual labels separately, i.e., they are local and transform the dataset into multiple single-label datasets (a dataset for each label) and then solve the multiple single-label tasks with off-the-shelf methods. The key observation here is that such approaches assume that labels are not related: If label relations exist, these approaches are not able to take advantage of their knowledge. Therefore, MLC approaches should be global and exploit potential relations between labels to produce more accurate models.

Notwithstanding, given a dataset, it is not clear which type of method one should use: a local or a global. There is no consensus on this issue [6]. On some datasets, it is preferable to use local, while on other global methods. Having this in mind, we believe that the best method should combine the advantages of both groups. We hence propose a method for MLC that randomly samples the output/label space and learns global models for the sampled label space. Furthermore, we combine the multiple models into an ensemble.

Output space selection and transformation methods already exist in the scope of MLC. One of the most well-known methods is Random k-Labelsets (RAkEL) [8]. It is a problem transformation method as it constructs an ensemble of ST classification models to solve the task of MLC. It does so by selecting random subset of labels (size is determined by the k parameter) for each base model. RAkEL then builds a powerset of the selected subset of labels and trains a ST classification model on it. This approach has been extended towards data-driven partitioning of the label space, which is achieved by using community detection algorithms from social networks [7]: These find better label subspaces as opposed to randomly selecting them. Another data-driven approach uses label hierarchies obtained by hierarchical clustering of flat label sets by using annotations that appear in the training data [5]. Finally, a dimensionality reduction method that uses random forests with Gaussian subspaces has been proposed [3]. This method also belongs to the algorithm adaptation group. It reduces the output space by making random projections of the output space into a new space which represents a highly compressed version of the original label space.

2   MLC Using Random Label Subset Selections

The proposed method is based on the predictive clustering (PC) framework. More specifically, we use predictive clustering trees (PCTs) that can be seen as a generalization of decision trees for the task of structured output prediction. The standard top-down induction of decision tree (TDIDT) algorithm is used to generate PCTs. The pseudo code for the randomized PCT induction algorithm (RPCT) is shown on the left side of Table 1 and it takes the following inputs: (i) a dataset S, (ii) a function $\delta_c(X)$ that randomly samples c descriptive variables from dataset X without replacements and (iii) a set of attributes R_t, that the learning process should use for supervision.

The RPCT algorithm first randomly samples from the pool of all available descriptive attributes for the current dataset. The sampled descriptive attributes, along with the target attributes R_t provided as input, are used to calculate the best possible split point (i.e., the best test) to use for partitioning the data instances. After the best test is found the data are split according to it. This process continues recursively until a stopping criterion is met and the prototype function is invoked. We use a prototype function that returns a vector of probabilities that an example belongs to the positive class for each target variable.

The test selection is handled by the *BestTest* function: It begins by removing the target attributes which should not be considered (Table 1, right, line 2). $\Pi(S, R_d, R_t)$ is a projection function that reduces the original dataset S to S_R by only considering descriptive and target attributes from sets R_d and R_t respectively. All possible tests on S_R are evaluated and the one that reduces the variance the most (w.r.t. S_R) is selected (Table 1, right, lines 3–9). The variance calculation function is also a parameter and can be instantiated based on the type of machine learning task we want to solve. In this paper, we focus on MLC so we calculate the variance as the sum of Gini indices over the individual target variables from the set $\Lambda = \{\lambda_1, \lambda_2, ..., \lambda_q\}$ as $Var(S) = \sum_{i=1}^{q} Gini(S, \lambda_i)$.

Ensembles combine the predictions of multiple predictive models to achieve better predictive performance. Predictions for new examples are made by querying base models and combining their predictions. In this section, we describe the process of generating ensembles, where the base models are not all learned from all available target attributes, but rather each model is learned from a (different) subset of them. For this, we will need the parameter R_t defined above. We named this ensemble method Random Output Selections (ROS).

Regular PCTs use the whole target space to calculate the heuristic score. The proposed ensemble approach introduces random selections in the output

Table 1. The top-down induction of randomized predictive clustering trees

Function $RPCT(S, \delta_c, R_t)$	**Function** $BestTest(S, R_d, R_t)$
Out: A predictive clustering tree	**Out:** Selected test t^*
	Out: Heuristic score h^* of test t^*
1: $R_d \leftarrow \delta_c(S)$	**Out:** Partitioning $\mathcal{P}^*$ induced by t^* on S
2: $(t^*, h^*, \mathcal{P}^*) \leftarrow \text{BestTest}(S, R_d, R_t)$	1: $(t^*, h^*, \mathcal{P}^*) \leftarrow (none, 0, \emptyset)$
3: **if** $t^* \neq none$ **then**	2: $S_R \leftarrow \Pi(S, R_d, R_t)$
4: **for each** $S_i \in \mathcal{P}^*$ **do**	3: **for each** possible test t in S_R **do**
5: $tree_i \leftarrow \text{RPCT}(S_i, \delta_c, R_t)$	4: $\mathcal{P} \leftarrow$ partitioning induced by t on S_R
6: **end for**	5: $h \leftarrow Var(R_t, S_R) - \sum_{S_i \in \mathcal{P}} \frac{|S_i|}{|S_R|} Var(R_t, S_i)$
7: **return** $node(t^*, \bigcup_i \{tree_i\})$	6: **if** $(h > h^*)$ **then**
8: **else**	7: $(t^*, h^*, \mathcal{P}^*) \leftarrow (t, h, \mathcal{P})$
9: **return** $leaf(Prototype(S))$	8: **end if**
10: **end if**	9: **end for**
	10: **return** $(t^*, h^*, \mathcal{P}^*)$

space, i.e., individual PCTs do not consider the whole target space anymore. Each base model (PCT) is consequently learned from only those targets that were included in the randomly generated partition R_t provided to it by the function Π. The output space partitions are generated before the induction of base models and are independent of the base model learning algorithm. The algorithm for construction of subspaces has the following parameters: (i) the number of base models b, (ii) a function $\theta_v(X)$ that samples uniformly at random without replacement v items from the set X and (iii) a set of target attributes (labels) T. ROS first creates a subspace which considers all target attributes, to make sure that every target attribute is considered by at least one base model. We generate the remaining $b-1$ subspaces with the θ_v function. We build ROS ensembles of PCTs by using the randomized PCT algorithm (RPCT). Each base model is learnt from different bootstrap replicate. Such perturbations of the learning set have been proven useful in cases, where unstable base models, such as decision trees, are used. RPCT introduces additional randomization while learning its individual base models by considering only a subset of descriptive attributes at each step, i.e., when selecting the best test at a given node by calling the function $\delta_c(X)$ just before. In addition, ROS randomly selects a subset of targets for each PCT in the ensemble (we refer to the method as RF-ROS).

Ensembles combine predictions of their base models. In this study, we use two different prediction-combining techniques, i.e., aggregation functions: (i) total averaging (i.e., the most commonly used voting technique) and (ii) subspace averaging. Total averaging combines votes of the individual base models using probability per-target distribution voting for all targets [1]. Subspace averaging does the same, but only the labels considered during learning of the respective base model participate in the voting.

3   Experimental Design

This section presents the experimental questions posed, benchmark datasets, the experimental setup and the evaluation measures used. We designed the experimental evaluation having the following research questions in mind:

1. What is the recommended label subspace size for RF-ROS ensembles?
2. Does it make sense to change the aggregation function, i.e., can subspace averaging improve the predictive performance of RF-ROS models?
3. Considering predictive performance, how do RF-ROS ensembles compare to other competing methods?

We use 13 publicly available benchmark datasets: Emotions, Scene, Yeast, Birds, TMC 2007, Genbase, Medical, Enron, Mediamill, Bibtex, Bookmarks, Corel 5k, and Delicious. The datasets vary in terms of number instances, descriptive and target attributes. More details about the datasets are available at the MULAN repository (http://mulan.sourceforge.net/datasets.html).

To evaluate the performance of the RF-ROS, we generated ensembles with different output space sizes: $v \in (\frac{q}{4}, \frac{q}{2}, \frac{3q}{4}, \sqrt{q}, \log q)$ with q the number of labels.

We also experimented with two aggregation functions: total and subspace averaging. We then compare the performance of RF-ROS with the performance of: (i) Random forests of standard PCTs (RF-PCT) [4], (ii) Random k-Labelsets (RAkEL) models [8] and (iii) Random forests with Gaussian subspaces (RF-Gauss) [3].

RF-PCT and RF-ROS ensembles used 100 PCTs (ensembles are typically saturated at that point) and descriptive space size $v = \lfloor 0.1 \cdot q \rfloor + 1$ [4]. The trees in the ensembles were not pruned [1]. For RAkEL models, the k parameter (size of labelset) was set to $q/2$ and the number of models to $min(2q, 100)$. A support vector machine (SVM) classifier was selected as a learning algorithm within RAkEL, with a linear kernel and a complexity constant $C = 1$. In RF-Gauss, the number of Gaussian subspace components was set to $\log q$. The other RF-Gauss parameters were set to $n_{min} = 1$ and $k = \sqrt{q}$ [3]. The statistical evaluation of the results was performed according to the guidelines of Demšar [2]. All statistical tests on the predictive performance values were conducted at the significance level $\alpha = 0.05$ (using three decimal places).

In order to determine the predictive performance of the induced models, we empirically evaluate them according to 12 different measures that belong to two groups: example and label based measures. The example based measures considered are: hamming loss, accuracy, precision, recall, F1, subset accuracy. The label based measures considered are: micro/macro precision, micro/macro recall, micro/macro F1 [6]. Results in terms of different measures lead to the same conclusions: In order to conserve space, we present only results for the example based measures F1 (more is better) and Hamming loss (less is better) in Table 2.

Table 2. The performance of the considered methods in terms of the example based measures F1 and Hamming loss. DNF (did not finish) denotes algorithms that did not produce results. The numbers in bold denote best performance on a dataset.

| Name | Example based F1 | | | | | Hamming loss | | | | |
| | | | RF-ROS | | | | | RF-ROS | | |
	RAkEL	RF-Gauss	RF-PCT	Tot-75	Sub-LOG	RAkEL	RF-Gauss	RF-PCT	Tot-75	Sub-LOG
Emotions	**0.637**	0.534	0.574	0.582	0.588	0.205	0.2	0.197	**0.196**	0.198
Scene	**0.681**	0.413	0.574	0.558	0.591	0.098	0.111	0.09	0.093	**0.088**
Yeast	**0.64**	0.573	0.587	0.583	0.602	0.2	0.199	**0.198**	**0.198**	0.199
Birds	**0.658**	0.51	0.566	0.556	0.579	0.05	0.048	0.044	0.044	**0.043**
TMC 2007	**0.81**	0.992	0.908	0.902	0.926	0.033	**0.001**	0.015	0.016	0.012
Genbase	**0.996**	0.991	0.981	0.981	0.986	**0.001**	**0.001**	0.002	0.002	**0.001**
Medical	**0.789**	0.515	0.673	0.669	0.683	**0.01**	0.016	0.013	0.013	0.012
Enron	**0.562**	0.508	0.527	0.518	0.559	0.049	0.047	0.046	0.046	**0.045**
Mediamill	DNF	0.545	**0.549**	0.547	0.541	DNF	**0.03**	**0.03**	**0.03**	0.032
Bibtex	DNF	0.173	0.211	0.209	**0.305**	DNF	0.014	**0.013**	**0.013**	**0.013**
Bookmarks	DNF	0.2	**0.206**	0.203	0.175	DNF	**0.009**	**0.009**	**0.009**	**0.009**
Corel	DNF	0.018	0.007	0.009	**0.089**	DNF	**0.009**	**0.009**	**0.009**	0.01
Delicious	DNF	**0.237**	0.194	0.193	0.202	DNF	**0.018**	**0.018**	**0.018**	0.021

4   Results

The proposed method has two degrees of freedom: target subspace size and aggregation function. Figure 1 shows the performance of RF-ROS on four datasets with various label and example counts. The plots for each dataset also show the point (total averaging, 100% target space, always the rightmost data point) which represents the performance of the RF-PCT model on that dataset. The results suggest that subspace averaging outperforms total averaging (especially for subset sizes below 50%). Moreover, the two aggregation functions exhibit inverse behavior w.r.t. the target subspace size. Total averaging performs better with larger target subspaces while subset averaging is better for smaller ones. When the target subspace size increases, both variants converge to a performance similar to that of the original RF-PCT method. This behavior is expected because larger subset size leads to larger overlap between the set of all target variables and its subsets.

We also observe that the performance of models with different aggregation functions converges at different rates. Although we observe convergence towards RF-PCT on all datasets, we speculate that the convergence rate is dataset dependent. For instance, on the Delicious dataset, both variants already converge with a target subspace size of 25%. On the Bibtex dataset, this number is a bit higher (50%) and on the Yeast and Scene datasets even higher (75%).

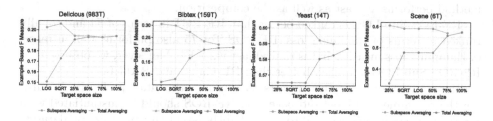

Fig. 1. Example based F1 results for Delicious, Bibtex, Yeast and Scene datasets.

Figure 2 shows average rank diagrams that confirm our speculations. Figures 2a and c show some statistically significant differences, so we recommend a larger subspace size ($v = \frac{3q}{4}$) with total averaging. Figures 2b and d do not show any statistically significant differences between the considered RF-ROS variants. Nevertheless, we recommend using the smallest evaluated subspace size ($v = \log q$) to be used with subspace averaging, as this is most efficient.

We compared the model performances of RF-ROS variants using these recommended parameters to the performance of RF-PCT, RAkEL and RF-Gauss (Fig. 3). The diagrams do not show any statistical significance in terms of F1. It is immediately visible that RAkEL performs very well. Although it did not finish on five datasets, it can still be considered a serious competitor on datasets with smaller label spaces. However, its predictive performance comes at a high

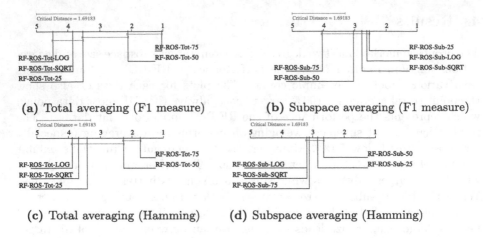

(a) Total averaging (F1 measure)   (b) Subspace averaging (F1 measure)

(c) Total averaging (Hamming)   (d) Subspace averaging (Hamming)

Fig. 2. Average rank diagrams of the RF-ROS variants (F1 and Hamming loss).

computational cost. This method is hindered by the fact that it uses label pow-ersets and SVMs to generate models which makes the running times of RAkEL substantially longer. RAkEL is not a clear winner w.r.t. the average rank dia-grams because the method was penalized for not finishing. If we take RAkEL out of consideration, the average rank diagrams in Fig. 3 suggest that the proposed method performs at least as well as the competition.

RF-ROS-Sub-LOG is ranked better than RF-PCT in terms of F1 and equally ranked in terms of Hamming loss. RF-ROS-Tot-75 also performs well in terms of Hamming loss measure but is ranked last w.r.t F1. Moreover, we observe that RF-ROS-Sub-LOG is ranked better than RF-Gauss and RAkEL.

Here, we summarize the answers to our experimental questions. Regard-ing the recommended label subspace size, RF-ROS should be instantiated with $v = \log q$. It could be beneficial to use a slightly larger subspace size on datasets with larger label spaces (i.e., $v \in (\sqrt{q}, \frac{q}{2})$). Next, subspace averaging should be preferred, because total averaging seems to degrade the predictive performance of the models and (with larger label subspace sizes) converges to the performance of the original method (RF-PCT). Note that even if we do not use the optimal value for the subspace size, the performance of RF-ROS is lower-bounded by

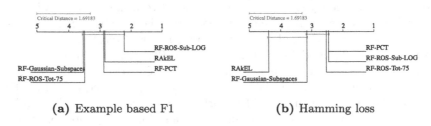

(a) Example based F1   (b) Hamming loss

Fig. 3. Average rank diagrams for RF-ROS and its competitors.

RF-PCT. Finally, RF-ROS ensembles perform well compared to the competition, which especially holds for the RF-ROS-Sub-LOG variant.

5   Conclusions and Future Work

We have proposed and evaluated a novel ensemble method for MLC, namedRF-ROS, that uses subsets of the label space to induce base models. We have experimented with different subspace sizes and two voting mechanisms, and found that the proposed method improves random forest models with PCTs as base learners. We have also shown that the proposed method generates models that performs equally well or better than the competition.

Future work is planned that will include evaluation against models generated by additional MLC methods. We will also add experiments on additional datasets. Next, we would like to try a new aggregation function where we would include predictions of the default model (i.e., predictions on the whole training set). We would also like to include out-of-bag errors to estimate the quality of individual base models and use this in conjunction with the mentioned aggregation function. Finally, a possible direction for future work is the extension of label subspace generation process that would work for hierarchies.

Acknowledgements. We acknowledge the financial support of the Slovenian Research Agency via the grants P2-0103,L2-7509, and a young researcher grant to MB, as well as the European Commission, through grants ICT-2013-612944 MAESTRA and ICT-2013-604102 HBP SGA1.

References

1. Bauer, E., Kohavi, R.: An empirical comparison of voting classification algorithms: Bagging, boosting, and variants. Mach. Learn. **36**(1), 105–139 (1999)
2. Demšar, J.: Statistical comparisons of classifiers over multiple data sets. J. Mach. Learn. Res. **7**, 1–30 (2006)
3. Joly, A., Geurts, P., Wehenkel, L.: Random forests with random projections of the output space for high dimensional multi-label classification. In: Calders, T., Esposito, F., Hüllermeier, E., Meo, R. (eds.) ECML PKDD 2014. LNCS, vol. 8724, pp. 607–622. Springer, Heidelberg (2014). doi:10.1007/978-3-662-44848-9_39
4. Kocev, D., Vens, C., Struyf, J., Džeroski, S.: Tree ensembles for predicting structured outputs. Pattern Recogn. **46**(3), 817–833 (2013)
5. Madjarov, G., Gjorgjevikj, D., Dimitrovski, I., Džeroski, S.: The use of data-derived label hierarchies in multi-label classification. J. Intel. Inf. Syst. **47**(1), 57–90 (2016)
6. Madjarov, G., Kocev, D., Gjorgjevikj, D., Džeroski, S.: An extensive experimental comparison of methods for multi-label learning. Pattern Recogn. **45**(9), 3084–3104 (2012)
7. Szymański, P., Kajdanowicz, T., Kersting, K.: How is a data-driven approach better than random choice in label space division for multi-label classification? Entropy **18**(8), 282 (2016)
8. Tsoumakas, G., Vlahavas, I.: Random k-labelsets: an ensemble method for multilabel classification. In: Kok, J.N., Koronacki, J., Mantaras, R.L., Matwin, S., Mladenič, D., Skowron, A. (eds.) ECML 2007. LNCS, vol. 4701, pp. 406–417. Springer, Heidelberg (2007). doi:10.1007/978-3-540-74958-5_38

Option Predictive Clustering Trees
for Hierarchical Multi-label Classification

Tomaž Stepišnik Perdih[1,2]([✉]), Aljaž Osojnik[1,2], Sašo Džeroski[1,2],
and Dragi Kocev[1,2]

[1] Department of Knowledge Technologies, Jožef Stefan Institute, Ljubljana, Slovenia
{tomaz.stepisnik,aljaz.osojnik,saso.dzeroski,dragi.kocev}@ijs.si
[2] Jožef Stefan International Postgraduate School, Ljubljana, Slovenia

Abstract. In this work, we address the task of hierarchical multi-label
classification (HMLC). HMLC is a variant of classification, where a single
example may belong to multiple classes at the same time and the classes
are organized in the form of a hierarchy. Many practically relevant problems can be presented as a HMLC task, such as predicting gene function,
habitat modelling, annotation of images and videos, etc. We propose to
extend the predictive clustering trees for HMLC – a generalization of
decision trees for HMLC – toward learning option predictive clustering
trees (OPCTs) for HMLC. OPCTs address the myopia of the standard
tree induction by considering alternative splits in the internal nodes of
the tree. An option tree can also be regarded as a condensed representation of an ensemble. We evaluate OPCTs on 12 benchmark HMLC
datasets from various domains. With the least restrictive parameter values, OPCTs are comparable to the state-of-the-art ensemble methods
of bagging and random forest of PCTs. Moreover, OPCTs statistically
significantly outperform PCTs.

1 Introduction

Supervised learning is one of the most widely researched areas of machine learning, where the goal is to learn, from a set of examples with known class, a function
that outputs a prediction for the class of a previously unseen example. The most
widely studied machine learning task is binary classification where the goal is to
classify the examples into two groups. The task where the examples can belong
to a single class from a given set of m classes ($m \geq 3$) is known as multi-class
classification. The case where the output is a real value is called regression.

In many real life problems of predictive modelling the target is structured
(e.g., the target is a vector of values with dependencies between them, or a time
series). In this work, we focus on the task of hierarchical multi-label classification
(HMLC). HMLC is a variant of classification, where a single example may belong
to multiple classes at the same time and the classes are organized in the form of
a hierarchy. An example that belongs to some class c automatically belongs to all
super-classes of c: This is called the hierarchical constraint. Problems of this kind
can be found in many domains including text classification, functional genomics,

A. Yamamoto et al. (Eds.): DS 2017, LNAI 10558, pp. 116–123, 2017.
DOI: 10.1007/978-3-319-67786-6_9

and object/scene classification. Silla and Freitas [19] give a detailed overview of the possible application areas and the different approaches to HMLC.

Decision tree based methods take a very notable place among approaches to HMLC. When used as base predictive models in an ensemble, they can yield a state-of-the-art performance [13,18]. A prominent global tree method for HMLC is a predictive clustering tree (PCT) for HMLC [20]. PCTs for HMLC inherit the properties of decision trees: they are interpretable models, but learning them is greedy. The performance of the trees is significantly improved when they are used in an ensemble setting [13]. However, the greediness of the tree construction process can lead to learning sub-optimal models. One way to alleviate this is to use a beam-search algorithm for tree induction [12], while another approach is to introduce option splits in the nodes [5,14].

In this work, we propose to extend predictive clustering trees (PCTs) for HMLC towards option trees, hence we propose to learn option predictive clustering trees (OPCTs). An option tree can be seen as a condensed representation of an ensemble of trees which share a common substructure. More specifically, the heuristic function for split selection can return multiple values that are close to each other within a predefined range. These splits are then used to construct an option node. For illustration, see Fig. 1.

The remainder of this paper is organized as follows. Section 2 proposes the algorithm for learning option PCTs for HMLC. Next, Sect. 3 outlines the design of the experimental evaluation. Section 4 continues with a discussion of the results. Finally, Sect. 5 concludes and provides directions for further work.

2   Option Predictive Clustering Trees

The predictive clustering trees framework views a decision tree as a hierarchy of clusters. The top-node corresponds to one cluster containing all data, which is recursively partitioned into smaller clusters while moving down the tree. The PCT framework is implemented in the CLUS system [1], which is available at http://clus.sourceforge.net.

Option predictive clustering trees (OPCT) extend the usual PCT framework, by introducing option nodes into the tree building procedure. Option decision trees were first introduced as classification trees by Buntine [5] and then analyzed in more detail by Kohavi and Kunz [14]. Ikonomovska et al. [10] analyzed regression option trees in the context of data streams. We also evaluated OPCTs for the multi-target regression task [16].

The major motivation for the introduction of option trees is to address the myopia of the *top-down induction of decision trees* (TDIDT) algorithm [4]. Viewed through the lens of the predictive clustering framework, a PCT is a non-overlapping hierarchical clustering of the whole input space. Each node/subtree corresponds to a clustering of a subspace and prediction functions are placed in the leaves, i.e., lowest clusters in the hierarchy. An OPCT, however, allows the construction of an overlapping hierarchical clustering. This means that, at each node of the tree several alternative hierarchical clusterings of the subspace can appear instead of a single one. When using TDIDT to construct a predictive clustering tree, and in particular when partitioning the data, all possible splits are evaluated by using a heuristic and the best one is selected. However, other splits may have very similar heuristic values. The best partition could be obtained with another split as a consequence of noise or of the sampling that generated the data. In this case, selecting a different split could be optimal. To address this concern, the use of option nodes was proposed [14].

The procedure of PCT learning for the HMLC task is presented in [13]. We modify it by introducing an option node into the tree when the best splits have similar heuristic values. Instead of selecting only the best split, we select several of them. Specifically, we select splits s, that satisfy the condition:

$$\frac{\text{Heur}(s)}{\text{Heur}(s_{best})} \geq 1 - e \cdot d^l, \tag{1}$$

where s_{best} is the best split, e determines how similar the heuristics must be, $d \in [0, 1]$ is a decay factor and l is the depth of the node we are attempting to split. E.g., when $e = 0.1$, we are selecting only splits whose heuristics are within 10% of the best split at the top level. We define the depth of a node to be the number of its ancestor nodes, excluding option nodes, as they do not split the data. The use of a decay factor makes the selection criterion more stringent in the lower nodes of the tree, where the impact of the split selection is also lower. After we have determined the candidate splits, we introduce an option node whose children are split nodes obtained by using the selected splits.

Introducing an option node with a large number of options is not advised [14] as it can lead to the explosion of model sizes. Therefore, we limit the maximum number of options for a single option node to 5 and also prohibit the induction of option nodes on depth 3 and greater.

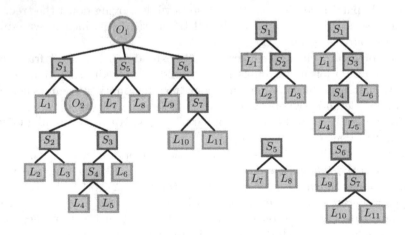

Fig. 1. An option tree (left) and the ensemble of its embedded trees (right). O_i are option nodes, S_j split nodes and L_k leaf nodes.

Once an OPCT is learned, we use it for prediction. In a regular PCT an example is sorted into a leaf (reached according to the tests in the nodes of the tree) where a prediction is made using a prototype function. Traversing an example through an OPCT is the same for split nodes and leaves. When we encounter an option node, however, we traverse the example down each of the options. This means that in an option node an example is sorted to multiple leaves, where multiple predictions are produced. To obtain a single prediction in an option node, we aggregate the obtained predictions.

An option tree is usually observed as a single tree, however, it can also be interpreted as a compact representation of an ensemble. We can extract *embedded trees* out of an option tree by replacing every option node with one of its options (Fig. 1). A given OPCT is also an extension of the PCT learned on the same data. By definition, whenever we introduce an option node, we include the best split. Consequently, the PCT is an embedded tree in the OPCT, resulting from replacing all option nodes with the best option.

3   Experimental Design

We evaluated the performance and efficiency of the proposed OPCT method with different parameter values and compared it to the standard PCTs and ensembles of PCTs. Evaluation was done on 12 datasets from biology, text classification and image annotation domains. They are described in Table 1. The datasets came pre-divided into training and testing sets and we used them in their original format, for easier comparison of the results.

OPCTs are evaluated for various values of parameters e and d. For e we consider values 0.1, 0.2, 0.5 and 1.0, while d takes values 0.5, 0.9 and 1.0. Notably, different selections of parameters can produce the same OPCT, if for a given dataset the same splits satisfy both criteria. Hereafter, the OPCT method with specific parameter values is denoted OPCT_eX_dY (e.g., for $e = 0.5, d = 0.9$, OPCT_e0.5_d0.9). The border case OPCT_e1_d1 always selects the 5 best options regardless of their heuristic score, making this setting similar to ensembles.

For PCTs and OPCTs we use the F-test as a pruning mechanism. Specifically, we check if a split results in a statistically significant improvement over the single node. If no split satisfies the F-test, the learning in the node stops. The significance level for the test was selected from the set of values $\{0.125, 0.1, 0.05, 0.01, 0.005, 0.001\}$ using internal 3-fold cross validation on the training set.

For ensembles, we considered bagging [2] and random forests [3]. For both methods we used 100 trees in the ensemble. Random forests algorithm also takes as input the size of the feature subset randomly selected at each node. For this we used the square root of the number of descriptive variables ($\lceil \sqrt{|D| + |C|} \rceil$).

Performance was measured using Area Under the Average Precision-Recall Curve ($AU\overline{PRC}$) [20]. For efficiency, we looked at the model size (number of leaves in a tree/ensemble). For statistical comparison of the methods we adopted the recommendations by Demšar [7]. Specifically, we used the Friedman test for statistical significance and Nemenyi post-hoc test to detect between which algorithms the significant differences occur. For both tests we selected confidence

Table 1. Descriptions of datasets used for the evaluation. The table shows the number of examples in the training and testing sets (N_{tr}/N_{te}), number of descriptive attributes (discrete/continuous, D/C), number of labels in the hierarchy ($|\mathcal{H}|$), maximal depth of the labels in the hierarchy ($\mathcal{H}_d$) and average number of labels per example ($\overline{\mathcal{L}}$).

	N_{tr}/N_{te}	$|D|/|C|$	$|\mathcal{H}|$	$\mathcal{H}_d$	$\overline{\mathcal{L}}$
Diatoms [9]	2065/1054	0/371	377	3.0	1.95
Enron [11]	988/660	0/1001	54	3.0	5.30
Expression–FunCat [6]	2494/1291	4/547	475	4.0	8.87
Exprindiv–FunCat [6]	2314/1182	1252	261	4.0	3.36
ImCLEF07A [8]	10000/1006	0/80	96	3.0	3.0
ImCLEF07D [8]	10000/1006	0/80	46	3.0	3.0
Interpro–FunCat [6]	2455/1264	2816	263	4.0	3.34
Reuters [15]	3000/3000	0/47236	100	4.0	3.20
SCOP-GO [6]	6507/3336	0/2003	523	5.5	6.26
Sequence-FunCat [6]	2455/1264	2/4448	244	4.0	3.35
WIPO [17]	1352/358	0/74435	183	4.0	4.0
Yeast-GO [6]	2310/1155	5588/342	133	6.3	5.74

level 0.05. The results of the statistical analysis are presented with *average ranking diagrams*. They plot the average ranks of the algorithms and connect those whose average ranks differ by less than the *critical distance*. The performance of the algorithms connected with a line is not statistically significantly different.

4   Results and Discussion

We present our experimental results as graphs with size on the horizontal axis and performance on the vertical axis. Figure 2 shows the results on four datasets. The remaining graphs are very similar and are omitted for brevity. Notably, the figures are on separate scales and on some figures the differences in performance between the different models are very small, e.g., on the SCOP-GO dataset.

Observing the points representing the results of OPCTs, the trade-off between size and performance is clearly visible. This trade-off is achieved as a consequence of different choices of the parameter values. The models' predictive performance generally rises with increasing model size, indicating that even the largest OPCTs do not overfit the training set, or possibly, different options overfit different parts of the input space. The increase in predictive performance in terms of increasing size also appears to saturate at the higher values of the observed parameter settings. This indicates that learning even larger less-restrictive OPCTs is not likely to provide a significant boost to predictive performance.

Compared to a PCT, OPCTs generally produce more accurate models that are mostly much larger. However, the increase in predictive performance is often

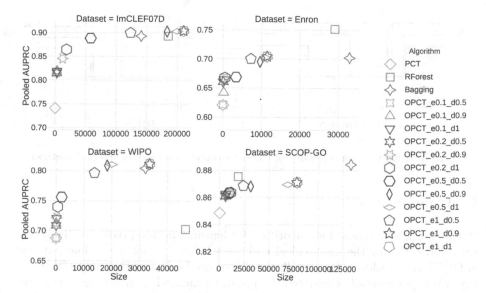

Fig. 2. Performances and sizes of models produced by different methods

noticeable even for the lowest parameter values when the difference in size is relatively small. The comparison between OPCTs and ensembles of PCTs is more varied. Bagging of PCTs is usually better than OPCTs (SCOP-GO), though often very slightly (Enron) and sometimes worse (IMCLEF07D). However, the size of a bagging ensemble can considerably surpass the size of even the largest OPCTs. On the Enron dataset, random forests of PCTs outperform all other methods by a solid margin. They also provide good performance on the SCOP-GO dataset with relatively small trees, however, on the WIPO dataset they produce the largest model which only outperforms a PCT.

We selected 3 parameter configurations as trade-off points between predictive performance and model size: OPCT_e1_d1, as it offers the best performance, OPCT_e1_d0.5, as its performance was similar to that of OPCT_e1_d1 but it often produced noticeably smaller models, and OPCT_e0.5_d0.5, as it consistently produced much smaller models than other two selected configurations, albeit at the cost of some performance.

We compared the performance and size of these three configurations to that of a PCT and their ensembles, using Friedman test to check if there is a significant difference between the algorithms and the Nemenyi post-hoc test to show where the differences occur. Results are presented in Fig. 3. The performance of a PCT and its size is significantly lower than that of ensembles of PCTs, OPCT_e1_d1 and OPCT_e1_d0.5. Additionally, the size of OPCT_e0.5_d0.5 is significantly lower than that of the four aforementioned methods, but its performance is not. We also observe that the average rank of OPCT_e1_d0.5 in performance is on par with ensembles of PCTs (it placed between bagging and random forests), while its average rank in size is noticeably better. As expected, a PCT always produced the smallest model with the worst performance.

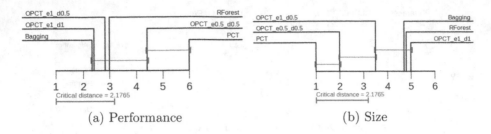

(a) Performance   (b) Size

Fig. 3. Average ranking diagrams of the performance and size of selected methods

5 Conclusions

In this work, we proposed an algorithm for learning option predictive clustering trees (OPCTs) for the hierarchical multi-label classification task. The purpose of OPCTs is to address the greediness of the standard algorithm for PCT learning. We experimentally evaluated the proposed method with various parameter values and compared it to PCTs and ensembles of PCTs (bagging and random forests). The results show that increasing the values of e and d increases the model performance and size compared to PCTs. At the highest parameter values of $e = 1$, $d = 1$, OPCTs are comparable to the state-of-the-art ensemble methods of bagging and random forest of PCTs.

We identified three interesting parameter selections for OPCTs and performed statistical comparison of these three methods and regular PCTs and their ensembles. The results show that regular PCTs have significantly lower performance and size than other methods with the exception of OPCT_e0.5_d0.5. Additionally, OPCT_e0.5_d0.5 produces significantly smaller models than bagging of PCTs, random forests of PCTs and OPCT_e1_d1. Average performance ranks of bagging, random forests, OPCT_e1_d1 and OPCT_e1_d0.5 are very similar, while average size rank of OPCT_e1_d0.5 is noticeably lower than that of the other three methods. Based on these results, we suggest the parameter values of $e \in \{0.5, 1\}$ and $d \in \{0.5, 1\}$ for future analyses.

There are several avenues for further work. Notably, the OPCT methodology described in this paper can be easily applied to the task of multi-label classification. In the future, we also plan to use the OPCT methodology as a part of a guided process to produce regular PCTs though either input from a domain expert, or through the use of additional validation data. Finally, we will investigate the use of OPCTs for performing feature ranking and selection for HMLC.

Acknowledgments. We acknowledge the financial support of the European Commission through the grants ICT-2013-612944 MAESTRA and ICT-2013-604102 HBP, as well as the support of the Slovenian Research Agency through young researcher grants and the program Knowledge Technologies (P2-0103).

References

1. Blockeel, H., Struyf, J.: Efficient algorithms for decision tree cross-validation. J. Mach. Learn. Res. **3**, 621–650 (2002)
2. Breiman, L.: Bagging predictors. Mach. Learn. **24**(2), 123–140 (1996)
3. Breiman, L.: Random forests. Mach. Learn. **45**(1), 5–32 (2001)
4. Breiman, L., Friedman, J., Olshen, R., Stone, C.J.: Classification and Regression Trees. Chapman & Hall/CRC, London (1984)
5. Buntine, W.: Learning classification trees. Stat. Comput. **2**(2), 63–73 (1992)
6. Clare, A.: Machine learning and data mining for yeast functional genomics. Ph.D. thesis, University of Wales Aberystwyth, Aberystwyth, Wales, UK (2003)
7. Demšar, J.: Statistical comparisons of classifiers over multiple data sets. J. Mach. Learn. Res. **7**, 1–30 (2006)
8. Dimitrovski, I., Kocev, D., Loskovska, S., Dzeroski, S.: Hierarchical annotation of medical images. Pattern Recogn. **44**(10–11), 2436–2449 (2011)
9. Dimitrovski, I., Kocev, D., Loskovska, S., Dzeroski, S.: Hierarchical classification of diatom images using ensembles of predictive clustering trees. Ecol. Inf. **7**(1), 19–29 (2012)
10. Ikonomovska, E., Gama, J., Zenko, B., Dzeroski, S.: Speeding-up hoeffding-based regression trees with options. In: Proceedings of the 28th International Conference on Machine Learning, ICML 2011, pp. 537–544 (2011)
11. Klimt, B., Yang, Y.: The enron corpus: a new dataset for email classification research. In: Boulicaut, J.-F., Esposito, F., Giannotti, F., Pedreschi, D. (eds.) ECML 2004. LNCS, vol. 3201, pp. 217–226. Springer, Heidelberg (2004). doi:10. 1007/978-3-540-30115-8_22
12. Kocev, D., Struyf, J., Džeroski, S.: Beam search induction and similarity constraints for predictive clustering trees. In: Džeroski, S., Struyf, J. (eds.) KDID 2006. LNCS, vol. 4747, pp. 134–151. Springer, Heidelberg (2007). doi:10.1007/ 978-3-540-75549-4_9
13. Kocev, D., Vens, C., Struyf, J., Džeroski, S.: Tree ensembles for predicting structured outputs. Pattern Recogn. **46**(3), 817–833 (2013)
14. Kohavi, R., Kunz, C.: Option decision trees with majority votes. In: Proceedings of the 14th International Conference on Machine Learning, ICML 1997, pp. 161–169. Morgan Kaufmann Publishers Inc., San Francisco (1997)
15. Lewis, D.D., Yang, Y., Rose, T.G., Li, F.: RCV1: A new benchmark collection for text categorization research. J. Mach. Learn. Res. **5**, 361–397 (2004)
16. Osojnik, A., Džeroski, S., Kocev, D.: Option predictive clustering trees for multi-target regression. In: Calders, T., Ceci, M., Malerba, D. (eds.) DS 2016. LNCS, vol. 9956, pp. 118–133. Springer, Cham (2016). doi:10.1007/978-3-319-46307-0_8
17. Rousu, J., Saunders, C., Szedmak, S., Shawe-Taylor, J.: Kernel-based learning of hierarchical multilabel classification models. J. Mach. Learn. Res. **7**, 1601–1626 (2006)
18. Schietgat, L., Vens, C., Struyf, J., Blockeel, H., Kocev, D., Džeroski, S.: Predicting gene function using hierarchical multi-label decision tree ensembles. BMC Bioinform. **11**(2), 1–14 (2010)
19. Silla, C., Freitas, A.: A survey of hierarchical classification across different application domains. Data Min. Knowl. Disc. **22**(1–2), 31–72 (2011)
20. Vens, C., Struyf, J., Schietgat, L., Džeroski, S., Blockeel, H.: Decision trees for hierarchical multi-label classification. Mach. Learn. **73**(2), 185–214 (2008)

Deep Learning

Re-training Deep Neural Networks to Facilitate Boolean Concept Extraction

Camila González$^{(\boxtimes)}$ ⓘD, Eneldo Loza Mencía, and Johannes Fürnkranz ⓘD

Knowledge Engineering Group, TU Darmstadt,
Hochschulstrasse 10, 64289 Darmstadt, Germany
camilag.bustillo@gmail.com, {eneldo,juffi}@ke.tu-darmstadt.de

Abstract. Deep neural networks are accurate predictors, but their decisions are difficult to interpret, which limits their applicability in various fields. Symbolic representations in the form of rule sets are one way to illustrate their behavior as a whole, as well as the hidden concepts they model in the intermediate layers. The main contribution of the paper is to demonstrate how to facilitate rule extraction from a deep neural network by retraining it in order to encourage sparseness in the weight matrices and make the hidden units be either maximally or minimally active. Instead of using datasets which combine the attributes in an unclear manner, we show the effectiveness of the methods on the task of reconstructing predefined Boolean concepts so it can later be assessed to what degree the patterns were captured in the rule sets. The evaluation shows that reducing the connectivity of the network in such a way significantly assists later rule extraction, and that when the neurons are either minimally or maximally active it suffices to consider one threshold per hidden unit.

Keywords: Deep neural networks · Inductive rule learning · Knowledge distillation

1 Introduction

Deep neural networks [10] achieve state of the art performance in a variety of different fields, like computer vision, speech recognition and machine translation. They can be leveraged both in supervised and unsupervised problem formulations, as they automatically learn insightful features out of unprocessed data. In the last few years, they have considerably risen in popularity as advancements in the training practices and availability of user friendly frameworks have made it much simpler to train accurate models, as long as sufficient data is available.

However, the fact that the models are governed by a high number of parameters makes tracing the path that led to a classification an arduous process, which is why they are often regarded as *black boxes*. This is a significant shortcoming, as it makes them unsuitable for safety critical applications and domains where there are juridical barriers which either explicitly forbid their use or implicitly discourage it by making the user liable for the model's decisions. Amongst the

© Springer International Publishing AG 2017
A. Yamamoto et al. (Eds.): DS 2017, LNAI 10558, pp. 127–143, 2017.
DOI: 10.1007/978-3-319-67786-6_10

legislation that aims for more comprehensible prediction models is the *General Data Protection Regulation (GDPR)*[1] planned to take effect in 2018. There is also an ongoing European legislative initiative on *Civil law rules on robotics*[2].

In fields such as health care and criminal sentencing, comprehensible models like decision lists or trees are favored because they provide understandable evidence to support their predictions [16,17]. *Decision support systems (DSS)* aim at integrating machine learning models into a human-centered decision process. Here, interpretability is of particular advantage because a justified decision is more likely to convince the human to support or disregard the machine's recommendation. Besides, the extent to which the model is used in practice depends heavily on how easily interpretable it is, as this is a relevant criteria for eliciting trust [14].

Compared to neural networks, *if-then* rule sets are a representation with a good trade-off between human and machine interpretability [9]. This is partly because they provide a symbolic representation which more closely resembles the way humans model logic. Also, each rule can be observed individually, so only a limited amount of information must be considered at any time. This advantage, together with the fact that they can be more flexibly pruned, sometimes makes them preferable over decision trees [7].

Consequently, researchers have looked into converting neural networks into a rule-based representation. One problem with such approaches is that much information is lost when the continuous range of activation levels of the internal neurons is mapped to a two-valued logical representation. In this paper, we investigate ways for retraining deep neural networks with the goal of encouraging sparse connectivity and minimally or maximally active hidden units, with the idea of facilitating a later extraction of rules from the network. We study the problem on the task of reconstructing Boolean functions, because there we can see whether the use of the network's structure really helps to recover the logical structure in the target function.

We will start our discussion with a brief general overview of prior work on rule extraction from neural networks (Sect. 2), with a particular focus on the DEEPRED algorithm, upon which forms the basis of our work (Sect. 3). The core contribution of this paper, an algorithm for retraining DNNs to extract better representations, is described in Sect. 4, and experimentally evaluated on the problem of reconstructing Boolean functions in Sect. 5.

2   Knowledge Distillation from Neural Networks

Much of the predictive strength of deep neural networks originates from their ability to form latent concepts in the hidden layers, and the high connectivity between these layers makes it difficult to distill the meaning of these concepts.

---

[1] EU Regulation 2016/679: http://eur-lex.europa.eu/eli/reg/2016/679/oj, http://www.eugdpr.org.

[2] http://www.europarl.europa.eu/oeil/popups/ficheprocedure.do?reference=2015/2103(INL).

One approach is to rely on visualization in order to analyze the behavior of the learned network (see, e.g., [29]). However, another line of research concentrates on ways of making the knowledge that is implicitly captured in these nodes explicit and amenable to human inspection. Typically this is done by transforming the neural network into more interpretable knowledge representations such as rules or decision trees. A prerequisite for such work is often to simplify the network by pruning unnecessary connections and neurons. We will briefly recapitulate work in these areas in Sects. 2.1 and 2.2, respectively.

2.1 Rule Extraction

Many approaches have been developed to extract symbolic representations from neural networks. However, most either do not consider the network's internal structure or are only applicable to shallow architectures. A distinction can be made between *pedagogical* methods, which regard the network as a black box and map relationships between the outputs and the inputs, *decompositional* ones that observe the contribution of individual parameters or neurons and *eclectic* methods which fall between the other two [4]. Other categorizations refer to the computational complexity of the approach, what data is used to build the model and whether a particular training regime is performed [2].

A first group is made up of *subset* approaches [8,27,28]. These are decompositional and typically assume a polarization of the activations and the use of exclusively binary inputs. They search the entire feature space and construct one expression per neuron of interest. Typically, a threshold is applied to the neuron's output to define an active and an inactive state. Rules are then learnt for the active state by finding combinations of the incoming weights that cause the bias of the hidden unit to be exceeded.

A shortcoming of these methods is that considering all subset combinations grows at an exponential rate with the number of incoming connections, which limits their applicability to larger networks. It also cannot be assumed that any network can be converted to one with only maximally or minimally active neurons while maintaining the initial accuracy. An even more difficult requirement to fulfill is that inputs should be discrete so they can be binarized without information loss.

Another problem that arises when sampling all possible inputs is that the way the network reacts to implausible instances may not be meaningful, so capturing this logic may result in an overly complicated rule model which is not better at classifying unseen, naturally occurring examples. Some methods thus focus primarily on the instances used to train the network when building the symbolic model.

Such is the case for the pedagogical TREPAN algorithm [5], which explains the outputs of the network with respect to the inputs by building decision trees directly between these layers. The tree building process makes use of queried instances, generated from the marginal distribution of each attribute, to avoid low amounts of data as the tree branches. However, in a comparison of different variants [18], that which did not generate new examples performed best.

CRED [20] also builds decision trees between network layers using the train data, but it works in a decompositional manner. First, a target condition is specified to discretize the class values, and decision trees are built to explain this output pattern with the hidden units as attributes, using the corresponding activation values of the training instances. The range of each hidden unit is partitioned in an online manner, so several thresholds may be considered per unit, and some units may be ignored. The trees are then converted into sets of decision rules. Redundant and unsatisfiable rules and terms are deleted, and rules are merged by forming their least general generalization (lgg) by selecting the most general condition of the attributes they share, and dropping all conditions of attributes they do not share. For instance, $a \leq 0.3 \wedge b > 4 \wedge c > 2 \rightarrow C_1$ and $a \leq 0.2 \wedge b > 3 \wedge d > 2 \rightarrow C_1$ would become $a \leq 0.3 \wedge b > 3 \rightarrow C_1$. Afterwards, analogous rule sets are built which explain each split value considered for a the hidden units with respect to the inputs, which now make up the attributes. Finally, the *total* rules are formed by substituting the hidden split values with these new rule sets.

2.2   Connection Pruning

Pruning connections or whole neurons of trained neural networks is a common way to adapt the topology of the network to the effective size of the problem, thus discouraging overfitting and increasing the generalization capabilities. It can also be leveraged to require less time and resources when making classifications [12,15,26].

A connection $w_{j,k}^l$ between two neurons $h_{l-1,k}$ and $h_{l,j}$ can be pruned by equaling the weight entry to zero. This is similar to applying dropout [25] but whereas dropout temporarily removes randomly chosen connections for one epoch at a time, pruning permanently removes selected connections from the network. Connections can be pruned iteratively by retraining the network after each pruning step, which allows to discard a considerably higher number [11]. Note that pruning connections can result in indirectly pruning whole inputs or hidden units, as a neuron without output connections is disconnected from the network.

In [23] a method is introduced to prune connections from shallow neural networks. First, the networks are trained with a weight-decay penalty. Connections $w_{p,j}^2$ between the hidden and output layers are then pruned if

$$\left| w_{p,j}^2 \right| \leq \eta, \tag{1}$$

and connections $w_{j,k}^1$ between the inputs and hidden units are pruned if

$$max_p \left| w_{p,j}^2 . w_{j,k}^1 \right| \leq \eta. \tag{2}$$

If no connection fulfills one of those conditions, then the entry $w_{j,k}^1$ for which the minimum of the maximum products is lowest is pruned. Afterwards, the network is retrained. If the final error falls below an acceptable level, the pruning step is repeated; otherwise the last acceptable parameters are restored and the process is stopped.

The author uses this approach extensively as a preprocessing step before applying rule extraction algorithms [21, 24]. The pruning phase is usually followed by a discretization of the hidden unit activations. In [22], the connectivity of the hidden units is further reduced by 'splitting' those with many input connections. Each new unit is treated as an output and a hidden layer is inserted in the middle between the inputs and the new output layer. The network is retrained and the new subnetwork pruned, and the process is repeated until each neuron only has a small number of inputs.

3   The DEEPRED Algorithm

In order to extract representations from deep neural networks which not only explain the network's predictive behavior but also uncover hidden features, we make use of the DEEPRED algorithm [30], which extends CRED (Sect. 2.1) to deep neural networks. It is scalable to large architectures, works in a decompositional manner and has been shown to be capable of extracting accurate models from deep neural networks. We extended DEEPRED with a post-pruning step (Sect. 3.3) meant to contain error propagation and reduce the complexity. This is carried out each time a rule set is generated from a decision tree, and between substitution steps when building the expression of the target with regards to the inputs.

3.1   Overview

The DEEPRED algorithm extracts rules between any two layers by building decision trees for layer h_l using the activations from layer h_{l-1} as attributes. The trees are then transformed to rule sets, and a merging step converts the intermediate representations into a single rule set connecting the inputs with the outputs. Redundant and unsatisfiable rules and conditions are deleted, but unlike in CRED no further pruning takes place. There is a version of the algorithm that performs a feature selection prior to rule extraction by considering the contribution of each input for correctly classifying the training data and removing inputs that do have a great impact. This proves to be very advantageous when the network is used for high-dimensional data.

Figure 1 exemplifies how DEEPRED would extract rule representations from a shallow neural network which emulates a Boolean function. The model is sampled to obtain activation values for each training instance. A first tree is built to predict under what activation settings of layer h_1 the target concept is fulfilled, namely that class C_1 has a higher probability than C_0. The tree is converted into a DNF representation and the expression is simplified. A tree is then built for each literal which appears in the simplified expression, using the input values as attribute data. Each of these expressions is extracted and simplified, and a last step substitutes the literals with regards to the hidden layer so that the expression which predicts class C_1 is in terms of the inputs.

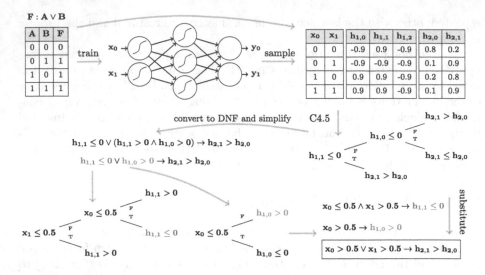

Fig. 1. The general workflow of DEEPRED.

3.2   Extraction of DNF Formulas from Trees

A rule can be regarded as a conjunction of terms, where a term is a condition indicating whether the activation state of a neuron falls or not above some threshold, and a rule set as an expression in disjunctive normal form (DNF). Each tree built by DEEPRED determines whether an input or the activation of a hidden unit fulfills one such term, using terms with respect of the adjacent shallower layer. For instance, a tree could determine if the value 0.5 is exceeded by the second neuron in the first hidden layer. Such a tree would have two possible classes, $h_{1,2} > 0.5$ and $h_{1,2} \leq 0.5$. A DNF for each of them can be obtained by joining the respective terms on all paths from the root to the corresponding leafs of the tree.

A separate DNF formula is maintained for each class, so there are two DNFs per split value, each of which fires as soon as one of its rules fires. A DNF formula for the event of neuron $h_{2,2}$ exceeding the threshold 0.5 may look like $(h_{1,1} > 0.5 \land h_{1,2} \leq 0.3) \lor (h_{1,2} > 0.3 \land h_{1,3} > 0.7) \rightarrow h_{2,2} > 0.5$.

The expressions for opposite class conditions, as would be those for $h_{2,2} > 0.5$ and $h_{2,2} \leq 0.5$, are complementary after being extracted for the tree, but this may no longer hold once pruning is applied. Thus, both may fire for a given example, or neither may do so. Usually, additional criteria such as a priority list for tie-breaking between multiple predictions or a default rule for the latter case are employed. However, in our case only one expression is maintained for the selected class, so the inconsistencies within intermediate expressions do not translate to ambiguities in the final class prediction.

3.3   Simplification and Post-pruning of Expressions

Transforming a decision tree into a rule set, as well as the process of building the expression of the target with regards to the inputs by sequentially substituting terms by DNFs, can result in expressions with redundant and unsatisfiable rules and redundant conditions. These are removed each time an expression is extracted from a decision tree and between substitution steps.

Very similar rules may still remain which do not provide more information than a simpler rule would. This affects the comprehensibility and can promote error propagation between layers. Yet, a too strong generalization of the intermediate concepts should be avoided as its repercussions on the final expression cannot be observed until the end. CRED (cf. Sect. 2.1) employs a pruning approach which is advantageous in shallow networks but proved in preliminary experiments to be too aggressive for deep networks. Instead, we use a method of reduced error pruning that only makes a change if this positively affects the accuracy with respect to the head of the rule.

Rules are ordered in terms of increasing precision. For each rule, the change in accuracy is calculated in case the rule is deleted, a term is removed from the rule (calculated for all terms) and the rule being merged with another one from the set (calculated for all remaining rules). The modification which leads to the highest accuracy is performed if it improves the accuracy over the current rule set. Unless the modification consists on removing the rule completely, the precision is calculated for the new rule, which is regarded as unseen. This is repeated until there are no unseen rules left.

4   Retraining DNNs to Extract Better Representations

One problem with rule extraction from neural networks is that the activations assume continuous values within some range, whereas a mapping to a decision tree or rule set reduces them to a discrete setting. The key idea of our work is that the transformation process may be supported by forcing the network weights to assume more extreme values. In this section, we therefore present methods for retraining a deep neural network in order to encourage sparseness in the weight matrices and make the hidden units be either maximally or minimally active. For this work, we consider the accuracy on the entire dataset for guiding the retraining, because our goal is to train the networks to exactly emulate predefined concepts. However, in a different setting it might be advisable to use a separate validation set, which is not used to optimize the parameters.

4.1   Weight Sparseness Pruning

We employ a connection pruning technique that is quite similar to that described in Sect. 2.2. In contrast to that method, ours can be applied to deep networks, and its aim is to sparsify all weight matrices so that the total number of connections between any two layers is reduced. This has the effect that single neurons are neither connected to a majority of the neurons of the following layer nor to a majority of those from the previous layer. The expectation is that, as observed by [22], rules extracted from minimally connected neurons will be simpler and more accurate.

The motivation for targeting such connections also comes from the performance of DEEPRED when applied to a network manually constructed to emulate the parity function with eight inputs [30]. The network constructed for this experiment has a recursive structure from the eight inputs to the output layer and is minimally connected. DEEPRED is not only able to extract the modeled DNF representation using a significantly lower number of instances than a pedagogical approach, but its intermediate rules also exactly replicate the recursive features.

In preliminary experiments, we could not repeat this effect on fully connected networks of the same topology trained with backpropagation, even if all combinations were used for training. When rules are extracted from such networks using a reduced set of instances, the majority of the logic is concentrated between the inputs and the first hidden layer. The rule sets extracted between these layers overfit the train data, and each depends on the majority of the inputs. Therefore, the accuracy on the unseen instances never exceeds fifty percent and actually decreases with increasing amounts of training data (a phenomenon that also affects C4.5). If, on the other hand, the number of connections is reduced, the network may be encouraged to learn a reduced amount of hidden features that are more abstract and apply to a greater percentage of examples.

General Methodology. A connection $w^l_{j,k}$ is represented by the index of the weight matrix l and the row and column indices j and k. The number of entries that have already been pruned in each row or column of each weight matrix is maintained in order to calculate the *neighborhood sparseness* of the remaining entries. This value is determined by the sum of entries that have been pruned in row j of matrix W^l plus those that have been pruned in column k of the same matrix.

On each step, a list of all existing connections is sorted in terms of increasing neighborhood sparseness, and it is attempted to prune the next head element, which is likely to be surrounded by unpruned entries. The target training accuracy that must be reached after retraining for the connection to remain pruned is the original train accuracy minus an allowed decrease. If the accuracy is satisfactory, the connection is pruned, the counts for column and row pruned entries are updated, and the list is re-sorted. Otherwise, the last accepted parameters are restored, and the next connection is removed from the list.

Iterations Used for Retraining. Preliminary experiments showed that often a small number of iterations suffice to determine whether a connection can be pruned, because the network gets stuck in a local minimum. On the other hand, some connections cause a steep decrease in accuracy when they are first removed, but the network later adapts. To allow the latter connections to be eliminated while not considerably increasing the retraining time, the retraining epochs are divided into smaller sets. If after a set the accuracy is equal or greater than the target accuracy, the connection is pruned, otherwise the retraining continues until either n retraining steps were performed from the time a connection was pruned or there were no improvements in the last m steps, with $n \geq m$.

Re-exploration of a Connection. It was also observed that if a connection could not be pruned at some stage, it was unlikely that it could be pruned later on, even if other connections had been pruned which affected the neurons it joined. Therefore, in the experiments outlined in Sect. 5 there was only one attempt of pruning per connection.

4.2   Activation Polarization

The activation range of the hidden units being continuous has several negative repercussions, such as making it more costly to classify new instances, so techniques have been developed for binarizing the parameters and activation values [1,3]. Yet, most networks are trained in such a way that the hidden units can take any value within the range.

As representing each neuron with only one expression is a more comprehensible way to illustrate that neuron's purpose in the network than if different expressions have to be regarded for an array of activation intervals, many decompositional rule extraction approaches reduce the possible states each neuron takes to being either at the bottom or the top of the activation range (cf. Sect. 2.1).

In order to extract rules which are true to the network while making this assumption, the networks must be trained in a way that the activations are polarized. There are several ways to achieve this. We propose a retraining step similar to that used in [19] to encourage sparse activations. The key idea is to penalize the loss function with a term based on the *KL divergence* between the mean absolute value of each activation

$$\hat{\rho}_{l,n} = \frac{1}{|D|} \sum_i |a_{l,n}^i| \tag{3}$$

and a ρ close to one, which results from the use of a hyperbolic tangent function. These terms are summed up over all hidden units, yielding

$$\sum_{l,n} KL(\rho \,\|\, \hat{\rho}_{l,n}) = \sum_{l,n} \rho \log \frac{\rho}{\hat{\rho}_{l,n}} + (1 - \rho) \log \frac{1 - \rho}{1 - \hat{\rho}_{l,n}}. \tag{4}$$

This term introduces the additional parameters ρ for the optimal activation average and β for the weighting of the penalty term. Instead of having to define β, the retraining method as implemented for this work starts by setting $\beta = 0$, and increases this value iteratively. Between each increment, a number of epochs are conducted, for as long as there is no decrease in accuracy, and the divergence stays above some threshold. The last weight and bias parameters are stored before each increase of β. If the process stops because the accuracy falls, the parameters which were saved last are restored.

5   Experiments

In order to show that we can derive meaningful conceptual descriptions from deep neural networks, we performed experiments on artificial datasets. Our goal was to demonstrate that our algorithms can reconstruct Boolean input functions from networks trained to model them. For this, we first made a quality assessment of the concepts extracted when the entire dataset is available. After exploring the limits of each approach, we compared the generalization abilities of the different variants by utilizing a subset of the combinations as training data and analyzing the accuracy on the remaining instances.

5.1   Experimental Setup

Data. We used twenty datasets constructed from Boolean functions with six to fourteen literals. Nineteen were generated by joining groups of randomly selected literals with alternating OR and AND operators and choosing to apply negation over each group with a 0.2 probability, and one was the parity function with eight inputs. The expressions were simplified with the ESPRESSO heuristic logic minimizer [13]. Each dataset was made up of all combinations of literals in the simplified expression.

Network Architecture and Training. The networks had three hidden layers, the first with twice as many neurons as inputs, the second with the average of that number and two and the third and output layers with two neurons. The hyperbolic tangent was used as activation function, and a softmax function was applied on the last layer. The networks were trained using the *TensorFlow* framework and cross-entropy as the loss function. They were trained with all input combinations until achieving a perfect accuracy. In this way, it was certain that they mimicked the logic of the predefined formulas.

Compared Algorithms. We compared several variants of our approach where (i) no retraining took place, (ii) weights sparseness pruning was performed, and (iii) a polarization of the activations was followed by weight sparseness pruning. Also, as observing one expression per hidden neuron of interest, which predicts when that unit is in an active state, is more comprehensible than having to consider an array of expressions, it was analyzed how the models would be affected if instead of allowing the online discretization of C4.5 to select thresholds of the activation range, only the midpoint of this range was considered.

Hyperparameter Setting. C4.5 was set to stop growing a tree when the percentage of the majority class exceeded 99% or only less than 1% of the original instances remained in a node. Only binary splits were allowed and the trees had a maximum depth of twenty nodes. For activation polarization, ρ was set to 0.99 and β was increased by 0.1 at a time. For weight sparseness pruning, a 1% decrease in accuracy was allowed. In both cases, each epoch set consisted of 1000 epochs. For the networks which were retrained in both manners, the penalty term from the activation polarization was added to the loss function used during connection pruning, multiplied by the last accepted value of β.

Evaluation Measures. The comprehensibility of the intermediate logic – which is to say that between subsequent layers – was assessed with the number of expressions and the total number of terms. The semantic quality was measured using the accuracy, which is to say the proportion of correctly classified instances among all classifications. Note that as the networks used perfectly replicate the Boolean functions, this corresponds to the fidelity of the extracted rules mimicking the network's behavior.

For determining whether observed performance differences between two classifiers were statistically significant, the sign and Wilcoxon signed ranks tests were used for a significance level of $p = 0.05$. Following [6], ties were distributed, and in the event of an uneven number of ties, the number N of datasets was reduced by one. Also subtracted from this number were comparisons which could not be performed because of uncompleted experiments. This occurred when the time or memory constraints for the extraction – set respectively at 24 h and 5000 MegaBytes – were surpassed or when no model could be built using one threshold per hidden unit. At least one experiment could not be completed for a total of three datasets, including that of the parity function.

5.2   Characteristics of the Trained Networks

After retraining the networks it was observed how many weights had been pruned, and how well the neurons could be polarized by calculating the deviation of the activations from zero averaged over all hidden units and examples. The results suggest that a trade-off takes place between the divergence from the range center and the number of pruned connections, as can be observed in Fig. 2. The pruning approach eliminated a great percentage of the connections, but at the cost of distributing the activation values more evenly across the range. When the networks were first polarized and the penalty term was maintained during the latter connection pruning, the activation values gathered even closer to the range boundaries, but considerably less connections were eliminated.

5.3   Reconstruction Using the Entire Dataset

There was a noticeable change in the number of intermediate expressions which were extracted – as well as in their complexity measured with the number of terms – when the networks were retrained under weight sparseness pruning. As can be observed in Fig. 3, models taken from less connected networks were much more compact.

How this reflects in the extracted models is exemplified in Fig. 4 for the expression $(x_3 \wedge x_1 \wedge x_5) \vee (\bar{x}_3 \wedge \bar{x}_0 \wedge \bar{x}_4 \wedge x_2 \wedge \bar{x}_5)$. The model extracted from

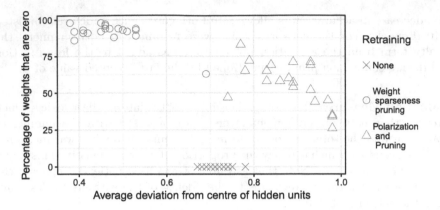

Fig. 2. Trade-off between deviation of the activations from the center of the range and weight entries that were pruned. Each point represents the result on one of the 20 datasets, either with no retraining, retraining after weight sparseness pruning or retraining after polarization and posterior weight sparseness pruning.

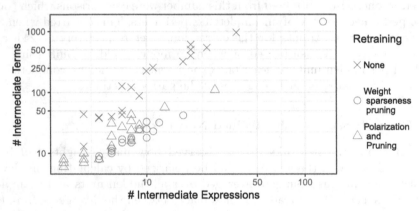

Fig. 3. Change in the complexity of the intermediate concepts when preceding extraction with weight sparseness pruning or polarization of the activations followed by pruning.

the original network finds an adequate representation for the second conjunction in $h_{1,6} > 0.32$, as well as for the first part of the first conjunction in $h_{1,0} \leq -0.26$, but fails to do so for x_5. The model extracted from the polarized and pruned network includes instead only expressions with only a couple of literals each and it therefore much simpler to trace.

Applying the different retraining methods on the networks did not cause a substantial change in accuracy when any threshold could be considered. Neither was a significant difference found when only zero was used to partition the activation ranges of the hidden units, except for when the networks had been subjected to weight sparseness pruning but no activation polarization. This effect

Model extracted from original network	
$h_{3,1} \leq 0.15$	$\rightarrow target$
$h_{2,6} > 0.32 \vee h_{2,4} > 0.56$	$\rightarrow h_{3,1} \leq 0.15$
$(h_{1,0} \leq -0.26 \wedge h_{1,7} \leq -0.65) \vee$ $(h_{1,0} \leq -0.26 \wedge h_{1,8} > 0.12)$	$\rightarrow h_{2,4} > 0.56$
$h_{1,6} > 0.32$	$\rightarrow h_{2,6} > 0.32$
$(x_4 \wedge x_5) \vee (x_1 \wedge \bar{x}_2 \wedge x_5) \vee$ $(x_1 \wedge \bar{x}_2 \wedge \bar{x}_3 \wedge x_4) \vee (x_1 \wedge \bar{x}_3 \wedge x_5)$	$\rightarrow h_{1,8} > 0.12$
$(\bar{x}_0 \wedge x_3 \wedge x_5) \vee (x_2 \wedge x_3 \wedge x_5) \vee$ $(\bar{x}_1 \wedge x_2 \wedge x_4 \wedge x_5) \vee$ $(\bar{x}_0 \wedge x_2 \wedge \bar{x}_3) \vee (x_3 \wedge x_4 \wedge x_5)$	$\rightarrow h_{1,7} \leq -0.65$
$x_1 \wedge x_3$	$\rightarrow h_{1,0} \leq -0.26$
$\bar{x}_0 \wedge x_2 \wedge \bar{x}_3 \wedge \bar{x}_4 \wedge \bar{x}_5$	$\rightarrow h_{1,6} > 0.32$

Model extracted from polarized and pruned network	
$h_{3,1} \leq -0.10$	$\rightarrow target$
$h_{2,6} > 0.88 \vee h_{2,6} \leq -0.99$	$\rightarrow h_{3,1} \leq -0.10$
$h_{1,0} \leq -0.01 \wedge h_{1,8} > -0.01$	$\rightarrow h_{2,6} \leq -0.99$
$h_{1,5} > -0.01 \wedge h_{1,8} \leq -0.01 \wedge h_{1,9} > -0.93$	$\rightarrow h_{2,6} > 0.88$
$\bar{x}_0 \wedge \bar{x}_4$	$\rightarrow h_{1,9} > -0.93$
$x_1 \wedge x_3$	$\rightarrow h_{1,0} \leq -0.01$
$\bar{x}_5$	$\rightarrow h_{1,8} \leq -0.01$
x_5	$\rightarrow h_{1,8} > -0.01$
$x_2 \wedge \bar{x}_3$	$\rightarrow h_{1,5} > -0.01$

Fig. 4. Effect of preceding DEEPRED with activation polarization and weight sparseness pruning in the model that reconstructs the expression $(x_3 \wedge x_1 \wedge x_5) \vee (\bar{x}_3 \wedge \bar{x}_0 \wedge \bar{x}_4 \wedge x_2 \wedge \bar{x}_5)$.

is shown in Fig. 5. The black circles, which refer to the models extracted from unpolarized and pruned networks which only include conditions of the hidden units with zero as threshold, illustrate a clear fall in accuracy when compared to the rest of the models.

These results reinforce the hypothesis that, when a high number of network connections are pruned and a retraining phase is performed between pruning steps, the logic modeled by the network is more heavily concentrated in the remaining neurons, thus needing to subdivide the neuron range into more intervals to describe it.

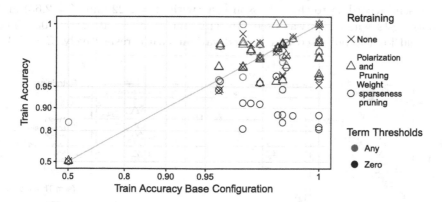

Fig. 5. Extent to which the concepts modeled by the neural networks were reconstructed when using all input combinations. Different configurations are compared to that where the network is not retrained and the term thresholds are selected by C4.5 in an online manner.

5.4   Reconstruction Using Part of the Dataset

For these experiments, the data was split into ten, four or two stratified folds and a cross-validation was performed. Also, the folds were inverted to observe the situation where lower data percentages were available. This resulted in ten experiments where 10% and 90% of the data was used, four with 25% and 75% being available, and two for the 50% case. The evaluation measure was averaged over all folds.

Again, the analysis focused on the effect of the different retraining methodologies. As models extracted from networks that had been pruned but not polarized using the range midpoint as sole threshold had a very low train accuracy, the effect of enforcing this constraint was only analyzed for the variant including both activation polarization and weight sparseness pruning.

Generally, the best performing models were those for which weight sparseness pruning had been performed, but significant differences were only found when less data was available. When more data was used to build the models, the predictive accuracy approached the accuracy on the train data. A special case was the parity function, for which none of the approaches extracted a well generalizing model, with the accuracies laying at 0.5 or below. Though the models extracted from pruned networks displayed slightly higher accuracies, we were eventually not able to resolve the issue for this very special case, which was our initial motivation for the pruning and re-training approaches (cf. Sect. 4.1), and leave further investigation for future work.

The results when using 50% of the data or less are illustrated in Fig. 6. The performances are shown in direct comparison to using plain C4.5 between the input and output data. Compared to C4.5, the variant which did not perform network pruning did not show any significant difference. That which only included weight sparseness pruning outperformed C4.5 when 50% and 10% of the data was present according to the Wilcoxon test (with $Z = 2.22$ and $Z = 2.63$) and for both tests when 25% was used (16 wins out of $N = 20$, $Z = 3.06$). The same held for the polarized and pruned variant (with, respectively, $Z = 2.31$,

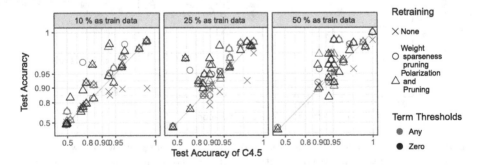

Fig. 6. Test accuracy of the models when using 10%, 25% and 50% of the dataset as train data and the remaining instances as test data. Different configurations are compared to the C4.5 algorithm, which disregards the internal structure.

$Z = 2.52$, 14 wins out of $N = 19$ and $Z = 2.75$). This variant also outperformed C4.5 using only one threshold per neuron according to both tests when 25% of the data was used (with 16 wins from $N = 19$ and $Z = 3.40$).

The unpolarized pruned variant outperformed that with no retraining according to both significance tests when 25% of the data was used (with 16 from $N = 19$ wins, $Z = 3.11$). The variant for which both retraining methods had been applied was deemed better by the Wilcoxon test when using 10% of the data ($Z = 2.07$).

6 Conclusion

Reducing the connectivity of the network proved to be a robust way for extracting simpler intermediate concepts, which were also better at classifying unseen instances. Yet it seems that encouraging low connectivity not only identifies irrelevant logic created from training too large architectures, but also concentrates the hidden features which are in fact relevant for the classification into fewer neurons. Thus a finer-grained partitioning of the activation ranges is required to regain the hidden patterns.

This was partly shown by an analysis of the characteristics of the network, which exposed a trade-off between the extent to which the activation values could be polarized and the percentage of connections that could be pruned. The negative consequences of this effect for rule extraction were confirmed by the dismal performance of models which combined connection pruning with only considering the center of the activation range as threshold.

However, when polarization of the activations was done jointly with connection pruning, the benefits of the latter could be leveraged while avoiding the undesired effect of concentrating more logic into less neurons. Although the number of connections which could be pruned in these networks was substantially lower, the intermediate models were not significantly more complex, and in terms of accuracy these approaches consistently performed within the highest.

Acknowledgements. We would like to thank the anonymous reviewers for their helpful suggestions. Computations for this research were conducted on the Lichtenberg high performance computer of the TU Darmstadt.

References

1. Aizenberg, I., Aizenberg, N.N., Vandewalle, J.P.: Multi-valued and Universal Binary Neurons: Theory, Learning and Applications. Springer, New York (2013). doi:10.1007/978-1-4757-3115-6
2. Andrews, R., Diederich, J., Tickle, A.B.: Survey and critique of techniques for extracting rules from trained artificial neural networks. Knowl. Based Syst. **8**(6), 373–389 (1995)
3. Courbariaux, M., Bengio, Y., David, J.: BinaryConnect: training deep neural networks with binary weights during propagations. In: Advances in Neural Information Processing Systems 28 (NIPS 2015), Montreal, Quebec, Canada, pp. 3123–3131 (2015)

4. Craven, M., Shavlik, J.W.: Using sampling and queries to extract rules from trained neural networks. In: Proceedings of the 11th International Conference on Machine Learning (ICML 1994), pp. 37–45. Morgan Kaufmann, New Brunswick (1994)
5. Craven, M., Shavlik, J.W.: Extracting tree-structured representations of trained networks. In: Advances in Neural Information Processing Systems 8 (NIPS 1995), pp. 24–30 (1995)
6. Demšar, J., Schuurmans, D.: Statistical comparisons of classifiers over multiple data sets. J. Mach. Learn. Res. **7**(1), 1–30 (2006)
7. Freitas, A.A.: Comprehensible classification models: a position paper. SIGKDD Explor. **15**(1), 1–10 (2013)
8. Fu, L.: Rule learning by searching on adapted nets. In: Proceedings of the 9th National Conference on Artificial Intelligence (AAAI 1991), Anaheim, CA, USA, vol. 2, pp. 590–595 (1991)
9. Fürnkranz, J., Gamberger, D., Lavrač, N.: Foundations of Rule Learning. Springer, Heidelberg (2012). doi:10.1007/978-3-540-75197-7
10. Goodfellow, I.J., Bengio, Y., Courville, A.C.: Deep Learning. Adaptive Computation and Machine Learning. MIT Press, Cambridge (2016)
11. Han, S., Pool, J., Tran, J., Dally, W.J.: Learning both weights and connections for efficient neural network. In: Advances in Neural Information Processing Systems 28 (NIPS 2015), Montreal, Quebec, Canada, pp. 1135–1143 (2015)
12. Hassibi, B., Stork, D.G.: Second order derivatives for network pruning: optimal brain surgeon. In: Advances in Neural Information Processing Systems 5 (NIPS 1992), pp. 164–171. Morgan Kaufmann, Denver (1992)
13. Hayes, J.P.: Digital Logic Design. Addison Wesley, Reading (1993)
14. Kayande, U., Bruyn, A.D., Lilien, G.L., Rangaswamy, A., van Bruggen, G.H.: How incorporating feedback mechanisms in a DSS affects DSS evaluations. Inf. Syst. Res. **20**(4), 527–546 (2009)
15. LeCun, Y., Denker, J.S., Solla, S.A.: Optimal brain damage. In: Advances in Neural Information Processing Systems 2 (NIPS 1990), Denver, Colorado, USA, pp. 598–605 (1989)
16. Liu, J., Li, M.: Finding cancer biomarkers from mass spectrometry data by decision lists. J. Comput. Biol. **12**(7), 971–979 (2005)
17. Malioutov, D.M., Varshney, K.R.: Exact rule learning via Boolean compressed sensing. In: Proceedings of the 30th International Conference on Machine Learning (ICML 2013), Atlanta, GA, USA, pp. 765–773 (2013)
18. Milaré, C.R., Carvalho, A.C.P.L.F., Monard, M.C.: Extracting knowledge from artificial neural networks: an empirical comparison of trepan and symbolic learning algorithms. In: Coello Coello, C.A., Albornoz, A., Sucar, L.E., Battistutti, O.C. (eds.) MICAI 2002. LNCS (LNAI), vol. 2313, pp. 272–281. Springer, Heidelberg (2002). doi:10.1007/3-540-46016-0_29
19. Ng, A.: Sparse autoencoder. CS294A Lecture Notes, Stanford University (2011)
20. Sato, M., Tsukimoto, H.: Rule extraction from neural networks via decision tree induction. In: Proceedings of the International Joint Conference on Neural Networks (IJCNN 2001), vol. 3, pp. 1870–1875. IEEE Press (2001)
21. Setiono, R.: Extracting rules from pruned neural networks for breast cancer diagnosis. Artif. Intell. Med. **8**(1), 37–51 (1996)
22. Setiono, R.: Extracting rules from neural networks by pruning and hidden-unit splitting. Neural Comput. **9**(1), 205–225 (1997)
23. Setiono, R.: A penalty-function approach for pruning feedforward neural networks. Neural Comput. **9**(1), 185–204 (1997)

24. Setiono, R., Liu, H.: Symbolic representation of neural networks. IEEE Comput. **29**(3), 71–77 (1996)
25. Srivastava, N., Hinton, G.E., Krizhevsky, A., Sutskever, I., Salakhutdinov, R.: Dropout: a simple way to prevent neural networks from overfitting. J. Mach. Learn. Res. **15**(1), 1929–1958 (2014)
26. Thodberg, H.H.: Improving generalization of neural networks through pruning. Int. J. Neural Syst. **1**(4), 317–326 (1991)
27. Towell, G.G., Shavlik, J.W.: Extracting refined rules from knowledge-based neural networks. Mach. Learn. **13**(1), 71–101 (1993)
28. Tsukimoto, H.: Extracting rules from trained neural networks. IEEE Trans. Neural Netw. Learn. Syst. **11**(2), 377–389 (2000)
29. Zeiler, M.D., Fergus, R.: Visualizing and understanding convolutional networks. In: Fleet, D., Pajdla, T., Schiele, B., Tuytelaars, T. (eds.) ECCV 2014. LNCS, vol. 8689, pp. 818–833. Springer, Cham (2014). doi:10.1007/978-3-319-10590-1_53
30. Zilke, J.R., Loza Mencía, E., Janssen, F.: DeepRED - rule extraction from deep neural networks. In: Proceedings of the 19th International Conference on Discovery Science (DS 2016), Bari, Italy, pp. 457–473 (2016)

An In-Depth Experimental Comparison of RNTNs and CNNs for Sentence Modeling

Zahra Ahmadi[1]([⊠]), Marcin Skowron[2], Aleksandrs Stier[1], and Stefan Kramer[1]

[1] Institut Für Informatik, Johannes Gutenberg-Universität, Mainz, Germany
zaahmadi@uni-mainz.de, stier@students.uni-mainz.de,
kramer@informatik.uni-mainz.de
[2] Austrian Research Institute for Artificial Intelligence, Vienna, Austria
marcin.skowron@ofai.at

Abstract. The goal of modeling sentences is to accurately represent their meaning for different tasks. A variety of deep learning architectures have been proposed to model sentences, however, little is known about their comparative performance on a common ground, across a variety of datasets, and on the same level of optimization. In this paper, we provide such a novel comparison for two popular architectures, Recursive Neural Tensor Networks (RNTNs) and Convolutional Neural Networks (CNNs). Although RNTNs have been shown to work well in many cases, they require intensive manual labeling due to the vanishing gradient problem. To enable an extensive comparison of the two architectures, this paper employs two methods to automatically label the internal nodes: a rule-based method and (this time as part of the RNTN method) a convolutional neural network. This enables us to compare these RNTN models to a relatively simple CNN architecture. Experiments conducted on a set of benchmark datasets demonstrate that the CNN outperforms the RNTNs based on automatic phrase labeling, whereas the RNTN based on manual labeling outperforms the CNN. The results corroborate that CNNs already offer good predictive performance and, at the same time, more research on RNTNs is needed to further exploit sentence structure.

1 Introduction

One aim of modeling sentences is to analyze and represent their semantic content for classification purposes. Neural network-based sentence modeling approaches have been increasingly considered for their significant advantages of removed requirements for feature engineering, and preservation of the order of words and syntactic structures, in contrast to the traditional bag-of-words model, where sentences are encoded as unordered collections of words. These neural network approaches range from basic Neural Bag-of-Words (NBoW), which ignores word ordering, to more representative compositional approaches such as Recursive Neural Networks (RecNNs) (e.g. [4]), Convolutional Neural Networks (CNNs) (e.g. [6]), Recurrent Neural Network (RNN) models (e.g. [9]), or LSTMs (which are outside the scope of this paper).

© Springer International Publishing AG 2017
A. Yamamoto et al. (Eds.): DS 2017, LNAI 10558, pp. 144–152, 2017.
DOI: 10.1007/978-3-319-67786-6_11

RecNNs work by feeding an external parse tree to the network. They are a generalization of classic sequence modeling networks to tree structures and have shown excellent abilities to model word combinations in a sentence. However, they depend on well-performing parsers to provide the topological structure, which are not available for many languages or do not perform well in noisy domains. Further, they often require labeling of all phrases in sentences to reduce the *vanishing gradient problem* [5]. Yet RecNNs implicitly model the interaction among input words, whereas Recursive Neural Tensor Networks (RNTNs) have been proposed to allow more explicit interactions [11]. On the other hand, CNNs are alternative models which apply one-dimensional convolution kernels in sequential order to extract local features. Each sentence is treated individually as a bag of n-grams, and long-range dependency information spanning multiple sliding windows is therefore lost. Another limitation of CNN models is their requirement for the exact specification of their architecture and hyperparameters [12].

We conducted extensive experiments over a range of benchmark datasets to compare the two network architectures: RNTNs and CNNs. Our goal is to provide an in-depth analysis of how these models perform across different settings. Such a comparison is missing in the peer-reviewed literature, likely because recursive networks often require labor-intensive manual labeling of phrases. Such annotations are unavailable for many benchmark datasets. In the next section, we propose two methods to label the internal phrases automatically. Later, we investigate whether there is an effect of using constituency parsing instead of dependency parsing in the RNTN model. In this way, we aim to contribute to a better understanding of the limitations of the two network models and provide a foundation for their further improvement.

2 Method

Recursive Neural Tensor Network Architecture. RNTNs [11] are a generalization of RecNNs where the interactions among input vectors are encoded in a single composition function (Fig. 1a). Here, we propose two methods for the automatic labeling of the phrases for RNTNs:

- **Rule-based method:** The RNTN model was first proposed for sentiment analysis purposes. Hence, our first approach uses a rule-based method to determine the valence of a phrase. We use four types of dictionaries: A dictionary of sentiments carrying terms (from unigrams to phrases consisting of n-gram words) with a corresponding sentiment score in the range of $[-k, +k]$, a negation dictionary, a dictionary of intensifier terms with a weight range of $[1, +k]$, and a dictionary of diminishers with a weight range of $[-k, -1]$. The analysis of a phrase is conducted from the end, backward to the beginning: If any sentiment term is found, we update the sentiment of the phrase from neutral to the value of the sentiment term in the dictionary. Then we search backwards for an intensifier or diminisher term. We increase or decrease the absolute value of the sentiment based on the weight of the intensifier/diminisher term and if required we adjust the score to a pre-defined range. In the next step, we adjust the score for a negation term. If one is found and there is no intensifier/diminisher before the sentiment term, the sentiment is reversed; otherwise if the phrase includes both the negation term and an intensifier/diminisher, the sentiment is set to weak negative. As an example, consider the terms *"not very good"* and *"not very bad"*, where both sentiments are weakly negative.

– **CNN-based method:** An alternative approach to labeling the phrases is to use a pre-trained CNN model. We use the architecture proposed here (see below for the description) to train a model on the sentence level, and use the resulting model to label the internal phrases for the RNTN. In this way, the RNTN can be applied to domains other than sentiment classification as well. The CNN model receives the complete sentences and their label as training data and will label the internal phrases in the test phase.

Convolutional Neural Network Architecture. Deep convolutional neural networks have led to a series of breakthrough results in image classification. Although recent evidence shows that network depth is of crucial importance to obtain better results [2,3], most of the models in the sentiment analysis and sentence modeling literature use a simple architecture, e.g. [6] uses a one-layer CNN. Inspired by the success of CNNs in image classification, our goal is to expand the convolution and Max-Pooling layers in order to achieve better performance by deepening the models and adding higher non-linearity to the structure. However, deeper models are also more difficult to train [3]. To reduce the computational complexity, we choose small filter sizes. In our experiments, we use a simple CNN model that consists of six layers (Fig. 1b): The first layer applies $1 \times d$ filters on the word vectors, where d is the word vector dimension. The essence of adding such a layer to the network is to derive more meaningful features from word vectors for every single word before feeding them to the rest of the network. This helps us achieving better performance since the original word vectors capture only sparse information about the words' labels. In contrast to our proposed layer, Kim uses a so-called *non-static* approach to modify the word vectors during the training phase [6].

The second layer of our CNN model is again a convolution layer with the filters of size $2 \times d$. The output of this layer is fed into a Max-Pooling layer with pooling size and stride 2. The reason for applying such a Max-Pooling layer in the middle layers of the network is to reduce the dimensionality and to speed up the training phase. This layer does not have notable effect on the accuracy of the resulting model. Next, on the fourth layer, convolving filters of size $2 \times d$ with a padding size 1 are again applied to the output of the previous layer. Padding preserves the original input size. The next layer applies Max-Pooling to the whole input at once. Using bigger pooling sizes leads to better results [12]. Finally, the last layer is a fully connected SoftMax layer which outputs the probability distribution over the labels.

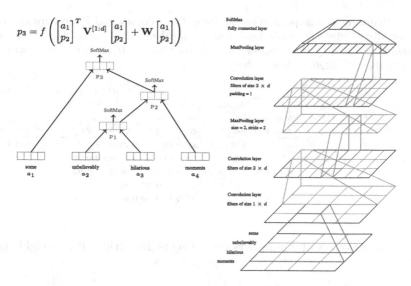

$$p_3 = f\left(\begin{bmatrix}a_1\\p_2\end{bmatrix}^T \mathbf{V}^{[1:d]}\begin{bmatrix}a_1\\p_2\end{bmatrix} + \mathbf{W}\begin{bmatrix}a_1\\p_2\end{bmatrix}\right)$$

Fig. 1. (a) An example of an RNTN architecture with word vector dimension of size 4 for sentiment classification of a given input sequence, which is parsed by a constituency parser. $\mathbf{V}$ and $\mathbf{W}$ are the tensor matrix and the recursive weight matrix, respectively. (b) Our proposed 6-layered CNN architecture. d is the dimension of the word vector.

3 Experiments

3.1 Experimental Settings

In our experiments, we use the pre-trained Glove [10] word vector models[1]: On the SemEval-2016 dataset, we use Twitter specific word vectors. On other datasets, we use the model trained on the web data from Common Crawl, which contains a case-sensitive vocabulary of size 2.2 million. Experiments show that RNTNs work best when the word vector dimension is set between 25 and 35 [11]. Hence, in all the experiments, the size of the word vector, the minibatch and the epochs were set to 25, 20 and 100, respectively. We use $f = tanh$ and a learning rate of 0.01 in all the RNTN models. In CNN models, the number of filters in the convolutional layers are set to 100, 200 and 300, respectively; and the maximum length of the sentences is 32. For shorter sentences, they are padded with zero vectors. In RNTN models which use constituency parsers, we use the Stanford parser [7]. For those models which use dependency parsers, we use the Tweebo parser [8] – a dependency parser specifically developed for Twitter data – for the SemEval-2016 dataset and on the rest of the datasets, we use the Stanford neural network dependency parser [1]. In rule-based methods, we use a dictionary of sentiments consisting of 6, 360 entries with maximum 2-gram words and a sentiment range of $[-3, +3]$, a negation dictionary consisting of 28 entries, a dictionary of intensifier terms consisting of 47 words with a weight range of

---

[1] http://nlp.stanford.edu/projects/glove/.

Table 1. Performance comparison on all datasets. Accuracy and F-measure are averaged over all the classes. n/a indicates non-defined cases as one of the classes was misclassified completely resulting in an undefined value. If an experiment was not applicable, the cell is left with a dash.

Dataset	RNTN										CNN		CNN (Kim model)		Rule-based	
	Constituency parser						Dependency parser									
	Rule		CNN		Manual		Rule		CNN							
	Acc.	F1	Acc.	F1	Acc.	F1	Acc.	F1	Acc.	F1	Acc.	F1	Acc.	F1	Acc.	F1
MR	0.63	0.63	0.70	0.70	-	-	0.50	0.50	0.49	0.49	**0.71**	**0.71**	**0.71**	**0.71**	0.64	0.64
SemEval-2016	0.53	0.45	0.52	0.51	-	-	0.52	0.45	0.50	0.49	0.56	0.56	**0.60**	**0.57**	0.53	0.52
SST-5	0.30	0.28	0.34	0.21	**0.41**	**0.32**	0.30	0.29	0.30	n/a	0.37	0.26	0.39	**0.32**	0.31	0.29
TREC	-	-	0.72	n/a	-	-	-	-	0.33	n/a	**0.86**	**0.86**	0.54	0.57	-	-
Subj	-	-	0.76	0.76	-	-	-	-	0.42	0.42	**0.89**	**0.89**	0.88	0.88	-	-

[1, 3], and a dictionary of diminishers consisting of 26 entries with a weight range of $[-3, -1]$.

3.2   Task 1: Sentiment Analysis

In the first task, we compare the models on a set of commonly used sentiment analysis benchmark datasets: The Movie Review (**MR**) dataset[2] that has positive or negative class, each contains 5331 instances. As the MR dataset does not have a separate test set, we use 10-fold cross-validation in the experiments. An extended version of the MR dataset relabeled by Socher et al. [11] in the Stanford Sentiment Treebank (**SST-5**)[3] has five fine-grained labels: negative, somewhat negative, neutral, somewhat positive and positive. **SST-5** contains 8544 training sentences, 1101 validation sentences and 2210 test sentences. The **SemEval-2016**[4] dataset is a set of tweets labeled as either of the three negative, neutral and positive labels. It has 12, 644 training tweets, 3001 validation tweets and 20, 632 test instances.

- **Comparison of automatic labeling methods:** We first use the manually labeled SST-5 dataset to test the effectiveness of our automatic labeling methods. We extract all the possible phrases of the whole dataset with respect to their parse trees and use our rule-based method to label them. In the next step we train the CNN model on the set of training instances and use the resulting model to label the phrases. The accuracy of the rule-based and the CNN labeling methods are 69% and 40%, respectively. As we see, the overall accuracy of the CNN-based model is significantly lower than that of the rule-based method. To have a better understanding of the classification performance, we look into their confusion matrices. We subtract the corresponding elements of the CNN-based confusion matrix from that of the rule-based variant and normalize them by dividing by the total number of phrases for each

---

[2] https://www.cs.cornell.edu/people/pabo/movie-review-data/.
[3] http://nlp.stanford.edu/sentiment/Data.
[4] http://alt.qcri.org/semeval2016/task4/.

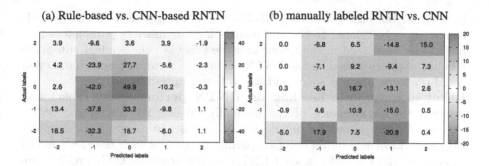

Fig. 2. (a) Heatmap of difference of rule-based RNTN and CNN-based RNTN confusion matrices on the SST-5 phrase set. The numbers are the percentage of normalized differences based on the total number of phrases for each label. (b) Heatmap of difference of the manually labeled RNTN and the CNN model confusion matrices on the SST-5 test set. The numbers in each cell indicate the percentage of normalized differences based on the total number of sentences for each label.

label (i.e. $\frac{conf_{rule}^{i,j} - conf_{cnn}^{i,j}}{total_i}$ where i and j are the actual and predicted labels, respectively). Figure 2a illustrates the resulting heatmap. Red color indicates cases where more phrases are predicted by the rule-based method than by the CNN-based method while the blue color indicates the opposite case. We observe that the CNN is a better model to correctly classify somewhat positive (1) and somewhat negative (−1) classes than the rule-based method. In turn, the rule-based method is superior in the classification of the neutral (0) and negative (−2) classes. To have a better interpretation of the numbers in the heatmap, it is beneficial to look at the distribution of labels in the whole population: 2.6%, 11.3%, 67.7%, 14.3% and 4.1% for −2 to +2 labels.

- **Constituency parser Vs. dependency parser:** The output of a dependency parser is a Directed Acyclic Graph (DAG). However, RNTNs accept a binary-branching parse tree as an input. Therefore, we have binarized the output of the dependency parser by starting from the word which does not point to any other word as its parent, and recursively binarize its children list by adding empty nodes when necessary. While analyzing the effect of using a dependency parser instead of a constituency parser in RNTNs (Table 1), a significant loss of performance is visible in some datasets (e.g. MR). This is particularly noticeable when the labeling method is CNN (e.g. 70% to 49% in MR). The reason for this could be the difference of the word order resulting from a dependency parser compared to the n-gram features extracted by the CNN.
- **RNTN Vs. CNN:** Table 1 shows a detailed comparison of the RNTN automatically labeled variants to the CNN model and the rule-based method. We have reported the average accuracy and F-measure over all classes. With the same settings of parameters, we see a better performance of the CNN model on the MR and SemEval-2016 datasets. The largest performance (in terms of F-measure) improvement can be observed on the SemEval-2016 dataset, 0.51

to 0.56, for the best performing RNTN and CNN approaches. The possible reasons may be related to the enormously large number of parameters that have to be optimized in the tensor and the effects of the applied automatic labeling of phrases used on the RNTN. Therefore, a future research direction could try to reduce this space and find a better initialization.

- **Effect of automatic labeling on RNTN performance:** Table 1 also presents the performance of the manually labeled RNTN on the SST-5 dataset. As we can see, automatic labeling results in a significant degradation of performance on SST-5. Comparing the results with the CNN model shows that the manually labeled RNTN outperforms the CNN architecture in terms of overall accuracy and F-measure. To have a closer look into the confusion matrix of both methods, we generate a heatmap similar to Fig. 2a, this time subtracting the CNN confusion matrix elements from that of the RNTN method (i.e. $\frac{conf_{rntn}^{i,j} - conf_{cnn}^{i,j}}{total_i}$). Blue color indicates more prediction of sentences by the CNN model than by the RNTN while the red color indicates the reverse case. Figure 2b indicates that the RNTN has a tendency to classify more instances into neutral (0) and positive (2) labels and it is better at correct prediction of somewhat negative (-1), neutral and positive labels while the CNN is better at classifying negative (-2) and somewhat positive (1) labels. Here, the distribution of sentences over labels is closer to the uniform distribution: 12.6%, 28.6%, 17.6%, 23.1% and 18.1% for -2 to $+2$ labels. Unfortunately, currently there is no other dataset that is manually labeled at the phrase level. A future direction includes further evaluation of the impact of the phrase labeling accuracy on various datasets.

3.3 Task 2: Sentence Categorization

We test this task on two datasets: The TREC[5] question dataset, where the goal is to classify a question into six coarse-grained question types (whether a question is about an entity, a person, a location, numeric information, abstract concepts or an abbreviation), and the Subj[6] dataset, where the goal is to classify a sentence as being objective or subjective. The TREC dataset has 5452 training instances and 500 test sentences. The Subj dataset contains 10,000 sentences in total but it does not have a separate test set, therefore we use 10-fold cross-validation. The results are reported in the bottom section of Table 1. In these experiments only CNN-based methods are applicable. We observe that the CNN model outperforms RNTN versions, and dependecy parsing drastically reduces the performance of the RNTN.

3.4 Comparison of CNN Architectures

In the next experiment, we compare our proposed deep CNN architecture to a one layer CNN to find out the cases where the deep structure is beneficial. The

---

[5] http://cogcomp.cs.illinois.edu/Data/QA/QC/.
[6] https://www.cs.cornell.edu/people/pabo/movie-review-data/.

one layer CNN architecture [6] has several parallel filters of different sizes and a max-pooling layer. In our experiments, we have used 100 filters of size 3, 4, and 5. Classification results (see next to last column of Table 1) indicate that the performance of the one layer architecture is comparable to the proposed deep architecture on the MR dataset and that it performs better on the rest of the sentiment datasets. The performance of Kim's architecture on the SST-5 dataset is comparable to the RNTN based on manual labeling. These results highlight the importance of keyphrase recognition in sentiment tasks, where applying larger filters is more beneficial than having several layers of small filters. However, on the other sentence categorization datasets, i.e. TREC and Subj, the proposed deep CNN outperforms the flat architecture.

4   Conclusions

In this paper we studied two well-known deep architectures, CNNs and enhanced versions of RNTNs, in the context of sentence modeling. In order to avoid the labor-intensive task of manually labeling the internal phrases for recursive networks, we proposed two methods to automatically label them for training and tuning phases: a rule-based method which is specifically used for sentiment prediction and a CNN based method for general purposes. Considering this part of study, the evaluation results on the SST-5 dataset indicate that the CNN method has a tendency to assign a positive or negative polarity to the phrases while the rule-based method classifies many of them as neutral. Based on the presented automatic labeling methods of internal nodes, we conducted an in-depth study of the RNTN model and compared the model to a relatively simple deep CNN architecture. Experimental results conducted on an extensive set of standard benchmark datasets demonstrate that the proposed CNN model outperforms the RNTN variants with automatic phrase labeling, whereas the RNTN with manual labeling (if available) outperforms the CNN. However, in that case, a one layer CNN with several filters of different sizes is comparable to the manually labeled RNTN. These results demonstrate that the syntactic structure of a sentence will help in the classification performance when it is possible to accurately label the internal nodes of a parse tree, otherwise CNNs can be more successful at representing the meaning of the sentence with respect to the task. The findings also show that there is still room for improvement of RNTN variants by determining tensor functions in a more informed manner.

Acknowledgements. The authors thank PRIME Research for supporting the first author during her research time. The second author is supported by the Austrian Science Fund (FWF): P27530-N15.

References

1. Chen, D., Manning, C.D.: A fast and accurate dependency parser using neural networks. In: Proceedings of Empirical Methods in Natural Language Processing, pp. 740–750 (2014)
2. Conneau, A., Schwenk, H., Barrault, L., Lecun, Y.: Very deep convolutional networks for text classification. In: Proceedings of the 15th Conference of the European Chapter of the Association for Computational Linguistics, pp. 1107–1116 (2017)
3. He, K., Zhang, X., Ren, S., Sun, J.: Deep residual learning for image recognition. In: IEEE Conference on Computer Vision and Pattern Recognition, pp. 770–778 (2016)
4. Irsoy, O., Cardie, C.: Deep recursive neural networks for compositionality in language. In: Advances in Neural Information Processing Systems, pp. 2096–2104 (2014)
5. Iyyer, M., Manjunatha, V., Boyd-Graber, J., Daumé III, H.: Deep unordered composition rivals syntactic methods for text classification. In: Proceedings of 53rd Annual Meeting of the Association for Computational Linguistics, pp. 1681–1691 (2015)
6. Kim, Y.: Convolutional neural networks for sentence classification. In: Proceedings of Empirical Methods in Natural Language Processing, pp. 1746–1751 (2014)
7. Klein, D., Manning, C.D.: Accurate unlexicalized parsing. In: Proceedings of the 41st Annual Meeting on Association for Computational Linguistics, pp. 423–430 (2003)
8. Kong, L., Schneider, N., Swayamdipta, S., Bhatia, A., Dyer, C., Smith, N.A.: A dependency parser for tweets. In: Proceedings of Empirical Methods in Natural Language Processing, pp. 1001–1012 (2014)
9. Li, J., Luong, M.T., Jurafsky, D., Hovy, E.: When are tree structures necessary for deep learning of representations? In: Proceedings of Empirical Methods in Natural Language Processing and Computational Natural Language Learning, pp. 2304–2314 (2015)
10. Pennington, J., Socher, R., Manning, C.D.: Glove: Global vectors for word representation. In: Proceedings of Empirical Methods in Natural Language Processing, pp. 1532–1543 (2014)
11. Socher, R., Perelygin, A., Wu, J.Y., Chuang, J., Manning, C.D., Ng, A.Y., Potts, C.P.: Recursive deep models for semantic compositionality over a sentiment treebank. In: Proceedings of Empirical Methods in Natural Language Processing, pp. 1631–1642 (2013)
12. Zhang, Y., Wallace, B.: A sensitivity analysis of (and practitioners' guide to) convolutional neural networks for sentence classification. CoRR abs/1510.03820 (2015)

Feature Selection

Improving Classification Accuracy
by Means of the Sliding Window Method
in Consistency-Based Feature Selection

Adrian Pino Angulo$^{(\boxtimes)}$ and Kilho Shin

Graduate School of Applied Informatics, University of Hyogo, Kobe, Hyogo, Japan
apinoa85@gmail.com, kilhoshin314@gmail.com

Abstract. In the digital era, collecting relevant information of a technological process has become increasingly cheaper and easier. However,
due to the huge available amount of data, supervised classification is
one of the most challenging tasks within the artificial intelligence field.
Feature selection solves this problem by removing irrelevant and redundant features from data. In this paper we propose a new feature selection
algorithm called SWCFS, which works well in high-dimensional and noisy
data. SWCFS can detect noisy features by leveraging the sliding window
method over the set of consecutive features ranked according to their
non-linear correlation with the class feature. The metric SWCFS uses to
evaluate sets of features, with respect to their relevance to the class label,
is the bayesian risk, which represents the theoretical upper error bound of
deterministic classification. Experiments reveal SWCFS is more accurate
than most of the state-of-the-art feature selection algorithms.

1 Introduction

Big data has been one of the most hottest trends for the last ten years. Supervised classification as a sub-field of machine learning, is increasingly gaining popularity among researchers due to its versatility and power of application at any
field where data is available. Among the most common examples of supervised
learning we can find: microarray problem classification [2], cancer diagnosis [3]
and network intruder detection [1]. Supervised classification in incredibly powerful to make predictions and suggestions by means of inferring a function from
labelled training data. The most basic structured data corresponds to a single
data matrix

$$\mathbf{D} = \begin{bmatrix} x_1^1 & \cdots & x_1^n & c_1 \\ \vdots & \ddots & \vdots & \vdots \\ x_m^1 & \cdots & x_m^n & c_m \end{bmatrix},$$

where every instance x_j is described by a row vector $[x_j^1, \ldots, x_j^n, c_j]$: x_j^i is a value
for the feature f_i; and c_j is a class label, which is a value for the class variable
C. The collected data have no utility unless useful information is discovered
from them. Supervised classification is a central issue in machine learning and

© Springer International Publishing AG 2017
A. Yamamoto et al. (Eds.): DS 2017, LNAI 10558, pp. 155–170, 2017.
DOI: 10.1007/978-3-319-67786-6_12

consists on finding a classification function $\ell : \mathbf{D} \to v(c)$ that is able to classify an arbitrary instance with unknown class from $v(c) \in C$. ℓ is built from analysing the relation between instances in $\mathbf{D}$. The performance of supervised classifiers is often measured in three directions: efficiency, representation complexity and accuracy. The efficiency refers to the time required to learn the classification function ℓ; while the representation complexity often refers to the number of bits used to represent the classification function. One of the most common metrics to measure the accuracy of a supervised classifier is the error rate defined as:

$$Err(\ell, \mathbf{D}) = \frac{1}{m} \sum_{j=1}^{m} \bar{\delta}(\ell(x_j), c_j),$$

where m is the number of instances in $\mathbf{D}$ and $\bar{\delta}$ is the complement of the Kronecker's delta function, which returns 0 if both arguments are equal and 1 otherwise. All these three factors can be strongly affected when there exist features in $\mathbf{D}$ that do not contain useful information to predict the class variable. Feature selection plays an essential role in supervised classification since its main goal is to identify and remove irrelevant and redundant features that do not contribute to minimize the error of a given classifier [4]. Basically, the advantages of feature selection include selecting a set of features $\tilde{F} = \{f_{i_1}, \ldots, f_{i_k}\} \subsetneq F$ with:

$$Err(\ell, \mathbf{D}_{\tilde{F}}) \leq Err(\ell, \mathbf{D}),$$

where $\mathbf{D}_{\tilde{F}}$ is the result of projecting $\tilde{F}$ over $\mathbf{D}$. The process of selecting features is composed of two basic components: an evaluation function and a search engine [5]. The evaluation function is a metric that evaluates quantitatively how good are a set of features to discriminate among class labels. On the other hand, the search engine is in charge of generating all the potential sets to be evaluated.

Feature selection algorithms can be divided into three broad categories: wrapper, filter and embedded methods. To evaluate a feature set $\tilde{F}$, wrapper methods use some accuracy score of a classifier after being trained in the dataset projected by $\tilde{F}$. Wrapper methods are very low in efficiency since training and testing the inferred function is required for each evaluation. Conversely, filters make use of explanatory analysis on data to assign a score to each feature set. Filters are usually less computationally expensive than wrappers, but they output a feature set that is not tuned to a specific type of predictive model. Embedded methods learn which features best contribute to the accuracy of the model while the model is being created. The most common type of embedded feature selection methods are regularization or penalization methods [6].

Many of filter algorithms evaluate relevance of individual features using statistical measures, and some of them also incorporate evaluation of mutual relationship among features into the result of feature selection. In particular, two sorts of inter-feature relation are known to harm the performance of feature selection, that is, redundancy and interaction. Two features are said to interact with each other if both individually can be considered irrelevant based on their correlation with the class; but when combined, they can become very relevant.

The ideal feature selection algorithm should be able to evaluate interacting features, if present, and could incorporate the results of evaluation into feature selection results. However, since detecting all of the interacting features is computationally expensive in high-dimensional environments, feature selection algorithms only focus on searching for: (i) relevant features, (ii) relevant features and non-redundant features, or (iii) interacting features.

To the best of our knowledge, SUPER-LCC [8] is one of the best feature selection algorithms proposed by Shin et al. [8], which better can find accurate sets. SUPER-LCC uses the backward search and the bayesian risk measure to detect interacting features with an extreme high-performance. (We will discuss in more detail this algorithm in Sect. 2.3). However, we have found that under some conditions, SUPER-LCC can not find features with high relevance score.

Assuming that there are several interacting set of features in the dataset that can equally predict the class, SUPER-LCC is not designed to select the one composed by features with the highest relevance. Pino and Shin [9] partially solve this problem by proposing the algorithm ASDCC [9] that uses a measure, which judge features according to their individual relevance score and their interaction rate. However, since ASDCC is based on the *Steepest Descent Search* [10], which needs $(|F| + |\tilde{F}|)(|F| + |\tilde{F}|)/2$ evaluations to output $\tilde{F}$, is not practical for high-dimensional data.

The main motivation of this paper is to improve both of the SUPER-LCC and ASDCC algorithms. Our approach is simple and is composed by two new gears. First, we use the *Steepest Descent Search*, but make it faster by using a sliding window method over F to only judge irrelevant features in the first iterations. Second, we use the *Binary Search* to detect and remove the non-interacting features with lower individual relevance before starting the search. We have found that this considerably reduce the search space. In the remaining of this paper we further analyse the most popular feature selection algorithms by giving concrete examples. Creating the ideal feature selection algorithm is a hard task. However in Sect. 3 we propose our new consistency-based algorithm namely *Sliding Window for Consistency-based Feature Selection* (SWCFS) that can find a feature set to approximately solve the optimization problem of maximizing interaction among relevant features and minimizing redundancy. In Sect. 4 we compare our algorithm with several state-of-the-art algorithms in 20 benchmark datasets.

2   Feature Selection Methods

Feature selection can be accomplished in a variety of ways depending on the characteristics of the data. In this section, we review most popular algorithms in the feature selection field and analyze their advantages and drawbacks.

2.1   Feature Ranking Methods

The individual relevance score $r(f_i; C)$ of a feature f_i is a common term that refers to the power of a single feature to predict the class feature C.

The individual relevance score can be used as a metric to select the features that better predicts the class under certain threshold. That is, features are ranked using their individual relevance score and then the top features are selected. These algorithms are called feature ranking methods and often use correlation, distance and information measures between a single feature and the class feature to find a set full of high-relevant features.

As an example, RELIEF [11] computes the relevance score of a feature f_i based on the capability of f_i to discriminate among instances of different classes. Assuming instance x_k with class c_+ is randomly sampled from the data, and H_k and M_k are two sets of instances (in the neighborhood of x_k) with class c_+ and c_- respectively, then a feature has high separability power if it has similar values in instances from H_k and different values in instances from M_k. RELIEFF is an extension of RELIEF that handle multiple classes by splitting the data into series of two-class data [12]. The individual relevance of each feature f_i in F is assessed by computing the average of its separability power in l instances randomly sampled. That is,

$$RF(f_i; C) = \frac{1}{|C|} \sum_{k=1}^{l} (-\frac{1}{|M_k|} \sum_{x_j \in M_k} d(x_k^i, x_j^i) + \sum_{c \neq c(x_k)} \frac{p(c)}{|H_k|(1 - p(c))} \sum_{x_j \in H_k} d(x_k^i, x_j^i)),$$

where $p(c)$ is the probability that an instance is labeled with class c and $d(x_k^i, x_j^i) = (x_k^i - x_j^i)/(max(f_i) - min(f_i))$, with $max(f_i)$ and $min(f_i)$ being the maximum and minimum value of feature f_i.

While RF requires numeric features, the Mutual Information measure accepts categorical features and can be used to measure correlation between a feature and the class:

$$MI(f_i; C) = \sum_{\substack{x_i \in V(f_i), \\ c \in V(C)}} Pr[f_i = x_i, C = c] \log \frac{Pr[f_i = x_i, C = c]}{Pr[f_i = x_i] Pr[C = c]}$$

Mutual Information is biased in favour of features with greater number of values and this is a problem when used for feature selection [16]. The Symmetrical Uncertainty measure deals with this problem by a normalizing function:

$$SU(f_i; C) = 2 \frac{MI(f_i; C)}{H(f_i) + H(C)}$$

The Symmetrical Uncertainty is the harmonic mean between $MI(f_i, C)/H(f_i)$ and $MI(f_i, C)/H(C)$. Therefore it is symmetrical and in the range of $[0, 1]$.

Although the ranking feature algorithms are usually simple and fast, they have two serious drawbacks that may affect the performance of supervised classifiers. First, redundant features are likely to be selected. Second, they usually can not detect interacting features.

2.2   Pairwise Evaluation Methods

Oppositely to the feature ranking algorithms, pairwise evaluation methods can detect and eliminate relevant features, but also are able to remove redundant features. Most of these algorithms use one of the measures mentioned in the section above. The way most of these algorithms operates is as follows. First, the relevance score $r(f_i, C)$ of each feature in $f_i \in F$ is computed and second, pairwise evaluations $r(f_i, f_j)$ between features are performed to detect features that are highly correlated to others.

As an example, the algorithm FCBF (*Fast Corelator based-Filter*) [16] first ranks all features $\{f_1, \ldots, f_n\}$ in the descending order of the Symmetrical Uncertainty scores. Then, starting from the best/first feature in the ranking f_1, it applies a redundancy filter to all of features f_j with $j > i$, and, if $SU(f_i; f_j) > SU(f_j; C)$ holds then it removes f_j. Since the overall complexity of algorithm FCBF is $O(mn \log n)$ where m is the number of instances in the data, this algorithm is scalable to large data.

Although feature ranking and pair-wise evaluation methods are quite fast and easy to implement, they are not able to detect interacting features. That's why in high-dimensional domains they may output low-quality sets.

To illustrate, consider the class target function $c = f_1 \oplus f_2$ where $\{f_1, f_2, \ldots, f_n\} \in F$ are binary features and $\oplus$ denotes the *xor* operator. Beforehand, we know $\{f_1, f_2\}$ won't be selected because both features by themselves are uncorrelated with c. If we consider that features in $F \backslash \{f_1, f_2\}$ can not accurately describe the class then we can not expect a good performance of the classifier after reducing F by any of the feature ranking or pairwise evaluation algorithms. Figure 1 depicts a numerical version of the aforementioned example.

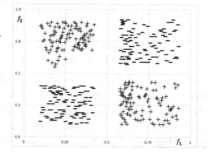

Fig. 1. Example of how non-relevant features can interact with each other to accurately discriminate between two classes.

Consistency-based measures are a successful choice to face this problem because they can detect high-order interacting features [20].

2.3   Consistency-Based Algorithms

Consistency-based algorithms can detect interacting features by collectively evaluating relevance (correlation) of a feature set to the class. Although exhaustive search of all possible feature sets is computationally too expensive, the result can be expected to be accurate.

We first introduce the *Bayesian risk* as a consistency measure example and then we define the consistency measure concept. To illustrate, for a dataset **D**, we view a feature of **D** as a random variable and a feature set $\tilde{F}$ as a joint variable. Then, we let $\Omega_{\tilde{F}}$ denote the sample space of $\tilde{F}$, C denotes a variable that describes classes and $\mathrm{Pr_D}$ denotes the empirical probability distribution of **D**. With these notations, the Bayesian risk is defined by

$$\mathfrak{Br}(\tilde{F}) = 1 - \sum_{x \in \Omega_{\tilde{F}}} \max\{ \Pr_{\mathbf{D}}[\tilde{F} = x, C = y] \mid y \in \Omega_C \}.$$

This function is also referred to as the *inconsistency rate* in [20]. The Bayesian risk has two important properties, that is, *determinacy* and *monotonicity*, and we first introduce the notion of *consistent feature sets* to explain the properties.

Definition 1. *For a dataset D described by F, a feature set $\tilde{F} \subseteq F$ is consistent, iff,* $\Pr_D[C = y \mid \tilde{F} = \boldsymbol{x}] = 0$ *or* 1 *for all* $x \in \Omega_{\tilde{F}}$ *and* $y \in \Omega_C$.

Then, the determinacy and monotonicity properties are described as follows.

Determinacy. $\mathfrak{Br}(\tilde{F}) = 0$, if, and only if, $\tilde{F}$ is consistent in **D**.
Monotonicity. $\mathfrak{Br}(\tilde{F}) \geq \mathfrak{Br}(G)$, if $\tilde{F} \subseteq G \subsetneq F$.

Formally, a consistency measure is defined as a function that returns real numbers on input of feature sets that has the determinacy and monotonicity properties. The *consistency-based feature selection*, on the other hand, is characterized by use of consistency measures as the evaluation function.

INTERACT [20] is the first instance of consistency-based feature selection algorithms that have practical performance in both time efficiency and prediction accuracy. It selects an answer from a small number of candidates, to be specific, $|\mathcal{F}|$ feature subsets. In the first step, INTERACT sorts the features in F into $(f_1, \ldots, f_{|\mathcal{F}|})$ in the increasing order of the symmetric uncertainty $\mathrm{SU}(f_i, C)$ and then sets $\tilde{F}$ to F. Initially, $\tilde{F}$ is equal F and then, Starting from $i = 1$, INTERACT lets $\tilde{F} = \tilde{F} \setminus \{f_i\}$ and computes $\mathfrak{Br}(\tilde{F} \setminus \{f_i\}) - \mathfrak{Br}(\tilde{F})$, which is non-negative by the monotonicity property; If $\mathfrak{Br}(\tilde{F}) \setminus \{f_i\} - \mathfrak{Br}(\tilde{F}) \leq \delta$, INTERACT judges that the feature f_i is not important and eliminates it from $\tilde{F}$; INTERACT repeats this procedure until all features are tested.

Although INTERACT presented good balance between accuracy and efficiency, Shin and Xu [21] have found that $\mathfrak{Br}(\tilde{F}) \setminus \{f_i\} - \mathfrak{Br}(\tilde{F})$ can accumulate, and consequently, INTERACT may output feature sets whose Bayesian risks are high. They also proposed a new algorithm, namely, *Linear Consistency Constrained* algorithm (LCC), that solves this problem [21]. The difference of LCC from INTERACT is slight: The criteria to eliminate f_i is on $\mathfrak{Br}(\tilde{F}) \setminus \{f_i\}$ instead of on $\mathfrak{Br}(\tilde{F}) \setminus \{f_i\} - \mathfrak{Br}(\tilde{F})$. Therefore, an output $\tilde{F}$ of LCC is *minimal* in the sense that both of $\mathfrak{Br}(\tilde{F}) \leq \delta$ and $\mathcal{G} \subsetneq \tilde{F} \Rightarrow \mathfrak{Br}(\mathcal{G}) > \delta$ hold.

Recently, the efficiency of LCC has been improved by conducting binary search instead of linear search [8]. This idea was materialized under the name of SUPER-LCC and works under the assumption that high-dimensional datasets are abundant in irrelevant features that can be removed in mass. By the first to the $(i-1)$-th iterations of the algorithm, the algorithm determines a sequence of indices of features $l_1 < l_2 < \cdots < l_{i-1}$, and defines $\tilde{F} = F \setminus \{f_1, \ldots, f_{l_{i-1}}\} \cup \{f_{l_1}, \ldots, f_{l_{i-1}}\}$. In the i-th iteration, the algorithm finds l_i such that

$$l_i = \underset{j = l_{i-1}+1, \ldots, n}{\mathrm{argmax}} \ \{\mathfrak{Br}(\tilde{F} \setminus \{f_{l_{i-1}+1}, \ldots, f_j\}) \leq \delta\}.$$

by *Binary Search* due to the monotonicity property of the bayesian risk. SUPER-LCC outputs the same set as LCC but on average has a computational complexity of $O(nm(\log n + \log m))$ where n is the number of features that describes the m

instances in D. To the best of our knowledge, SUPER-LCC is the algorithm with better practical performance in both of efficiency and accuracy. According to their authors, for data with more than hundred thousand features, SUPER-LCC needs some seconds to give a response in an ordinary personal computer [8].

On the other hand, *Steepest Descent Consistency Constrained* algorithm (SDCC) [10] is further stemmed from LCC and aims to improve the prediction performance of LCC by expanding the search range of LCC by leveraging a steepest descent search instead of a linear search. That is, when $\tilde{F}$ is the current best feature subset, SDCC asks the evaluation function to calculate the Bayesian risk scores of all of the subsets that are obtained by eliminating a single feature from $\tilde{F}$. If the minimum of the Bayesian risks computed is no greater than δ, SDCC updates $\tilde{F}$ with one of the subsets that yield the minimum. The outputs of SDCC are minimal in the same sense as stated above. Hence, if $\tilde{F}$ is the final output, SDCC evaluates $(|F| + |\tilde{F}|)(|F| - |\tilde{F}| + 1)/2$ feature subsets. Since SDCC is not practical for high-dimensional domains Pino and Shin developed a new version of the SDCC algorithm namely, *Accurate Steepest-Descent-Consistency-Constrained* (ASDCC) [9]. ASDCC introduces two rules to early detect and remove non-interacting features and scores the features according to their individual relevance and their bayesian risks.

In the remaining of this paper, we further analyse ASDCC and also the intrinsic characteristics and drawbacks of SDCC and SUPER-LCC to achieve the goal of creating a novel feature selection algorithm as fast as SUPER-LCC and at least as accurate as SDCC.

3 Our Proposal

Steepest-descent is a first-order optimization algorithm that finds a local minimum of a given function by stepping the solution in the direction where the function decreases most quickly [10]. The main advantage of SDCC over LCC can be justified as follows. LCC eliminates the first feature f_i that satisfies $\mathfrak{Br}(\tilde{F} \setminus \{f_i\}; C) \leq \delta$ from $\tilde{F}$, while SDCC tests all $f_i \in \tilde{F}$ and eliminates f_i that minimizes $\mathfrak{Br}(\tilde{F} \setminus \{f_i\}; C)$ such that $\mathfrak{Br}(\tilde{F} \setminus \{f_i\}; C) \leq \delta$. We consider $\tilde{F}$ as a point in the space of subsets of the entire features of $\mathbf{D}$. The neighbours of $\tilde{F}$ are determined by $\tilde{F} \setminus \{f_i\}$ for $f_i \in \tilde{F}$; and the distance between $\tilde{F}$ and $\tilde{F} \setminus \{f_i\}$ is given by $\mathfrak{Br}(\tilde{F}\{f_i\}; C) - \mathfrak{Br}(\tilde{F}; C)$. When a function f over feature subsets is $f(\tilde{F}) = |Ft|$, the gradient from $\tilde{F}$ to $\tilde{F} \setminus \{f_i\}$ is $1/(\mathfrak{Br}(\tilde{F} \setminus \{f_i\}; C) - \mathfrak{Br}(\tilde{F}; C))$. Therefore, an increase of the inconsistency score by eliminating a single feature for SDCC is at least equal than by eliminating a single feature for LCC. This also means that SDCC can eliminate more features than LCC.

Although it is known that SDCC significantly beats LCC in terms of the inconsistency score, SDCC performs $(|F| + |\tilde{F}|)(|F| - |\tilde{F}| + 1)/2$ evaluations to output $\tilde{F}$. Furthermore, we have detected that SDCC removes a lot of features highly-correlated with the class variable, which may affect the performance of classifiers. In the remainder of this section we discuss some efficiency and effectiveness issues of SDCC. Moreover, we propose a new algorithm to solve these issues.

3.1   Defieciencies of *steepest-descent* Search

To set the scene of this section, consider two performance-related issues in the
SDCC algorithm that are revealed in the following example. Note that **Problem
1** and **Problem 2** are solved by ASDCC algorithm. Therefore, the contribution
of this paper is related to solving **Problem 3**.

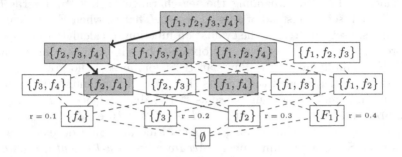

Fig. 2. An example of search paths by *steepest-descent*. r stands for the individual
relevance of a feature.

Figure 2 is the Hasse diagram of $F = \{f_1, f_2, f_3, f_4\}$, and the gray nodes
represent the feature subsets whose *inconsistency* is zero. With $\delta = 0$, the solid
lines represent an example of the paths that SDCC can track. In the first iteration,
SDCC investigates the four feature subsets of $\{f_2, f_3, f_4\}$, $\{f_1, f_3, f_4\}$, $\{f_1, f_2, f_4\}$
and $\{f_1, f_2, f_3\}$. The *inconsistency* of three of them are zero, and SDCC chooses
$\{f_2, f_3, f_4\}$. In the same way, in the second iteration, SDCC investigates $\{f_3, f_4\}$,
$\{f_2, f_4\}$ and $\{f_2, f_3\}$ and chooses $\{f_2, f_4\}$. In the last iteration, SDCC investigates
$\{f_4\}$ and $\{f_2\}$ and then terminates.

Problem 1: Small Total Relevance Score: In Fig. 2, $\{f_2, f_4\}$ and $\{f_1, f_4\}$
are the two candidates that SDCC can select, because they are minimal in the
inclusion relation among the feature subsets in F with $\mathfrak{Br}(F; C) \leq 0$. Although
the Sdcc selects one of $\{f_2, f_4\}$ and $\{f_1, f_4\}$ arbitrarily, $\{f_1, f_4\}$ is likely to be
a better answer than $\{f_2, f_4\}$, because $r(f_1, C) + r(f_4, C) = 0.5 > r(f_2, C) +
r(f_4, C) = 0.4$. In general, provided all the minimal sets G in F with $\mathfrak{Br}(G; C) \leq
\delta$, SDCC arbitrary selects any set G regardless any other information.

Solution to Problem 1. The Individual relevance insensitivity problem occurs
because the individual relevance of features has no meaning in the *steepest-
descent* algorithm. That is, the *steepest-descent* arbitrarily removes any feature
f^- such that $f^- \in \operatorname{argmin}_{f_i \in \tilde{F}}\{\mathfrak{Br}(\tilde{F} \setminus \{f_i\}; C) \mid \mathfrak{Br}(\tilde{F} \setminus \{f_i\}; C) \leq \delta\}$.

A straightforward way to deal with this problem is by removing the feature
f^- with the smallest individual relevance. That is, in each iteration remove
feature f^-, such that

$$f^- \in \operatorname{argmin}\{r(f; C) \mid f \in \operatorname{argmin}\{\mathfrak{Br}(\tilde{F} \setminus \{f_i\}) \mid f_i \in \tilde{F}, \mathfrak{Br}(\tilde{F} \setminus \{f_i\}) \leq \delta\}\}. \tag{1}$$

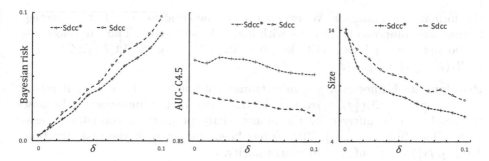

Fig. 3. Comparison between the original SDCC [10] and its corrected version that searches features based on Eq. (1) in terms of the bayesian risk, the AUC-ROC by $C4.5$ classifier and, the number of features selected.

To validate the effect of this solution, we have compared SDCC and the corrected that searches features by Eq. 1 version using 50 datasets chosen from the UCI machine learning repository [22]. As we expected the corrected version significantly outperforms the original version in terms of the AUC-ROC, the bayesian risk and the number of features selected. Figure 3 depicts the averages of the bayesian risk, AUC-ROC when $C4.5$ is used as a classifier and the number of features selected across the 50 datasets. The threshold parameter δ varies in the interval $[0, 0.1]$ with an increment of 0.01.

Although these results are quite good, maximization of the average of the individual features (collective relevance) can not be guaranteed because the individual relevance of features is measured back stage. This means that until now the process of removing a feature is composed by two sequential steps and the individual relevance score is only used in the second one. In many cases, this unbalanced trade-off between the bayesian risk and the collective relevance of a set, may lead to undesirable results as stated in Sect. 2.3. We now consider the individual relevance of features as a crucial factor to judge the quality of a feature set, by proposing the *interelevance score* measure defined as follows.

$$IR(\tilde{F}; f_i; C) = (1 - \alpha)A(\tilde{F}; f_i; C) + \alpha B(f_i; C)$$

$$\text{with } A(\tilde{F}; f_i; C) = \begin{cases} \frac{\mathfrak{Br}(\tilde{F}\setminus\{f_i\};C) - \mathfrak{Br}(F;C)}{\delta - \mathfrak{Br}(F;C)}, & \text{if } \mathfrak{Br}(F;C) \leq \delta \\ \mathfrak{Br}(\tilde{F} \setminus \{f_i\}; C) - \mathfrak{Br}(F;C), & \text{if } \mathfrak{Br}(F;C) = \delta \end{cases}$$

$$B = \begin{cases} \frac{r(f_i;C) - r^-}{r^+ - r^-} & \text{if } r^+ > r^- \\ 0 & \text{if } r^+ = r^- \end{cases}$$

where $r^+ = \max_{f_i \in F} r(f_i; C)$, $r^- = \min_{f_i \in F} r(f_i; C)$ and α satisfies $0 \leq \alpha \leq 1$. The *interelevance score* IR is normalization function that evaluates how good is a given feature f_i for the current feature set $\tilde{F}$. IR measures: (i) how relevant is f_i and (ii) the effect of removing f_i from $\tilde{F}$ from the consistency point of view. Function A normalize the *bayesian risk* obtained by removing feature f_i from $\tilde{F}$. $\mathfrak{Br}(F; C)$ and δ are taken as the minimum and maximum value respectively in the normalization function. We expect that IR metric allows to select interacting feature sets composed by features with high relevance score. Thus, to select f^-, we do not use Eq. 1 as a criterion, but use $f^- \in \text{argmin}\{\text{IR}(\tilde{F}; f_i; C) \mid f_i \in \tilde{F}, \mathfrak{Br}(\tilde{F} \setminus \{f_i\}; C) \leq \delta\}$.

Problem 2: Unnecessary evaluations: Although in the second iteration SDCC computes $\mathfrak{Br}(\{f_2, f_3\}; C) > 0$, this operation is unnecessary because the result can be inferred by the monotonicity property of consistency measures. Since $\mathfrak{Br}(\{f_1, f_2, f_3\}; C) > 0$ has been computed in the first iteration, $\mathfrak{Br}(\{f_2, f_3\}) > 0$ is inferred by monotonicity.

Solution to Problem 2. In order to avoid unnecessary evaluations, if $\mathfrak{Br}(\tilde{F}; C \setminus \{f_i\}) > \delta$ holds with $f_i \in \tilde{F}$ then feature f_i is not evaluated anymore and never removed. Furthermore, the *interelevance* score has a property that allows saving evaluations when *steepest-descent* is run over a ranked feature set.

Proposition 1. *Let $\tilde{F} = \{f_1, \ldots, f_{k-1}, f_k\}$ be in a increasing order of $r(f_i; C)$ where r is a relevant measure. If $\mathfrak{Br}(\tilde{F} \setminus \{f_p\}; C) = \mathfrak{Br}(\tilde{F}; C)$ holds then, there does not exist a feature $f_j \in \tilde{F}$ with $j < p$ such that $IR(\tilde{F}; f_j; C) < IR(\tilde{F}; f_p; C)$ holds.*

Proof. This is easy to see because $\mathfrak{Br}(\tilde{F} \setminus \{f_p\}; C)$ is the minimum by the monotonicity property and $r(f_p; C) \leq r(f_j; C)$ always holds.   □

Proposition 1 is essential to turn our *steepest-descent* algorithm faster since when f_p is found, it can be immediately removed without affecting the final solution and $p - 1$ evaluations of B are saved in each iteration. In the ideal scenario, where $p = k$ holds in each iteration, the number of evaluations performed by our *steepest-descent* algorithm is linear respect to the number of features in F. Oppositely, in the worst case scenario where $p = 1$ holds in each iteration, the number of evaluation performed by the new version of SDCC is the same as the SDCC.

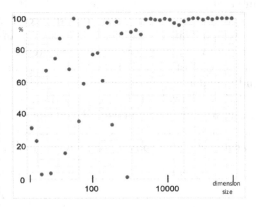

Fig. 4. Percentage of the first consecutive features $\{f_1, \ldots, f_l\}$ such that $\mathfrak{Br}(F; C) = \mathfrak{Br}(F \setminus \{f_1, \ldots, f_l\}; C)$ to the entire feature set F.

Problem 3: Low scalability to high-dimensional data: Our new version of SDCC may be still slow in datasets with large number of interacting features. Although it is well known high-dimensional datasets are rich in non-interacting features, we do not assume their class variable can be described by a small number of features. Therefore, we now describe two mechanism to reduce even more the number of evaluation of our proposal.

Solution 1 to Problem 3: Eliminating the big mass of irrelevant features by Super-Lcc. High-dimensional datasets are likely to be abundant in irrelevant and non-interacting features. Assuming $|F|$ is very large, we can expect $\mathfrak{Br}(F \setminus F'; C) = \mathfrak{Br}(F; C)$ with $F' = \{f_1, \ldots, f_l\}$ for a large value of l. To make sure this expectation is true, we randomly picked 44 datasets from the *UCI machine learning repository* and determine l.

The experiments were conducted in small ($|F| \leq 100$), medium ($100 < |F| \leq 10000$) and high-dimensional data ($10000 < |F|$) using $\delta = \mathfrak{Br}(F; C)$. Figure 4 depicts the results, and we see that values of l are very close to the numbers of the entire features $|F|$, when the dataset is high-dimensional. This means that for these high-dimensional datasets our *steepest-descent* algorithm will remove a huge number of consecutive features one by one, which is not so efficient. However, recently Shin et al. in [8] have found that l can be determined efficiently by means of *binary search*. In fact, $\{f_1, \ldots, f_l\}$ are removed by the first iteration of SUPER-LCC. This finding broke the premise that consistency-based algorithms were computationally too expensive to apply to high-dimensional data. We use their finding to efficiently remove F' with only a few iterations (see Eq. 1). We use the first iteration of SUPER-LCC to eliminate the largest $\{f_1, \ldots, f_l\}$ such that $\mathfrak{Br}(F \setminus \{f_1, \ldots, f_l\}; C) \leq \mathfrak{Br}(F; C) + \delta$ and then apply *steepest descent* to the remainder of the features, that is, $\{f_{l+1}, \ldots, f_n\}$.

Solution 2 to Problem 3: Windowing the search. When feature selection is performed using consistency measures, in each iteration of the search we can categorize features as: indispensable, useless and potential features. Being $\tilde{F}$ the current feature set, indispensable features must remain in $\tilde{F}$ in order to keep the *bayesian risk* under the threshold. That is, a feature $f_x \in \tilde{F}$ is indispensable if $\mathfrak{Br}(\tilde{F} \setminus \{f_x\}; C) > \delta$ holds. On the contrary, useless features can be safely removed without degrading the *bayesian score* of $\tilde{F}$. A feature f_y is said to be useless when $\mathfrak{Br}(\tilde{F} \setminus \{f_y\}; C) = \mathfrak{Br}(\tilde{F}; C)$ holds. On the other hand, if a feature is neither of indispensable nor useless then it is a potential feature. That is, for potential feature f, $\mathfrak{Br}(\tilde{F} \setminus \{f\}; C) \leq \mathfrak{Br}(\tilde{F}; C) + \delta$ holds. Potential features are the most interesting type of features: they necessarily become indispensable or useless at any moment of the search and must be evaluated in the next iteration. Speaking about efficiency, the worst case scenario, in a given iteration, is that all features are potential. This means that our version of the *steepest-descent* algorithm needs $|\tilde{F}|$ evaluations to remove the one that minimize IR. To overcome this drawback we propose to limit the search in each iteration to a portion of the features in the current set. This can be done by applying a mobile window search. Let $\bar{d}$ be the average of the differences of the individual relevance of consecutive features in F

$$\bar{d} = \frac{1}{n-1} \sum_{i=1}^{n-1} \Big(r(f_{i+1}; C) - r(f_i; C) \Big) = \frac{1}{n-1}(r^+ - r^-), \qquad (2)$$

we define the window search w_k in the k-th iteration as:

$$w_1 = r^- + \omega(r^+ - r^-) \tag{3}$$

$$w_k = w_{k-1} + \lambda\bar{d}, \text{ with } k > 1, \tag{4}$$

where $\omega = (0,1]$ and $\lambda \in \mathbb{R}^+$ are predefined parameters that influence the initial size of the window search w_1 and the acceleration of the expansion of the window search w_k in the k-th iteration respectively. If the relevance score of a feature falls into the region of the window $[r^-, w_k)$ then will be evaluated in the k-th iteration. The number of features evaluated in each iteration is not only determined by the position of useless features but also by the size of the window search. This may significantly improve the efficiency of our *steepest-descent* version in datasets abundant in potential features.

Let F be the entire feature set and δ be the upper bound of the permissible *bayesian risk* of the output sets. We combine all the solutions given above as follows.

1. The relevance $r(f_i; C)$ of each feature $f_i \in F$ is computed using the *Symmetrical Uncertainty* measure, and F is mapped to $\tilde{F}$ by sorting the features in incremental order of $SU(f_i; C)$.

2. The maximum set of consecutive useless features $\{f_1, \ldots, f_l\}$ is identified and removed by using the *binary search* (see Eq. 1)

3. The window size is computed in each iteration.

Algorithm 1: SWCFS Algorithm

Input: D: the dataset
δ: inconsistency score threshold
ω: initial size of the window search
λ: windows size coefficient
Output: $\tilde{F}$ suboptimal set

1 Rank features in F in incremental order according to SU
2 Fix $\tilde{F} = F$
3 Find the maximum l such that $\mathfrak{Br}(\tilde{F} \setminus \{f_1, \ldots, f_l\}; C) = \mathfrak{Br}(\tilde{F}; C)$
4 Update $\tilde{F} = \tilde{F} \setminus \{f_1, \ldots, f_l\}$
5 Compute $r^+ = \max_{f_i \in \tilde{F}} r(f_i; C)$ and $w_1 = \omega r^+$
6 Let $\bar{d}$ be the average of the difference between $SU(f_i; C)$ and $SU(f_{i-1}; C)$ for $f_i \in \tilde{F}$, $k = 1$ and $IR^- = \inf$
7 **repeat**
8 $\quad$ $f^- = Null$
9 $\quad$ **foreach** $f_i \in \tilde{F}$ **do**
10 $\quad\quad$ **if** $SU(f_i; C) \leq w_k$ **then**
11 $\quad\quad\quad$ $\delta[f_i] = \mathfrak{Br}(\tilde{F} \setminus \{f_i\}; C)$
12 $\quad\quad\quad$ **if** $\delta[f_i] > \delta$ **then** continue;
13 $\quad\quad\quad$ **if** $\delta[f_i] = \mathfrak{Br}(\tilde{F}; C)$ **then** $f^- = f_i$, **break**;
14 $\quad\quad\quad$ **if** $IR(\tilde{F}; f_i; C) \leq IR^-$ **then** $f^- = f_i$, $IR^- = IR(\tilde{F}; f_i; C)$;
15 $\quad\quad$ **end**
16 $\quad$ **end**
17 $\quad$ **if** $f^- = Null$ **then break**;
18 $\quad$ $\tilde{F} = \tilde{F} \setminus \{f^-\}$
19 $\quad$ $k = k + 1$
20 $\quad$ $w_k = w_{k-1} + \lambda\bar{d}$
21 **until** *True*;

Fig. 5. The algorithm of SWCFS

The *steepest-descent* algorithm is performed using the interelevance score IR by evaluating only the features included in the current window and taking into account the following rules with $f_i \in \tilde{F}$:

Rule 1. If f_i is an useless feature then it is immediately removed from $\tilde{F}$ (line 13).

Rule 2. Else if f_i is indispensable then f_i is not evaluated anymore and never will be removed from $\tilde{F}$ (line 12).

Rule 3. Otherwise the feature f_i that minimize IR is removed from $\tilde{F}$ if $IR(\tilde{F}; f; C) > IR(\tilde{F}; \varnothing; C)$ holds. The algorithm stops when all features have been tested and none of the features can be removed. Figure 5 depicts the entire algorithm.

4   Experiments

We empirically evaluate the performance of the proposed algorithm and make comparisons with some state-of-the-art feature selection algorithms: RELIEFF (RF) [11], CFS [18], FCBF [16] and SUPER-LCC [8] and ASDCC [9]. We exclude from comparison algorithms SUPER-CWC [8] and FSDCC [9] because we verify they output similar results to SUPER-LCC and ASDCC respectively.

The configuration of the experiments is as follows. First, we run the feature selection algorithms over the datasets and obtain selected feature subsets for respective algorithms. To evaluate the classification capability of the selected feature sets, we run ten-fold cross validation on the reduced data using two classifiers: NAIVE BAYES and

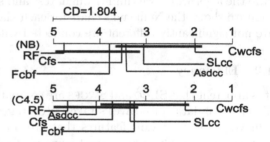

Fig. 6. Nemenyi test with $\alpha = 0.05$

C4.5. The *bayesian risk* parameter δ of SUPER-LCC and SWCFS algorithms was fixed to 0.01. We report results about the AUC-ROC values of both classifiers and the number of features selected by each algorithm. Before running experiments we run SWCFS across many datasets with different values of α and verified that $\alpha = 0.5$ works well. Table 1 shows the AUC-ROC values after running the classifiers on the reduced data and the number of features selected by each algorithm.

4.1   Numbers of Features Selected and Auc-Roc Scores

Speaking about the size of the output, SWCFS outputs smaller or equal when compared with SLCC. Furthermore, when compared with all the algorithms it turns out that SWCFS is ranked top for a half of the datasets. Speaking about AUC-ROC scores, SWCFS is ranked top for more than the 68% and 62% of the

Table 1. Results of Auc-Roc values for the reduced data and number of features selected by each algorithm

data	NB-AUC values RF	Cfs	Fcbf	SLcc	ASdcc	Swcfs	C4.5-AUC values RF	Cfs	Fcbf	SLcc	Asdcc	Swcfs	size RF	Cfs	Fcbf	SLcc	Asdcc	Swcfs
OPT.	.945	.967	.966	.966	.967	.968	.858	.924	.929	.933	.928	.935	30	38	21	9	10	8
ARR.	.468	.850	.854	.848	.850	.848	.464	.738	.737	.733	.733	.733	1	25	12	21	28	21
MAD.	.523	.644	.646	.647	.646	.647	.500	.770	.613	.811	.814	.811	1	6	4	15	12	15
MFE	.966	.973	.985	.986	.970	.991	.972	.968	.961	.952	.954	.964	360	85	136	7	8	6
SEM	.983	.956	.952	.955	.958	.956	.877	.881	.876	.865	.879	.885	175	74	30	31	45	27
AUD	.946	.939	.905	.962	.923	.952	.907	.905	.924	.921	.905	.924	10	6	16	12	9	12
KRV	.969	.930	.968	.972	.970	.983	.972	.930	.959	.997	.995	.997	5	3	7	21	18	15
MF1	.922	.948	.947	.977	.981	.981	.923	.908	.925	.916	.914	.911	90	67	38	8	9	7
MF2	.961	.969	.968	.969	.968	.970	.903	.905	.899	.906	.905	.910	15	12	37	11	13	11
MF3	.979	.986	.986	.981	.984	.984	.907	.915	.907	.920	.907	.924	7	26	57	7	7	7
MF4	.949	.950	.945	.950	.949	.950	.918	.919	.918	.922	.918	.922	3	4	2	5	4	5
MF5	.964	.965	.969	.967	.937	.969	.903	.904	.911	.901	.904	.906	196	103	27	21	17	17
MF6	.925	.955	.955	.957	.955	.957	.859	.880	.884	.871	.871	.871	7	25	14	12	14	12
PEN	.977	.963	.963	.963	.964	.964	.973	.973	.974	.975	.970	.974	16	11	11	7	10	7
SPL	.981	.984	.993	.990	.984	.989	.967	.969	.970	.969	.969	.970	19	6	22	9	8	9
WAV	.510	.945	.932	.938	.945	.946	.500	.858	.882	.877	.876	.884	1	15	6	10	8	9
AVG.	.873	.933	.933	.939	.934	.941	.838	.897	.892	.904	.903	.908	58.5	31.6	27.5	12.9	13.8	11.8

datasets for NAIVE BAYES and C4.5 classifiers respectively. To statistically compare the algorithms, we run Friedman test and statistical differences were found. Figure 6 shows the Nemenyi's chart for each classifier. Group of algorithms that are not significantly different are connected with a thick line.

4.2 Efficiency

It is apparent that SLCC and SWCFS are compatible in terms of efficiency in high-dimensional data since SWCFS takes advantage of the first iteration of SLCC to remove the less relevant features that are not necessary to create consistency sets. In the case where only small number of features are eliminated in the first step of SWCFS, the numbers of evaluations depends on the size of the sliding window. However, if the sliding window is reasonably small then the number of evaluation can be comparable with the number of evaluations of LCC algorithm. Nevertheless, as Fig. 4 shows, it turns out that high-dimensional data are prone to be rich in irrelevant features that can be removed in the first iteration of SWCFS.

5 Conclusion and Future Works

In this paper we propose a new feature selection algorithm based on consistency measures and individual feature scoring functions. The search strategy used by the algorithm is the SDCC, which has quadratic order. However, we modify SDCC by leveraging the *binary search* that allows to remove, in many cases, more than 95% of non-relevant features with small number of evaluations. In addition, a sliding window was added to SDCC to avoid unnecessary evaluations. Experiments reveal that the new proposal is very accurate when compared with several state-of-the-art algorithm. In the future we will evaluate the new algorithm in high-dimensional data and make a further analysis about the optimal values for its parameters.

References

1. Rohrmair, G., Lowe, G.: Using data-independence in the analysis of intrusion detection systems. Theor. Comput. Sci. **340**(1), 82–101 (2005)
2. Angeleska, A., Jonoska, N., Saito, M.: Rewriting rule chains modeling DNA rearrangement pathways. Theor. Comput. Sci. **454**, 5–22 (2012)
3. De Maria, E., Fages, F., Rizk, A., Soliman, S.: Design, optimization, and predictions of a coupled model of the cell cycle, circadian clock, DNA repair system, irinotecan metabolism and exposure control under temporal logic constraints. Theor. Comput. Sci. **412**(21), 2108–2127 (2011)
4. Kohavi, R., John, G.H.: Wrappers for feature subset selection. Artif. Intell. **97**, 273–324 (1997)
5. Molina, L.C., Belanche, L., Nebot, A.: Feature selection algorithms: a survey and experimental evaluations. In: Proceedings of the 2002 IEEE International Conference on Data Mining (ICDM 2002), 9–12 December 2002, Maebashi City (2002)
6. Hodorog, M., Schicho, J.: A regularization approach for estimating the type of a plane curve singularity. Theor. Comput. Sci. **479**, 99–119 (2013)
7. John, G.H., Kohavi, R., Pfleger, K.: Irrelevant feature and the subset selection problem. In: ICML (1994)
8. Shin, K., Kuboyama, T., Hashimoto, T., Shepard, D.: Super-CWC and super-LCC: super fast feature selection algorithms. In: Proceedings of 2015 IEEE International Conference on Big Data (Big Data), pp. 1–7 (2015)
9. Pino Angulo, A., Shin, K.: Fast and accurate steepest-descent consistency-constrained algorithms for feature selection. In: Pardalos, P., Pavone, M., Farinella, G.M., Cutello, V. (eds.) MOD 2015. LNCS, vol. 9432, pp. 293–305. Springer, Cham (2015). doi:10.1007/978-3-319-27926-8_26
10. Shin, K., Xu, X.M.: A consistency-constrained feature selection algorithm with the steepest descent method. In: Torra, V., Narukawa, Y., Inuiguchi, M. (eds.) MDAI 2009. LNCS, vol. 5861, pp. 338–350. Springer, Heidelberg (2009). doi:10.1007/978-3-642-04820-3_31
11. Kira, K., Rendell, L.A.: A practical approach to feature selection. In: Proceedings of the Ninth International Workshop on Machine Learning, pp. 249–256. Morgan Kaufman Publishers Inc. (1992)
12. Kononenko, I.: Estimating attributes: analysis and extensions of RELIEF. In: Bergadano, F., De Raedt, L. (eds.) ECML 1994. LNCS, vol. 784, pp. 171–182. Springer, Heidelberg (1994). doi:10.1007/3-540-57868-4_57
13. Xiaofei, H., Deng, C., Partha, N.: Laplacian score for feature selection. In: Proceedings of the 18th International Conference on Neural Information Processing Systems (NIPS 2005), pp. 507–514 (2005)
14. Zhu, L., Miao, L., Zhang, D.: Iterative Laplacian score for feature selection. In: Liu, C.-L., Zhang, C., Wang, L. (eds.) CCPR 2012. CCIS, vol. 321, pp. 80–87. Springer, Heidelberg (2012). doi:10.1007/978-3-642-33506-8_11
15. Quanquan, G., Zhenhui, L., Jiawei, H.: Generalized Fisher score for feature selection. In: Proceedings of the Twenty-Seventh Conference on Uncertainty in Artificial Intelligence (UAI 2011), pp. 266–273 (2011)

16. Yu, L., Liu, H.: Feature selection for high-dimensional data: a fast correlation-based filter solution. In: Proceedings of the Twentieth International Conference on Machine Learning (ICML-2003) (2003)

17. Guyon, I., Weston, J., Barnhill, S.: Gene selection for cancer classification using support vector machines. Mach. Learn. **46**, 389 (2002)

18. Hall, M.A., Smith, L.A.: Feature selection for machine learning: comparing a correlation-based filter approach to the wrapper. In: Proceedings of the Twelfth International, pp. 235–239. AAAI Press (1999)

19. Ding, C., Peng, H.: Minimum redundancy feature selection from microarray gene expression data. In: Proceedings of the IEEE Computer Society Conference on Bioinformatics (CSB 2003) (2003)

20. Zhao, Z., Liu, H.: Searching for interacting features. In: Proceedings of the 20th International Joint Conference on Artifical Intelligence (IJCAI 2007) (2007)

21. Shin, K., Xu, X.M.: Consistency-based feature selection. In: Velásquez, J.D., Ríos, S.A., Howlett, R.J., Jain, L.C. (eds.) KES 2009. LNCS, vol. 5711, pp. 342–350. Springer, Heidelberg (2009). doi:10.1007/978-3-642-04595-0_42

22. Lichman, M.: UCI machine learning repository, School of Information and Computer Science, University of California, Irvine (2013). http://archive.ics.uci.edu/ml

Feature Ranking for Multi-target Regression with Tree Ensemble Methods

Matej Petković[1,2]([✉]), Sašo Džeroski[1,2], and Dragi Kocev[1,2]

[1] Department of Knowledge Technologies, Jožef Stefan Institute, Ljubljana, Slovenia
[2] Jožef Stefan International Postgraduate School, Ljubljana, Slovenia
{matej.petkovic,saso.dzeroski,dragi.kocev}@ijs.si

Abstract. In this work, we address the task of feature ranking for multi-target regression (MTR). The task of MTR concerns problems where there are multiple continuous dependent variables and the goal is to learn a model for predicting all of the targets simultaneously. This task is receiving an increasing attention from the research community. However, performing feature ranking in the context of MTR has not been studied. Here, we propose three feature ranking methods for MTR: Symbolic, Genie3 and Random Forest. These methods are then coupled with three types of ensemble methods: Bagging, Random Forest, and Extremely Randomized Trees. All of the ensemble methods use predictive clustering trees (PCTs) as base predictive models. PCTs are a generalization of decision trees capable of MTR. In total, we consider eight different ensemble-ranking pairs. We extensively evaluate these pairs on 26 benchmark MTR datasets. The results reveal that all of the methods produce relevant feature rankings and that the best performing method is Genie3 ranking used with Random Forests of PCTs.

Keywords: Multi-target Regression · Feature ranking · Feature importance · Ensembles · Predictive Clustering Trees

1 Introduction

Single target regression (STR) is a subfield of predictive modelling, where the goal is to learn a model able to predict the values of a single numeric target variable. STR can be generalized to multi-target regression (MTR), where the goal is to learn a model that predicts $T \geq 2$ targets. The STR and MTR tasks can be formalized as described below.

We are given a set of examples $\boldsymbol{x}$ from the input domain $\mathcal{X} \subseteq \mathcal{X}_1 \times \cdots \times \mathcal{X}_D$, $D \geq 1$ being the number of descriptive attributes (features). We assume that the domain $\mathcal{X}_i$ of the i-th descriptive attribute x_i is either a subset of $\mathbb{R}$ or an arbitrary finite set, i.e., x_i is either numeric or nominal. Each example $\boldsymbol{x}$ is associated with a target value $\boldsymbol{y}(\boldsymbol{x})$ from the target domain $\mathcal{Y} \subseteq \mathcal{Y}_1 \times \cdots \times \mathcal{Y}_T \subseteq \mathbb{R}^T$, T being the number of target attributes (targets). STR considers domains where $T = 1$, while MTR considers domains with $T \geq 2$. In the latter case, the j-th component of the target vector $\boldsymbol{y}(\boldsymbol{x})$ is denoted by $y_j(\boldsymbol{x})$. The true

© Springer International Publishing AG 2017
A. Yamamoto et al. (Eds.): DS 2017, LNAI 10558, pp. 171–185, 2017.
DOI: 10.1007/978-3-319-67786-6_13

mapping $y : \boldsymbol{x} \mapsto y(\boldsymbol{x})$ (STR) or $\boldsymbol{y} : \boldsymbol{x} \mapsto \boldsymbol{y}(\boldsymbol{x})$ (MTR) is unknown and the goal of regression is to find its approximation, given a dataset $\mathscr{D} \subseteq \mathcal{X} \times \mathcal{Y}$.

STR is a well established research topic, while MTR is recently attracting interest in the research community [22, 23]. MTR is a structured output prediction task with applications in a wide range of real life problems where we are interested in simultaneously predicting multiple continuous variables. Prominent examples come from ecology and include predicting the abundance of different species living in the same habitat [12] and predicting properties of forests [21].

A possible way to approach a MTR problem is problem transformation, which transforms one MTR problem to several STR problems and build one predictive model for each target separately. Another way to approach the problem is by algorithm adaptation, i.e., to change STR methods in such a way they are able to exploit the potential relatedness between the multiple targets. For example, regression trees can be generalized so that the heuristic function considers the multiple targets and the leaves make predictions for all targets. For an overview of the different MTR, we refer the reader to Borchani et al. [5].

Another important task in machine learning is feature ranking, which is typically seen as a data preprocessing step. Here, the importances of descriptive attributes (features) are estimated and an ordering (or ranking) of the features is made, based on the estimated importances. There are two main reasons for doing this. First, we may want to reduce the dimensionality of the input space, so that only the features that contain the most information about target(s) are kept in the dataset. By doing this, we decrease the amount of memory/time needed to build a predictive model, while the performance of the model is not degraded. Second, dimensionality reduction typically results in models that are easier to understand, which comes in handy when a machine learning expert works in collaboration with a domain expert. Predictive models, such as decision trees, are easier to interpret when a small number of the most relevant features are used to learn them.

There is a plethora of feature ranking methods for the machine learning tasks of single target regression and classification. For an overview, see Stanczyk and Jain [24]. However, in the case of MTR, the task of feature ranking has not been studied to a great extent. To the best of our knowledge, there is no previous work from the machine learning community.

In the field of statistics, a few such methods can be found. Their main drawback is that they allow only for numeric features, since they typically assume a (generalized) linear model $\boldsymbol{y} = A\boldsymbol{x} + \boldsymbol{e}$, where A is a $T \times D$ matrix and $\boldsymbol{e}$ is a random noise vector. One such method is forward selection. It starts with a constant model $\boldsymbol{y}(\boldsymbol{x}) = c \in \mathbb{R}$, and repeatedly adds the most significant feature that improves the model. The sooner the feature is included in the model, the greater the importance. For on overview of these methods, see Brobbey [9].

In this work, we propose three feature ranking methods based on ensembles of predictive clustering trees (PCTs) [4, 22]. PCTs are generalization of decision trees able to handle various types of structured output prediction tasks, including MTR. The proposed feature ranking methods can handle both numeric and nominal features.

The proposed methods exploit different properties of the ensemble learning mechanism to estimate feature importances. More specifically, two of the methods are adaptations of the feature importance measures already known from the single target regression task: Genie3 [18] and Random Forest ranking [7]. Genie3 uses the variance reduction at each tree node as a proxy for the importance of the feature that is used in the test at a given node. Random Forest ranking permutes the values of a feature on the out-of-bag set of data to estimate how much worse performance this will yield as compared to the original data. This decrease of predictive performance is then taken as a proxy for the feature's importance. Finally, the third method named Symbolic counts how often a feature appears in the nodes of the trees from an ensemble. These appearances can be also weighted with the nodes' depth at which a given feature appears. This is a general method that is applicable to an arbitrary machine learning task for which tree-based models can be learned.

Furthermore, these three ranking methods can be coupled with three ensemble learning methods: Bagging [6], Random Forests [7] and Extremely randomized trees [14]. Note that Random Forests ranking cannot be coupled with Extremely randomized trees because the latter do not perform bootstrapping of the examples. This yields in total 8 pairs of ensemble learning method and feature ranking method.

We extensively evaluate the proposed methods on 26 benchmark MTR datasets. The evaluation is performed by comparing the performance of the standard 5NN (5 nearest neighbors) prediction method with a 5NN prediction method that uses the obtained feature importances as weights during distance calculation. The experiments investigate the relevance of the obtained feature rankings and look for the optimal combination of ensemble learning and feature ranking method.

The remainder of this paper is organized as follows. Section 2 presents the PCT approach to MTR. Ensembles of PCTs for MTR and feature ranking methods based on these are described in Sect. 3. Section 4 outlines the experimental design, while Sect. 5 discusses the results of the experimental evaluation. Finally, the conclusions and a summary are given in Sect. 6.

2   Predictive Clustering Trees for Multi-target Regression

PCTs generalize decision trees and can be used for a variety of learning tasks, including clustering and different types of prediction. The PCT framework views a decision tree as a hierarchy of clusters: The root of a PCT corresponds to one cluster containing all data, which is recursively partitioned into smaller clusters while moving down the tree. The leaves represent the clusters at the lowest level of the hierarchy and each leaf is labeled with its cluster's prototype (prediction).

PCTs are induced with the standard top-down induction of decision trees algorithm presented in Table 1 [8]. It takes as input a set of examples E and outputs a tree. The heuristic h that is used for selecting the tests is the reduction of variance caused by partitioning the instances (see line 4 of the *BestTest*

Table 1. The top-down induction algorithm for PCTs.

procedure PCT(E) **returns** tree	**procedure** BestTest(E)
1: $(t^*, h^*, \mathcal{P}^*) = \text{BestTest}(E)$	1: $(t^*, h^*, \mathcal{P}^*) = (none, 0, \emptyset)$
2: **if** $t^* \neq none$ **then**	2: **for each** candidate test t **do**
3: **for each** $E_i \in \mathcal{P}^*$ **do**	3: $\mathcal{P} = $ partition induced by t on E
4: $tree_i = \text{PCT}(E_i)$	4: $h = Var(E) - \sum_{E_i \in \mathcal{P}} \frac{|E_i|}{|E|} Var(E_i)$
5: **return** node(t^*, $\bigcup_i \{tree_i\}$)	5: **if** $(h > h^*) \wedge \text{Acceptable}(t, \mathcal{P})$ **then**
6: **else**	6: $(t^*, h^*, \mathcal{P}^*) = (t, h, \mathcal{P})$
7: **return** leaf(Prototype(E))	7: **return** $(t^*, h^*, \mathcal{P}^*)$

procedure in Table 1). By maximizing the variance reduction, the cluster homogeneity is maximized: The algorithm is thus guided towards small trees with good predictive performance. If no acceptable test can be found (line 6 of the PCT procedure), i.e., no test reduces the variance significantly, then a leaf is created and the prototype of the instances belonging to that leaf is computed.

The main difference between the algorithm for learning PCTs and other algorithms for learning decision trees is that the former considers the variance function and the prototype function (that computes predictions in leaves) as parameters that can be instantiated for a given learning task. In this work, we focus on the task of MTR and define the variance function as follows. First, we define the average $\overline{y}_j$ and variance of the target y_j over subset $E \subseteq \mathscr{D}_{\text{TRAIN}}$ as

$$\overline{y}_j(E) = \frac{1}{|E|} \sum_{x \in E} y_j(x) \quad \text{and} \quad Var_j(E) = \frac{1}{|E|} \sum_{x \in E} (y_j(x) - \overline{y}_j(E))^2. \tag{1}$$

We then compute the weights $w_j = Var_j(\mathscr{D}_{\text{TRAIN}})$ and use them as normalization factors in the definition of variance function:

$$Var(E) = \frac{1}{T} \sum_{j=1}^{T} \frac{1}{w_j} Var_j(E).$$

In a leaf L, the prototype function returns a vector $(\overline{y}_1(E_L), \ldots, \overline{y}_T(E_L))$, where E_L denotes the set of all examples that fall into the leaf L. For a detailed description of PCTs for MTR, we refer the reader to Blockeel [4] and Kocev [22]. The PCT framework is implemented in the CLUS system (available at http://clus.sourceforge.net).

3   Feature Ranking via Ensembles of PCTs

We use PCTs as the base models in three types of ensembles [22] that are constructed to calculate the variable importance, i.e., the feature ranking. In the following, we first present the ensemble methods and then describe the feature ranking methods.

3.1 Ensembles of PCTs

An ensemble is a set of base predictive models constructed with a given algorithm. The prediction for each new example is made by combining the predictions of every model from the ensemble. This can be done by taking the average in regression tasks, and the majority or probability distribution vote in classification tasks [6]. For the task of MTR, we consider ensembles of PCTs [22], where the predictions are the average values for each target.

A necessary condition for an ensemble to be more accurate than any of its individual members, is that the members are accurate and diverse models [16]. This means that they perform better than random guessing. On the other hand, it means that they make different errors on new examples.

There are several ways to introduce diversity among the base predictive models in an ensemble. We describe how this is done in Random Forests [7], Bagging [6] and Extra Trees ensembles [14].

Random Forest (RF) and Bagging. A Random Forest is an ensemble of trees where diversity among the predictive models is obtained in two ways. First, instead of being learned from the original dataset $\mathscr{D}_{\text{TRAIN}}$, each tree is built from a different bootstrap replicate $\mathcal{B}$. The chosen examples from such a replicate form a so called bag $\mathcal{B}$, while the rest are called out-of-bag examples (OOB). Hence, we perform a call PCT($\mathcal{B}$) rather than PCT($\mathscr{D}_{\text{TRAIN}}$) as we would do in the case when a single PCT is to be grown.

Additionally, we modify the line 2 of the *BestTest* procedure (see Table 1), to change the feature set during learning. More precisely, at each node in a decision tree, a random subset of the input attributes is taken, and the best test is selected from the splits defined on these. The number of attributes that are retained is given as a function of the total number of descriptive attributes D, e.g., $\lceil\sqrt{D}\rceil$, $\lceil\log_2(D)\rceil$, $D/4$, etc. In the special case when we keep all attributes, we obtain the Bagging procedure, where the only source of diversity is the difference in the bootstrap replicates of the data.

Extra trees ensemble (ET). The source of diversity in ET comes from the extreme randomization of the tree learning procedure. Here, at each node all attributes are considered (as in Bagging), but we do not evaluate all tests that the attributes yield. Rather, we choose randomly only one per attribute. Among these D tests, we choose the best one, hence the only difference compared to standard top-down PCT induction, is that a modified line 2 of the *BestTest* procedure is used. Note that ET uses the initial dataset $\mathscr{D}_{\text{TRAIN}}$ for learning the base predictive models and does not make bootstrap replicates.

3.2 Ensemble Feature Ranking Methods

Feature ranking of the descriptive variables can be obtained either by exploiting the ensemble structure of the learning algorithm or the mechanism of Random Forests. For its simplicity, we first describe symbolic ranking. Then, we discuss Genie3 and Random Forest ranking.

In the following, we denote a tree by $\mathcal{T}$, whereas $\mathcal{N} \in \mathcal{T}$ denotes a node. Trees form a forest $\mathcal{F}$. Its size (the number of trees in the forest) is denoted by $|\mathcal{F}|$. The set of all internal nodes of a tree $\mathcal{T}$ in which the attribute x_i appears as part of the test, is denoted by $\mathcal{T}(x_i)$.

Symbolic ranking (Symb). Let $d(\mathcal{N})$ denote the *depth* of $\mathcal{N} \in \mathcal{T}$. The depth is defined recursively: if $\mathcal{N}$ is the root of $\mathcal{T}$, then $d(\mathcal{N}) = 0$. Otherwise, $d(\mathcal{N}) = 1 + d(\text{parent}(\mathcal{N}))$. In the basic version of symbolic ranking, we simply count how many times a given attribute occurs in the tests in the internal nodes of the trees in the forest. Since the attributes that appear at lower depths (i.e., closer to the root) are intuitively more important than those that appear deeper in the trees, we introduce the parameter $w \in (0, 1]$ and define the importance of the attribute x_i as

$$importance_{\text{SYMB}}(x_i) = \frac{1}{|\mathcal{F}|} \sum_{\mathcal{T} \in \mathcal{F}} \sum_{\mathcal{N} \in \mathcal{T}(x_i)} w^{d(\mathcal{N})}. \tag{2}$$

Note that symbolic ranking is applicable to all three ensemble methods that we use, and that the basic version of the ranking corresponds to the choice $w = 1$.

Genie3. The main motivation for Genie3 ranking is that splitting the current subset $E \subseteq \mathscr{D}_{\text{TRAIN}}$, according to a test where an important attribute appears, should result in high variance reduction. As in the symbolic ranking case, greater emphasis is put on the attributes higher in the tree, i.e., on the splits where $|E|$ is larger. The Genie3 importance of the attribute x_i is defined as

$$importance_{\text{GENIE3}}(x_i) = \frac{1}{|\mathcal{F}|} \sum_{\mathcal{T} \in \mathcal{F}} \sum_{\mathcal{N} \in \mathcal{T}(x_i)} |E(\mathcal{N})| h^*(\mathcal{N}),$$

where $E(\mathcal{N})$ is the set of examples that come to the node $\mathcal{N}$, and $h^*(\mathcal{N})$ is the value of the variance reduction function described in the *BestTest* procedure. Genie3 ranking is applicable to all three ensemble methods that we use.

Random Forest (RF). This feature ranking method tests how much does noise in a given descriptive attribute decrease the predictive performance of the trees in the forest. The greater the performance degradation, the more important the attribute. This feature ranking algorithm uses the internal out-of-bag estimates of the error, therefore it cannot be used with ensembles of extra trees.

Once a tree $\mathcal{T}$ is grown, the algorithm evaluates the performance of the tree by using the corresponding $\text{OOB}_{\mathcal{T}}$ examples. This results in the predictive error $Err(\text{OOB}_{\mathcal{T}}) \geq 0$. Here, we assume that lower error value corresponds to better predictions. To assess the importance of the attribute x_i for the tree $\mathcal{T}$, we randomly permute the values of this attribute in the set $\text{OOB}_{\mathcal{T}}$ and obtain the set $\text{OOB}_{\mathcal{T}}^i$. Then, the error $Err(\text{OOB}_{\mathcal{T}}^i)$ is computed and the importance of the attribute x_i for the tree $\mathcal{T}$ is defined as the relative increase of error after noising the attribute. The Random Forest importance of the attribute is the average of these values across all trees in the forest, namely

$$importance_{\text{RF}}(x_i) = \frac{1}{|\mathcal{F}|} \sum_{\mathcal{T} \in \mathcal{F}} \frac{Err(\text{OOB}_{\mathcal{T}}^i) - Err(\text{OOB}_{\mathcal{T}})}{Err(\text{OOB}_{\mathcal{T}})}.$$

Note that $Err(\text{OOB}_T^i) = Err(\text{OOB}_T)$ if the attribute x_i does not appear in T. This can speed up the computation of $importance_{\text{RF}}$, but this feature ranking method is still the most time consuming. While the time complexity of the first two is negligible as compared to the one of growing the forest, this one has an additional linear factor: the number of examples in the dataset.

4 Experimental Design

In this section, we present the experimental design used to evaluate the performance of the proposed feature ranking methods. We begin by stating the main experimental questions and then briefly summarize the MTR datasets used in this study. We next describe the evaluation procedure and give the specific parameter instantiations of the methods.

4.1 Experimental Questions

The main focus of this study is to answer the following questions:

1. Can additional knowledge from feature importances lead to better predictive performance of a regressor, i.e., are the obtained feature rankings relevant?
2. Which ranking method is the most appropriate for a given ensemble method?
3. Which ensemble method is the most appropriate for a given ranking algorithm?
4. Which *ensemble-ranking* pair is the best overall?

For answering these questions, we design several experiments and comparisons of performance. We learn different feature rankings by considering combinations of ensemble learning methods and feature ranking methods. More specifically, we construct 8 different feature rankings: *Random Forest Symb*, *Random Forest Genie3*, *Random Forest RF*, *Bagging Symb*, *Bagging Genie3*, *Bagging RF*, *Extra trees Symb*, and *Extra trees Genie3*. We then use the obtained feature importances as weights in k nearest neighbor predictor (kNN) and compare the different rankings to address the questions outlined above. Finally, based on the obtained results, we identify the method that yields the best feature ranking.

4.2 Data Description

We use 26 MTR benchmark problems. Table 2 presents the basic statistics of the datasets: The number of features per dataset ranges from 6 to 576 and features are mainly numeric. The number of targets ranges from 2 to 16, while the number of examples takes values between 42 and 60607.

The datasets come from different domains: *andro*, *ENB* and *water quality* originate from studies of water quality; the *ATP* datasets concern the prediction of airline tickets prices; *collembola*, the *Forestry* datasets, *soil quality* and *vegetation condition* describe soil and vegetation conditions; *EDM* stands for electrical discharge machining; *jura* contains measurements of heavy metals concentrations; *metal-data* is about meta learning; *OES* stands for occupational employment survey; *osales* (online product sales) and *scpf* (see-click-predict fix) originate from two Kaggle competitions; *RF1* and *RF2* describe river flows; *SCM1d* and *SCM20d* were derived from a competition in supply chain management; *sigmeareal* and *sigmeasim* deal with cross-pollination between conventional and GM crops, and *slump* concerns the prediction of concrete slump.

Table 2. Description of the benchmark problems in terms of the number of nominal and numeric descriptive attributes, the number of targets, and the number of examples.

Dataset	Nominal	Numeric	Targets	Examples
andro [17]	0	30	6	49
ATP1d [23]	0	411	6	337
ATP7d [23]	0	411	6	296
collembola [19]	8	39	3	393
EDM [20]	0	16	2	154
ENB [28]	0	8	2	768
Forestry Kras [26]	0	160	11	60607
Forestry LIDAR IRS [25]	0	29	2	2730
Forestry LIDAR Landsat [25]	0	150	2	6218
Forestry LIDAR Spot [25]	0	49	2	2730
jura [15]	0	15	3	359
metal-data [27]	0	53	10	42
OES10 [23]	0	298	16	403
OES97 [23]	0	263	16	334
osales [1]	0	401	12	639
RF1 [23]	0	64	8	9125
RF2 [23]	0	576	8	9125
SCM1d [23]	0	280	16	9803
SCM20d [23]	0	61	16	8966
scpf [2]	0	23	3	1137
sigmeareal [11]	0	6	2	817
sigmeasim [11]	2	9	2	10368
slump [29]	0	7	3	103
soil quality [12]	0	156	3	1944
vegetation condition [21]	1	39	7	16967
water quality [13]	0	16	14	1060

4.3 Evaluation Methodology

We adopted the following evaluation methodology to properly assess the performance of the proposed methods. First, we randomly divide each dataset $\mathscr{D}$ into 2/3 for the training part $\mathscr{D}_{\text{TRAIN}}$ and 1/3 for the testing part $\mathscr{D}_{\text{TEST}}$. A ranking is computed from an ensemble that is built on the training part only. This procedure is repeated 10 times and the performance measures are averaged.

The quality of the ranking is assessed by using the kNN algorithm. Instead of the standard Euclidean distance, its weighted version was used in kNN. For two input vectors $\boldsymbol{x}^1$ in $\boldsymbol{x}^2$, the distance d between them is defined as

$$d(\boldsymbol{x}^1, \boldsymbol{x}^2) = \sqrt{\sum_{i=1}^{D} w_i d_i^2(\boldsymbol{x}_i^1, \boldsymbol{x}_i^2)}, \tag{3}$$

where the distance $d_i : \mathcal{X}_i \times \mathcal{X}_i \to [0, 1]$ is defined as

$$d_i(\boldsymbol{x}^1, \boldsymbol{x}^2) = \begin{cases} \mathbf{1}[\boldsymbol{x}_i^1 \neq \boldsymbol{x}_i^2] & : \mathcal{X}_i \text{ nominal} \\ \frac{|\boldsymbol{x}_i^1 - \boldsymbol{x}_i^2|}{\max_{\boldsymbol{x}} \boldsymbol{x}_i - \min_{\boldsymbol{x}} \boldsymbol{x}_i} & : \mathcal{X}_i \subseteq \mathbb{R} \end{cases},$$

where max and min go over the known examples $\boldsymbol{x}$. The weights are set to $w_i = \max\{importance(x_i), 0\}$ and are equal to the feature importances obtained by Symb and Genie3 ranking. They need to be made non-negative for RF ranking. In this way, we ensure that d is well defined and ignore the attributes that are of lower importance than a randomly generated attribute.

The evaluation through a kNN predictive model was chosen for two main reasons. First, this is a distance based model, which can easily use the information contained in the feature importances in the learning phase. The second reason is kNN's simplicity: its only parameter is the number of neighbors k, which we set to 5. In the prediction stage, the neighbors' contributions to the predicted value are equally weighted, so we do not introduce additional parameters that would influence the performance.

The rationale for using kNN as an evaluation model is as follows. If a feature ranking is meaningful, then the feature importances used as weights in the calculation of distances should yield better predictive power as compared to not using these weights [10].

We assess the predictive performance with the average relative root mean squared error $\overline{\text{RRMSE}}$. If we denote the predicted value of the target y_j by $\hat{y}_j(\boldsymbol{x})$, the RRMSE for this target is defined as

$$\text{RRMSE}(y_j) = \sqrt{\frac{\frac{1}{|\mathscr{D}_{\text{TEST}}|} \sum_{\boldsymbol{x} \in \mathscr{D}_{\text{TEST}}} (y_j(\boldsymbol{x}) - \hat{y}_j(\boldsymbol{x}))^2}{Var_j(\mathscr{D}_{\text{TRAIN}})}},$$

and $\overline{\text{RRMSE}}$ can be expressed as $\overline{\text{RRMSE}} = \sqrt{\frac{1}{T} \sum_{j=1}^{T} \text{RRMSE}^2(y_j)}$.

4.4   Statistical Analysis of the Results

For comparing two algorithms, we use the Wilcoxon's test, and for comparing more than two algorithms, we use the Friedman's test [12]. In both cases, the null hypothesis H_0 is that all considered algorithms have the same performance. If H_0 is rejected by the Friedman's test, we additionally apply Nemenyi's post-hoc test [12] to investigate where the statistically significant differences occur. Finally, to control the false discovery rate, the Benjamini-Hochberg procedure [3] was applied: let p_i be the i-th smallest among the obtained p-values, and m the number of tests. Let i_0 be the largest i, such that $p_i \leq \frac{i}{m}\alpha =: \hat{\alpha}_i$. Then, we can reject the hypotheses that correspond to p-values p_i, for $1 \leq i \leq i_0$.

The results of the Nemenyi's tests are presented by average ranks diagrams. Each diagram shows the average rank of each algorithm over the considered datasets, and the critical distance, i.e., the distance by which the average ranks of two algorithms must differ to be considered statistically significantly different. Additionally, the groups of algorithms among which no statistically significant differences occur are connected with a red line. In the analysis, the significance level was set to $\alpha = 0.05$.

4.5   Parameter Instantiation

The algorithm for inducing an ensemble of PCTs for MTR takes as input the following parameters: the number of base predictive models in the forest (all ensemble types), minimal number of examples in a leaf of a tree (all ensemble types), and the feature subset size (Random Forest only). In all cases, we grow 100 trees, whose leaves must contain at least two examples each. Additionally, the feature subset size in the case of Random Forests is set to $\lceil\sqrt{D}\rceil$.

Next, recall that the symbolic ranking requires selecting a value for w. In a preliminary study, we investigate the influence of several values of w, i.e., $w \in \{0.25, 0.5, 0.75, 1\}$, on the performance of feature ranking. We perform Friedman's test with the null hypothesis H_0 that the four symbolic rankings perform equally well. It turns out that the differences among the rankings are not statistically significant in the case of Bagging (p-value is 0.418) and ET (p-value is 0.230), whereas in the case of RF, they are (p-value is 0.000697). In the RF case, we can proceed to Nemenyi's test, whose results are shown in Fig. 1.

The diagram reveals that only the symbolic ranking with weight $w = 1.0$ is statistically significantly worse than the rankings with weights $w = 0.5$ and $w = 0.25$. This can be explained by Eq. 2: this value for the weight is the only one where the depth of the node where an attribute appears is not taken into account when computing the relevance.

Since the average ranks of the ranking methods with $w = 0.25$, $w = 0.5$, $w = 0.75$ and $w = 1.0$ are respectively 2.38, 2.25, 2.54 and 2.83 for Bagging, and 2.42, 2.38, 2.25 and 2.94 for ET, the ranking Symb50 is a reasonable choice for all three ensembles, since it is always ranked at least second. The reason for this is less obvious but we hypothesize that it could be an artifact of the algorithm for inducing ensembles (see Table 1). Namely, splits in the ensemble trees are

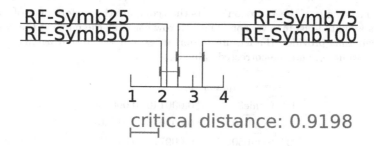

Fig. 1. The average ranks diagram for Nemenyi's test, performed for the symbolic ranking methods with the Random Forest ensemble at significance level of $\alpha = 0.05$.

binary. If we assume that the best test in an internal node $\mathcal{N} \in \mathcal{T}$ partitions $E(\mathcal{N}) \subseteq \mathscr{D}$ approximately in half, then the attribute in $\mathcal{N}$'s test influences one half of the instances that arrive to its parent; hence, the parent should receive twice as large a reward as each of its two children.

5 Results and Discussion

In this section, we present the results from the experimental evaluation, responding to the experimental questions posed above. The baseline kNN is denoted as 5NN, while the weighted 5NN is denoted by the combination of ensemble and ranking method (*ensemble-ranking*).

5.1 Are the Obtained Feature Rankings Relevant?

The investigation of whether a given feature ranking is relevant has a pivotal role in this work. More specifically, we investigate whether 5NN prediction can benefit from using the additional information from the feature importances. To this end, we compare the performance of 5NN without and with feature importances. We use the Wilcoxon's test to assess the statistical significance of the differences in performance between the two 5NN methods.

Table 3 gives the results of the statistical evaluation. It shows that all hypotheses are rejected, since we have $p_i < \hat{\alpha}_i$ for all i. Therefore, we can conclude that using feature ranking is clearly beneficial, i.e., the obtained feature rankings (more precisely, feature importances) are relevant and meaningful. Considering that the answer to the first question is positive, we now proceed with discussing the remaining experimental questions.

5.2 Comparison of the Different Ranking Methods

We compare the performance of the different ranking methods when coupled with a given ensemble learning method. In other words, we perform three analyses, where the ensemble method is fixed in each analysis. The results of this analysis

Table 3. The results of the Wilcoxon's tests that compare the performance of standard 5NN to its weighted-distance version. The i-th row contains the name of the *ensemble-ranking* pair that provided the feature importances and was tested against standard 5NN; the p-value p_i; and the corrected value $\hat{\alpha}_i$.

ensemble-ranking	p_i	$\hat{\alpha}_i$
RF-Genie3	0.000146	0.006250
RF-Symb50	0.001938	0.012500
ET-Symb50	0.009776	0.018750
Bagging-RF	0.017592	0.025000
ET-Genie3	0.020120	0.031250
Bagging-Genie3	0.028920	0.037500
Bagging-Symb50	0.031316	0.043750
RF-RF	0.035411	0.050000

for Random Forests and Bagging are given in Fig. 2. For Random Forests, the Friedman test found that there are statistically significant differences among the three ranking methods with $p = 0.00316$. The follow-up Nemenyi test reveals that the performance of a feature ranking obtained with Genie3 is statistically significantly better than the Random Forest ranking.

The differences between the different rankings for the Bagging ensemble method are not statistically significant ($p = 0.347$) and we can note that the three ranking methods have close average ranks. Furthermore, the Wilcoxon test for the Extra Trees ensemble revealed that there is no statistically significant difference ($p = 0.191$) between the two rankings, but Genie3 has a better sum of ranks than Symb50. Finally, the Random Forest ranking should be avoided: it has the worst computational complexity and it is consistently the worst performing ranking method (both for the Bagging and RF ensemble method).

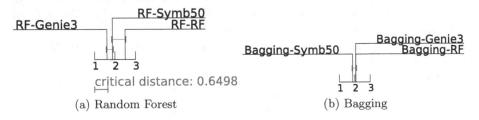

(a) Random Forest   (b) Bagging

Fig. 2. The average ranks diagrams for Nemenyi's post-hoc test at a significance level of $\alpha = 0.05$, performed for the rankings RF, Genie3 and Symb50, for (a) RF ensemble method and (b) Bagging ensemble.

5.3 Comparison of the Different Ensemble Methods

We also compare the performance of the different ensemble methods when used together with a given ranking method. In other words, we perform three analyses where the ranking method is fixed in each analysis. The Friedman test for Genie3 and Symb50 and the Wilcoxon test for Random Forest ranking did not rejected the null hypotheses with p-values 0.403, 0.346 and 0.176, respectively. Nevertheless, top ranked ensemble methods for the given ranking methods are: Random Forests for Genie3, and Bagging for Symb50 and Random Forests ranking.

5.4 Selecting the Best Ensemble-Ranking Pair

Finally, one of the goals of this paper is to select the best ensemble-ranking pair of methods. For this purpose, we evaluate the performance of the 8 ensemble-ranking pairs by performing a Friedman test. It reveals that there are statistically significant differences in performance among the methods and the results of the post hoc Nemenyi test are shown in Fig. 3. The average ranks diagram shows that the best performing pair is the Random Forest ensemble method coupled with the Genie3 ranking method. Moreover, the best performing method pair is statistically significantly better than the worst performing method pair (Random Forest ensembles coupled with Random Forest ranking).

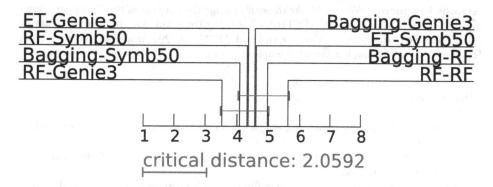

Fig. 3. The average ranks diagram from the Nemenyi's post-hoc test, performed for all *ensemble-ranking* pairs, at a significance level of $\alpha = 0.05$.

6 Conclusions

In this work, we proposed three base feature ranking methods that can be coupled with three ensemble learning methods. We investigated and evaluated eight *ensemble-ranking* options. These are the first methods that can address the task of feature ranking in the case of MTR with numeric and nominal attributes. More specifically, we extend Genie3, Random Forest and Symbolic ranking towards the task of MTR. We then coupled these rankings with the following ensemble learning methods: Bagging, Random Forests and Extra trees – all of which use predictive clustering trees as base predictive models.

We perform an extensive experimental evaluation of the proposed feature ranking methods using 26 benchmark MTR datasets. The evaluation is based on a 5NN predictive model that uses the obtained feature importances as weights.

The results show that all of the proposed eight methods yield a relevant feature ranking, i.e., the 5NN predictive models that use the feature importances as weights statistically significantly outperform the standard 5NN. Next, the best values for the weight parameter of the Symbolic ranking is 0.5. Furthermore, the best performing method is the one that uses Random Forests for learning the ensemble and Genie3 for calculating the feature importances. Moreover, this method is also computationally efficient: Random Forests are among the most efficient ensemble learning methods and Genie3 adds just a small computational cost of a single traversal of each tree in the ensemble.

We plan to extend this work along three major directions. First, we will compare the proposed methods to methods that use the data transformation approach, i.e., transform a MTR problem to a set of STR problems, coupled with a feature ranking algorithm for STR. Second, we will extend the proposed method to other structured output prediction tasks, such as multi-label classification, and hierarchical multi-label classification. Finally, we will investigate the influence of the ensemble size on the produced feature rankings.

Acknowledgments. We would like to acknowledge the support of the European Commission through the project MAESTRA - Learning from Massive, Incompletely annotated, and Structured Data (Grant number ICT-2013-612944), and of the Slovenian Research Agency through a young researcher grant.

References

1. Kaggle: Online product sales. https://www.kaggle.com/c/online-sales. Accessed 05 May 2017
2. Kaggle: See click predict fix. https://www.kaggle.com/c/see-click-predict-fix. Accessed 05 May 2017
3. Benjamini, Y., Hochberg, Y.: Controlling the false discovery rate: a practical and powerful approach to multiple testing. J. R. Stat. Soc. Ser. B (Methodol.) **57**(1), 289–300 (1995)
4. Blockeel, H.: Top-down induction of first order logical decision trees. Ph.D. thesis, Katholieke Universiteit Leuven, Leuven, Belgium (1998)
5. Borchani, H., Varando, G., Bielza, C., Larrañaga, P.: A survey on multi-output regression. Wiley Interdiscip. Rev. Data Min. Knowl. Discov. **5**(5), 216–233 (2015)
6. Breiman, L.: Bagging predictors. Mach. Learn. **24**(2), 123–140 (1996)
7. Breiman, L.: Random forests. Mach. Learn. **45**(1), 5–32 (2001)
8. Breiman, L., Friedman, J., Olshen, R., Stone, C.J.: Classification and Regression Trees. Chapman & Hall/CRC, Boca Raton (1984)
9. Brobbey, A.: Variable Selection in Multivariate Multiple Regression. Master's thesis, Department of Mathematics and Statistics, Memorial University, Newfoundland and Labrador, Canada (2015)

10. Cunningham, P., Delany, S.J.: k-Nearest Neighbour Classifiers. Technical report 2, University College Dublin (2007)
11. Demšar, D., Debeljak, M., Džeroski, S., Lavigne, C.: Modelling pollen dispersal of genetically modified oilseed rape within the field. In: Proceedings of the 9th Annual Meeting of the Ecological Society of America. p. 152 (2005)
12. Demšar, J.: Statistical comparisons of classifiers over multiple data sets. J. Mach. Learn. Res. **7**, 1–30 (2006)
13. Džeroski, S., Demšar, D., Grbović, J.: Predicting chemical parameters of river water quality from bioindicator data. Appl. Intell. **1**(13), 7–17 (2000)
14. Geurts, P., Erns, D., Wehenkel, L.: Extremely randomized trees. Mach. Learn. **36**(1), 3–42 (2006)
15. Goovaerts, P.: Geostatistics for Natural Resources Evaluation. Oxford University Press, New York (1997)
16. Hansen, L.K., Salamon, P.: Neural network ensembles. IEEE Trans. Pattern Anal. Mach. Intell. **12**(10), 993–1001 (1990)
17. Hatzikos, E.V., Tsoumakas, G., Tzanis, G., Nick, B., Vlahavas, I.P.: An empirical study on sea water quality prediction. Knowl. Based Syst. **21**(6), 471–478 (2008)
18. Huynh-Thu, V.A., Irrthum, A., Wehenkel, L., Geurts, P.: Inferring regulatory networks from expression data using tree-based methods. PLoS One **5**(9), 1–10 (2010)
19. Kampichler, C., Džeroski, S., Wieland, R.: Application of machine learning techniques to the analysis of soil ecological data bases: relationships between habitat features and Collembolan community characteristics. Soil Biol. Biochem. **32**(2), 197–209 (2000)
20. Karalič, A., Bratko, I.: First order regression. Mach. Learn. **26**(2–3), 147–176 (1997)
21. Kocev, D., Džeroski, S., White, M., Newell, G., Griffioen, P.: Using single- and multi-target regression trees and ensembles to model a compound index of vegetation condition. Ecol. Model. **220**(8), 1159–1168 (2009)
22. Kocev, D., Vens, C., Struyf, J., Džeroski, S.: Tree ensembles for predicting structured outputs. Pattern Recognit. **46**(3), 817–833 (2013)
23. Spyromitros-Xioufis, E., Tsoumakas, G., Groves, W., Vlahavas, I.: Multi-target regression via input space expansion: treating targets as inputs. Mach. Learn. **104**(1), 55–98 (2016)
24. Stańczyk, U., Jain, L.C. (eds.): Feature Selection for Data and Pattern Recognition. Studies in Computational Intelligence. Springer, Heidelberg (2015)
25. Stojanova, D.: Estimating Forest Properties from Remotely Sensed Data by using Machine Learning. Master's thesis, Jožef Stefan International Postgraduate School, Ljubljana, Slovenia (2009)
26. Stojanova, D., Panov, P., Gjorgjioski, V., Kobler, A., Džeroski, S.: Estimating vegetation height and canopy cover from remotely sensed data with machine learning. Ecol. Inform. **5**(4), 256–266 (2000)
27. Todorovski, L., Blockeel, H., Dzeroski, S.: Ranking with predictive clustering trees. In: Elomaa, T., Mannila, H., Toivonen, H. (eds.) ECML 2002. LNCS (LNAI), vol. 2430, pp. 444–455. Springer, Heidelberg (2002). doi:10.1007/3-540-36755-1_37
28. Tsanas, A., Xifara, A.: Accurate quantitative estimation of energy performance of residential buildings using statistical machine learning tools. Energy Build. **49**, 560–567 (2012)
29. Yeh, I.C.: Modeling slump flow of concrete using second-order regressions and artificial neural networks. Cem. Concr. Compos. **29**, 474–480 (2007)

Recommendation System

Recommending Collaborative Filtering Algorithms Using Subsampling Landmarkers

Tiago Cunha[1(✉)], Carlos Soares[1], and André C.P.L.F. de Carvalho[2]

[1] INESC-TEC/FEUP, Porto, Portugal
{tiagodscunha,csoares}@fe.up.pt
[2] ICMC - USP, São Carlos, São Paulo, Brazil
andre@icmc.usp.br

Abstract. Recommender Systems have become increasingly popular, propelling the emergence of several algorithms. As the number of algorithms grows, the selection of the most suitable algorithm for a new task becomes more complex. The development of new Recommender Systems would benefit from tools to support the selection of the most suitable algorithm. Metalearning has been used for similar purposes in other tasks, such as classification and regression. It learns predictive models to map characteristics of a dataset with the predictive performance obtained by a set of algorithms. For such, different types of characteristics have been proposed: statistical and/or information-theoretical, model-based and landmarkers. Recent studies argue that landmarkers are successful in selecting algorithms for different tasks. We propose a set of landmarkers for a Metalearning approach to the selection of Collaborative Filtering algorithms. The performance is compared with a state of the art systematic metafeatures approach using statistical and/or information-theoretical metafeatures. The results show that the metalevel accuracy performance using landmarkers is not statistically significantly better than the metafeatures obtained with a more traditional approach. Furthermore, the baselevel results obtained with the algorithms recommended using landmarkers are worse than the ones obtained with the other metafeatures. In summary, our results show that, contrary to the results obtained in other tasks, these landmarkers are not necessarily the best metafeatures for algorithm selection in Collaborative Filtering.

Keywords: Metalearning · Subsampling landmarkers · Collaborative filtering

1 Introduction

Recommender Systems (RSs) recommend potentially interesting items to users in order to deal with the information overload problem [1]. Collaborative Filtering (CF) is the most popular of the available recommendation strategies. Despite the large amount of research dedicated to this topic, there are still several challenges that need to be addressed. One of them is how to choose the best CF algorithm

© Springer International Publishing AG 2017
A. Yamamoto et al. (Eds.): DS 2017, LNAI 10558, pp. 189–203, 2017.
DOI: 10.1007/978-3-319-67786-6_14

for a given dataset. Since training and evaluating all algorithms for the new dataset requires a prohibitive amount of time and resources, automatic solutions based on prior knowledge are of the utmost importance. Metalearning (MtL) is an approach useful for that purpose [7].

MtL is concerned with discovering patterns in data and understanding the effect on the behavior of algorithms [30]. It has been extensively used for algorithm selection [6,27,28]. MtL casts the algorithm selection problem as a learning task. For such, it uses a metadataset, where each meta-example corresponds to a problem. For each meta-example, the predictive features are characteristics (metafeatures) extracted from the corresponding problem and the target represents the performance of algorithms when applied to the problem (metatarget) [5].

Metafeatures are regarded as the most important element in a MtL task [5]. It is essential for them to be representative of the problem at hand. The metafeatures used must contain information that discriminates the performance of different algorithms in such a way that the patterns found are useful for future applications. However, this is not a trivial task. The research in this topic has originated several different types of metafeatures, such as statistical and/or information-theoretical, model-based and landmarkers, which are related to the dataset, model and performance properties, respectively [29,30].

The algorithm selection task for CF has received considerable attention recently [2,7,10,14,23]. Related work has investigated the effect of different statistical and information-theoretical metafeatures with positive performances. However, none has investigated the merits of landmarkers as metafeatures. Since these metafeatures use simple estimates of performance to predict the actual performance of algorithms, its efficacy in solving the algorithm selection problem is not only expected but has been demonstrated in various other tasks [3,11,17,18,20,21,25]. Therefore, it is important to understand if their effect is similarly positive in selecting CF algorithms.

Hence, the main contribution of this paper is the proposal of several subsampling landmarkers and their experimental validation in terms of their merits to select CF algorithms. To do so, this paper provides an extensive collection of baselevel datasets, algorithms and evaluation measures similarly to the ones found in the state of the art [7]. The subsampling landmarkers are proposed and analyzed as relative landmarkers. Such landmarkers look not only towards the absolute performance estimations, but also to the relative performance between landmarkers. Our motivation lies in ensuring a proper exploration of the landmarkers concept for the CF scope. All different metafeatures are compared to the state of the art approach in statistical and information-theoretical metafeatures [7] in terms of metalevel accuracy and impact on the baselevel performance. The results show that landmarkers are not statistical significantly better than the statistical and/or information-theoretical metafeatures.

This document is organized as follows: Sect. 2 presents related work on CF, MtL and algorithm selection for CF; Sect. 3 presents the approach used for subsampling landmarkers and relative landmarkers and explains the experimental setup. In Sect. 4, several aspects of the proposed approach are evaluated and discussed. Section 5 presents the conclusions and directions for future work.

2   Related Work

2.1   Collaborative Filtering

RSs were proposed to complement Information Retrieval systems, providing an alternative to solve the problem of information overload and recommend potentially interesting items to users [4]. RSs are inspired by human social behavior, where it is common to take into account the tastes, opinions and experiences of acquaintances when making decisions [4]. Several strategies are used in such systems, such as: (1) recommend items that similar users find relevant, (2) recommend items with similar characteristics, (3) recommend items depending on the user's context, (4) recommend items based on social relationships and (5) recommend items using knowledge about the user's behavior. From the several strategies available, Collaborative Filtering (CF) is the most popular.

CF recommendations are based on the premise that a user will probably like the items favored by a similar user. CF employs the feedback from each individual user to recommend items to similar users [33]. The feedback is a numeric value, proportional to the user's appreciation of an item. Most feedback is based on a rating scale, although other variants such as like/dislike actions and clickstream are also suitable. The data structure used in CF is named rating matrix R. It is usually described as $R^{U \times I}$, representing a set of users U, where $u \in \{1, ..., N\}$ and a set of items I, where $i \in \{1, ..., M\}$. Each element of this matrix (R_{ui}) is the feedback provided by user u for item i. Figure 1 presents such matrix.

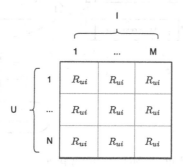

Fig. 1. Rating matrix.

CF algorithms can be organized in two major groups: memory-based and model-based [4]. Memory-based algorithms apply heuristics to a rating matrix to compute recommendations, whereas model-based algorithms induce a model from this matrix. Most memory-based algorithms adopt Nearest Neighbor strategies, while the model-based ones are mostly based on Matrix Factorization [33].

The evaluation of RSs is usually performed by procedures that split the dataset into training and testing subsets (using sampling strategies, such as k-fold cross-validation [16]) and assesses the performance of the trained model on the testing dataset. Different evaluation metrics exist [22]: for rating accuracy, error measures such as Mean Absolute Error (MAE) or Root Mean Squared Error (RMSE); for classification accuracy, one uses Precision/Recall or Area Under the Curve (AUC); for ranking accuracy, common measures are Normalized Discounted Cumulative Gain (NDCG) and Mean Reciprocal Rank (MRR).

2.2 Metalearning

MtL addresses the algorithm selection problem similarly to a traditional learning process (see Fig. 2). First, the problems are characterized by a set of measurable characteristics (i.e., metafeatures) and the compared algorithms are evaluated according to their performance in the learning task. This creates a meta-dataset, where each meta-example has as predictive attributes the characteristics extracted for the problem and the target attribute is usually the algorithm that obtained the best performance in the specific dataset. Next, a learning algorithm is trained using the metadataset. The trained model represents patterns in the data that relate the metafeatures with the best performing algorithms. Hence, it can be used to predict the best algorithm for a new problem [29].

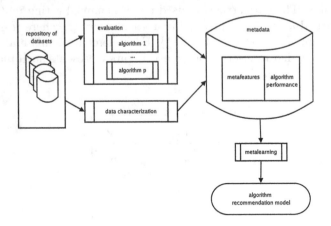

Fig. 2. Metalearning process [5].

As in any other learning problem, the success of a MtL approach depends on the information contained in the independent variables, i.e. the metafeatures. The MtL literature divides metafeatures into three main groups [5,29,30]: statistical and/or information-theoretical, model-based and landmarkers.

Statistical and/or information-theoretical metafeatures describe the dataset characteristics using a set of measures from statistics and information theory.

These metafeatures assume that there are patterns in the data which can be related to the best algorithms. Examples include simple measures such as the number of examples and features in the dataset to more advanced measures such as entropy and kurtosis of features and even correlation between features [5].

Model-based characteristics are properties extracted from models induced from the dataset. They refer, for instance, to the number of leaf nodes in a decision tree [5]. The rationale is that there is a relationship between model characteristics and algorithm performance which are dataset-independent. Then, it is expected that these characteristics are able to discriminate among algorithms.

Finally, landmarkers are fast estimates of the algorithm performance on the dataset. There are two different types of landmarkers: those obtained from the application of fast and simple algorithms on complete datasets and those which are achieved by using complete models for samples of datasets, also known as subsampling landmarkers [5]. Such metafeatures rely on the assumption that by estimating the performance of fast and simple models or by using samples of the data, the performance estimates will correlate well with the best algorithms, hence enabling future predictions. In fact, these metafeatures have proven successful on the selection of algorithms for various tasks [3,11,17,18,20,21,25].

2.3   Algorithm Selection for CF

Related work in algorithm selection for CF has studied the problem using only statistical and/or information-theoretical metafeatures. These have focused on different aspects of the data distributions [2,10,14,23], the matrix structure [23] and neighborhood statistics [14]. A more recent work has combined the majority of the metafeatures used previously in a single framework [7]. This extensive set of metafeatures (referred to here as Systematic) are used in our experimental study in order to properly compare statistical and/or information-theoretical metafeatures with the set of subsampling landmarkers proposed here.

In order to understand the systematic metafeatures, one must consider first the framework used to generate them. It requires three main elements: object o, function f and a post-function pf. The framework applies the function f to the object o and, afterwards the post-function pf to the outcome of the function f in order to derive the final metafeature. Thus, any metafeature can be represented using the following notation: $\{o.f.pf\}$ [26]. For instance, the metafeature *column.maximum.mean* refers to the mean value of all the maximum values in all columns in the dataset.

Consider now a rating matrix R, with rows (i.e., users) U and columns (i.e., items) I. The objects to be used in the framework are R, U and I. The functions f considered to characterize these objects are: original ratings (ratings), count the number of elements (count), mean value (mean) and sum of values (sum). The post-functions pf are maximum, minimum, mean, standard deviation (sd), median, mode, entropy, Gini index, skewness and kurtosis. Additionally, we consider 4 simple metafeatures: number of users, items, ratings and matrix sparsity. This results in 74 metafeatures which were reduced by correlation feature selection, ending up with the following set: $D.ratings.kurtosis$, $D.ratings.sd$, $I.count.kurtosis$, $I.count.minimum$, $I.mean.entropy$, $I.sum.skewness$, $nusers$, $sparsity$, $U.mean.minimum$, $U.sum.kurtosis$, $U.mean.skewness$ and $U.sum.entropy$.

3   Subsampling Landmarkers for Collaborative Filtering

This section presents our proposal of subsampling landmarkers for the selection of CF algorithms and the experimental procedure used to validate them. Our motivation for using landmarkers is that, although they have been successfully applied to the algorithm selection problem in other learning tasks [3,11,17,18, 20,21,25], they were never adapted for selecting CF algorithms. Since there are no fast/simple CF algorithms, which can be used as traditional landmarkers, we have followed the alternative approach of developing subsampling landmarkers, i.e. applying the complete CF algorithms on samples of the data.

3.1   Subsampling Landmarkers

Subsampling landmarkers are based on the estimation of the performance of algorithms on random samples from the original datasets. This means that for each CF dataset, random samples are extracted. Then, CF algorithms are trained on these samples and their performance assessed using different metrics. The outcome is a subsampling landmarker for each pair algorithm/evaluation measure. In order to properly validate the impact of subsampling landmarkers, we recur to different ways to take advantage of these metafeatures, also known as relative landmarkers [11]:

- Absolute: this is the most straightforward approach since it does not operate any transformation on the subsampling landmarkers. It uses the estimated performance values as the metafeature.
- Ranking: this approach is based on the ranking of the landmarkers $L = \{l_1, l_2, ..., l_n\}$. Therefore, the metafeatures are now the rank of the landmarker, where 1 indicates the best landmarker and n the worst.
- Pairwise: this approach performs pairwise comparison for all pairs of landmarkers. Consider two landmarkers l_i and l_j. If the performance of l_i is greater, equal or worse than l_j, then the final metafeature values are 1, 0 or -1, respectively. Such comparisons are performed for all pairs of landmarkers. Thus $n \times (n - 1)$ new metafeatures are added for each evaluation measure.
- Ratio: this approach also performs pairwise comparisons. However, it does so by using the ratios of the performances instead of assigning 1, 0 or -1 values. Given two landmarkers l_i and l_j, a metafeature with the value l_i/l_j is created.

As an example, let us consider two CF algorithms, A and B, and the NMAE performance measure. Given a data sample, they are applied to it and the corresponding NMAE score is computed. Table 1 illustrates such values and all the

Table 1. Example of relative landmarkers.

Algorithm	NMAE	Absolute	Ranking	Pairwise	Ratio
A	0.73	0.73	1	1	0.839
B	0.87	0.87	2	−1	1.192

corresponding subsampling landmarkers. Notice Absolute is equal to the original NMAE, Ranking assigns the ranking of the algorithms, Pairwise assigns 1 to the best algorithm and −1 to the worst and Ratio presents the ratios of NMAE. It should be noted that the process is repeated for each evaluation measure.

3.2 Experimental Procedure

The experimental setup used in this work is divided into baselevel and metalevel, referring, respectively, to the CF and classification stages of the process.

Baselevel. The baselevel setup is concerned with the CF datasets, algorithms and measures used to evaluate the performance of CF algorithms on those datasets. The 38 datasets used come from different domains, namely Amazon Reviews [24], BookCrossing [36], Flixter [35], Jester [13], MovieLens [15], MovieTweetings [9], Tripadvisor [31], Yahoo! [32] and Yelp [34]. It is important to observe that each domain can contain more than one dataset.

The experiments were carried out with MyMediaLite, a software library for recommender systems [12]. Two CF tasks were addressed: Rating Prediction (RP) and Item Recommendation (IR). While RP aims to predict the rating an user would assign to a new instance, in IR the goal is to recommend a ranked list of items in terms of user preference. Since the tasks are different, so are the algorithms and evaluation measures. The following CF algorithms were used for RP: Matrix Factorization (MF), Biased MF (BMF), Latent Feature Log Linear Model (LFLLM), SVD++, 3 variants of Sigmoid Asymmetric Factor Model (SIAFM, SUAFM and SCAFM), User Item Baseline (UIB) and Global Average (GA). Regarding IR, the algorithms used are BPRMF, Weighted BPRMF (WBPRMF), Soft Margin Ranking MF (SMRMF), WRMF and Most Popular (MP). In IR, the algorithms are evaluated using NDCG, while in RP the algorithms are evaluated using NMAE. All experiments use 10-fold cross-validation.

Metalevel. The metalevel is first characterized by the construction of the metafeatures. This work applies the statistical and/or information-theoretical metafeatures (described in Sect. 2.2) to all 38 CF datasets to extract the metafeatures for the Systematic approach. In order to extract the subsampling landmarkers (see Sect. 3.1), random samples of 10% for each of the original 38 CF datasets are extracted. Next, all algorithms are trained on said samples and their performance assessed via suitable evaluation metrics. This allows the extraction of what are referred as the Original relative landmarkers. Afterwards, the remaining relative landmarkers (Ranking, Pairwise and Ratio) are computed based on the values for the Original relative landmarker, as explained previously in Sect. 3.1. The entire process creates 5 different sets of metafeatures.

Two baselevel measures (NMAE and NDCG) are used to create two separate metatargets. The best algorithm, and consequently the target variable for each dataset, depends on the evaluation measures. For each pair dataset/evaluation measure, the best algorithm is chosen as the target variable. Hence, we study the algorithm selection problem for 2 different metatargets. The final metadatabases, consisting of combinations of all different metafeatures and metatargets, are the experimental basis for the algorithm selection problem addressed here.

Since the model selection problem is approached here as a classification task, 11 classification algorithms from the `caret` package [19] representing several biases were chosen to address it: ctree, C4.5, C5.0, kNN, LDA, Naive Bayes, SVM (linear, polynomial and radial kernels), random forest and a baseline algorithm: Majority Vote. The Majority Vote does not take into account any metafeatures and always predicts the class which appears more often. Since the metadatasets have a reduced number of examples, the accuracy of the metalevel algorithms was estimated using a leave one out strategy.

Meta-level performance is measured in two ways. First, the accuracy of the meta-level prediction is assessed, i.e. whether the best algorithm is selected or not. However, in MtL it is also important to understand the impact on the baselevel performance of the meta-level prediction. It assesses how the algorithms recommended by the metamodels actually affect the baselevel performance. It is based on the comparison of baselevel performance between the algorithm selected by the metamodel and the best possible algorithm. The goal is to understand what is the actual cost of failing in the prediction of the best algorithm in terms of baselevel performance.

Consider a dataset D and the performance of n algorithms on D, $P_D = \{p_1, p_2, ..., p_n\}$, according to a specific evaluation measure. It is possible to create a ranking $R_D = \{a_1, a_2, ..., a_n\}$ in decreasing order of those performance values. This means that a_1 is the best algorithm on D, with a performance of p_1. Consider now that $\hat{a} = a_q$ is the algorithm predicted by a metamodel for dataset D, $q \in \{1, ..., n\}$. The impact at the baselevel of using the metamodel for algorithm selection is assessed by comparing p_q, the performance of the selected algorithm, with p_1, the performance of the best algorithm. In this work, this comparison is done in three ways: performance (PE), error (ER) and ranking (RK), which are given by: $PE(\hat{a}, D) = p_q$, $ER(\hat{a}, D) = p_1 - p_q$ and $RK(\hat{a}, D) = q$.

The three measures are computed for all datasets and averaged. The comparisons average performance (AP), average error (AE) and average rankings (AR) for a set of M datasets are defined as follows:

$$AP(\hat{a}_i) = \frac{\sum_{i=1}^{M} PE(\hat{a}_i, D_i)}{M}$$

$$AE(\hat{a}_i) = \frac{\sum_{i=1}^{M} ER(\hat{a}_i, D_i)}{M}$$

$$AR(\hat{a}_i) = \frac{\sum_{i=1}^{M} RK(\hat{a}_i, D_i)}{M} \tag{1}$$

where $\hat{a}_i$ is the algorithm selected for dataset D_i.

4   Results and Discussion

4.1   Metalevel Evaluation

The metalevel accuracy performance for all strategies evaluated in this experimental study can be seen in Fig. 3. For readability purposes, only the performance of the best metamodel is presented. After manual inspection, the choice fell on SVM with polynomial kernel. Two baseline methods are included for fair comparison. The Majority Vote baseline assesses if the MtL approach is finding any useful patterns. The Systematic metafeatures baseline assesses if there is any advantage in using the proposed subsampling landmarkers in the CF scenario.

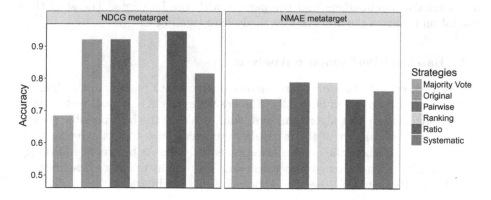

Fig. 3. Metalevel accuracy for all relative landmarkers and baselines.

Several observations can be made:

– Most landmarkers outperform the Majority Vote baseline. The exceptions are the Original and Ratio relative landmarkers in the NMAE metatarget.
– Landmarkers are better than the Systematic metafeatures in the NDCG metatarget.
– Landmarkers have slightly better performance than the Systematic metafeatures in the NMAE metatarget: this happens for the Ranking and Pairwise relative landmarkers.

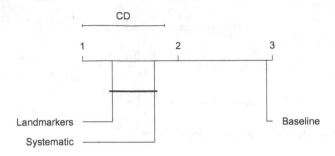

Fig. 4. CD diagram for the comparison of metafeature strategies.

The observations seem to indicate that (1) the metafeatures proposed are better than the baseline in terms of metalevel accuracy and (2) they seem to have slightly better performances than the Systematic metafeatures. To validate this assessment, we employ statistical significance tests using Critical Difference (CD) diagrams [8]. CD diagrams plot the average rank for each strategy and calculate the CD interval. Strategies connected by a CD line cannot be considered to perform differently. On the other hand, if two strategies are not connected by a CD line, they obtain, in fact, different performance, i.e. one strategy is ranked higher than the other. To apply this framework, we combine the performances of all relative landmarkers and compare it with the baselines. The statistical validation confirms the observations made here (see Fig. 4).

4.2   Baselevel Performance Analysis

Figure 5 presents the baselevel performance analysis with regards to the Average Performance (discussed in Sect. 3.2). The oracle represents an ideal system that always predicts the best algorithm, and, thus, achieves the best possible performance. The performance of the methods were scaled such that it is represented as a percentage, where the oracle corresponds to 100%. As before, the Majority Vote and MtL with Systematic metafeatures are used as reference baselines.

The results show that the MtL approach using landmarkers:

– outperforms the Majority Voting baseline on the NDCG metatarget, but not on the NMAE metatarget.
– never beats the Systematic approach on either metatarget.

The results on the baselevel performance show that, although the landmarkers perform better in terms of metalevel accuracy, the same is not true for the baselevel performance analysis in terms of Average Performance. This shows that in spite of correctly predicting the best algorithm more often, the performance of the selected algorithms in terms of the baselevel evaluation measure is worse, on average. Thus, when the landmarkers fail to predict the correct best algorithm, they usually choose an algorithm with worse performance than when the systematic metafeatures fail to predict the best algorithm.

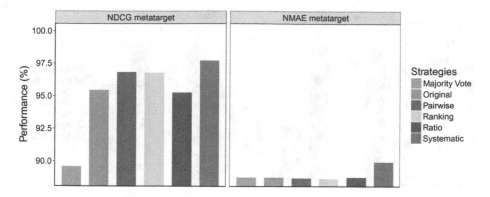

Fig. 5. Baselevel performance analysis regarding Average Performance.

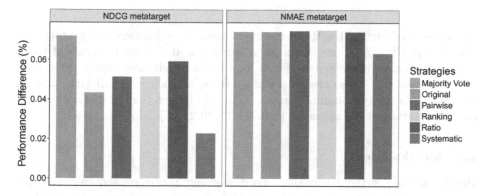

Fig. 6. Baselevel performance analysis regarding Average Error.

To validate this analysis, we performed the baselevel performance analysis, based on the Average Error (discussed in Sect. 3.2). The results are presented in Fig. 6. It shows that the error obtained by the Systematic approach has a smaller difference to the best error on both metatargets, hence confirming our previous observation.

In another analysis, we looked towards the Average Ranking (discussed in Sect. 3.2). The results are presented in Fig. 7. The baselines Majority Vote and Systematic are included for comparison with the landmarkers. The following observations regarding the landmarkers can be made:

- They rarely outperform the baseline Majority Voting: this only happens in 3 relative landmarkers in the NMAE metatarget.
- They are always worse than the Systematic metafeatures.

This analysis confirms the reason for the poor performance of landmarkers in terms of baselevel performance: the average ranking for the predicted CF algorithms is always higher than the Systematic approach. This means that

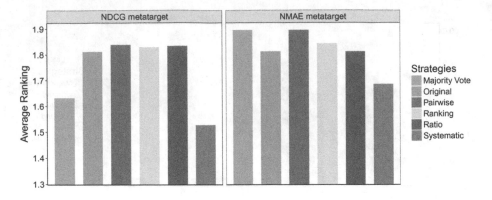

Fig. 7. Baselevel performance analysis regarding Average Ranking.

the metamodels trained with landmarkers tend to recommend on average the second best CF algorithm. When we consider the difference in terms of baselevel performance presented in Fig. 6, one understands how costly these misclassifications are. These are surprising results, as they contradict the results in other tasks, where landmarkers are typically better than statistical and/or information-theoretical measures [3,11,17,18,20,21,25].

4.3   Metaknowledge

Metaknowledge is the knowledge about learning processes acquired through experience with past learning episodes [30]. It explains how specific metafeatures influence which one is the best algorithm. Such knowledge is typically embedded in the metamodels built and sometimes it is difficult to access and/or interpret. Furthermore, considering the vast amount of metamodels built and analyzed so far, it is difficult to discuss all the knowledge potentially obtained with this study. Here, we address this problem simply by analyzing metafeature importance.

We analyze all different strategies in terms of feature importance across all metatargets studied. To do so, we build Random Forest models and take advantage of its inbuilt mechanism for feature importance. We use the implementation available in the `caret` package [19], which computes an importance score for each feature. We average the importance percentages across all models which share the same metafeatures and present the results in Fig. 8. Features with average importance below 10% were discarded.

The results show that the Systematic strategy contains the most influential metafeatures throughout. Special attention goes to the number of users and the skewness of the distribution of the sum of ratings per item. The remaining metafeatures focus on the kurtosis and entropy of the distribution of the sum of ratings of users. In terms of landmarkers, the Original relative landmarker highlights the importance of NMAE for SCAFM and LFLLM, while in the Ranking

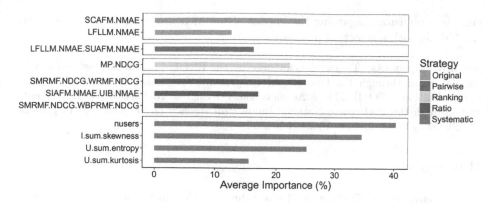

Fig. 8. Feature importance.

relative landmarker, the NDCG for MP is essential. In terms of relative land-markers which focus on the comparison of landmarkers, the results show that the comparison of NMAE performances of LFLLM and SUAFM algorithms are quite important among the Pairwise relative landmarkers. In the Ratio relative landmarkers, the ratios in terms of NDCG performance between SMRMF and both WRMF and WBPRMF and the ratio of NMAE between SIAFM and UIB are the most important ones. Although this analysis lacks some depth in terms of patterns found in the metamodels, it highlights two very important issues: (1) which are the most influential metafeatures and (2) since we are using land-markers, which algorithms and evaluation measures are essential for the problem. Both are essential for future CF algorithm selection works.

5   Conclusions and Future Work

Landmarkers have been reported as a successful way to characterize problems in Metalearning approaches to algorithm selection in several tasks. In this work, we propose a set of subsampling landmarkers for Collaborative Filtering (CF) methods. The landmarkers were compared with the state of the art systematic metafeatures, based on statistical and/or information-theoretical measures, both in terms of metalevel accuracy and baselevel performance analysis. Somewhat surprisingly, in our experiments, their performance was not statistical signifi-cantly better than the systematic approach, in terms of metalevel accuracy. Fur-thermore, the impact on the baselevel performance produces worse results when using landmarkers in terms of average performance, average error and average rankings. Thus, the major contributions of this work are: (1) to propose subsam-pling landmarkers for CF tasks and (2) showing that the widely accepted assump-tion that landmarkers are better than statistical and/or information-theoretical metafeatures may not be true in CF. Future work includes the adaptation of other types of landmarkers for CF, using for instance different sampling strate-gies and the extension of the experimental procedures in order to allow more generic conclusions regarding the impact of metafeatures of different natures on the CF algorithm selection problem.

Acknowledgments. This work is financed by the ERDF – European Regional Development Fund through the Operational Programme for Competitiveness and Internationalisation - COMPETE 2020 under the Portugal 2020 Partnership Agreement, and through the Portuguese National Innovation Agency (ANI) as a part of project «FASCOM | POCI-01-0247-FEDER-003506».

References

1. Adomavicius, G., Tuzhilin, A.: Toward the Next Generation of Recommender Systems: A Survey of the State-of-the-Art and Possible Extensions. IEEE Trans. on Knowl. and Data Eng. **17**(6), 734–749 (2005)
2. Adomavicius, G., Zhang, J.: Impact of data characteristics on recommender systems performance. ACM Trans. Manag. Inf. Syst. **3**(1), 1–17 (2012)
3. Bensusan, H., Kalousis, A.: Estimating the Predictive Accuracy of a Classifier. In: European Conference on Machine Learning, pp. 25–36 (2001)
4. Bobadilla, J., Ortega, F., Hernando, A., Gutiérrez, A.: Recommender systems survey. Knowl.-Based Syst. **46**, 109–132 (2013)
5. Brazdil, P., Giraud-Carrier, C., Soares, C., Vilalta, R.: Metalearning: Applications to Data Mining, 1st edn. Springer, Heidelberg (2009). doi:10.1007/978-3-540-73263-1
6. Brazdil, P., Soares, C., da Costa, J.: Ranking Learning Algorithms : Using IBL and Meta-Learning on Accuracy and Time Results. Mach. Learn. **50**(3), 251–277 (2003)
7. Cunha, T., Soares, C., de Carvalho, A.C.: Selecting Collaborative Filtering algorithms using Metalearning. In: European Conference on Machine Learning and Knowledge Discovery in Databases, pp. 393–409 (2016)
8. Demšar, J.: Statistical comparisons of classifiers over multiple data sets. J. Mach. Learn. Res. **7**, 1–30 (2006)
9. Dooms, S., De Pessemier, T., Martens, L.: MovieTweetings: a Movie Rating Dataset Collected From Twitter. In: CrowdRec at RecSys 2013 (2013)
10. Ekstrand, M., Riedl, J.: When recommenders fail: predicting recommender failure for algorithm selection and combination. In: ACM Conference on Recommender Systems, pp. 233–236 (2012)
11. Fürnkranz, J., Petrak, J., Bradzil, P., Soares, C.: On the use of fast subsampling estimates for algorithm recommendation. Technical report (2002)
12. Gantner, Z., Rendle, S., Freudenthaler, C., Schmidt-Thieme, L.: MyMediaLite: a free recommender system library. In: ACM Conference on Recommender Systems, pp. 305–308 (2011)
13. Goldberg, K., Roeder, T., Gupta, D., Perkins, C.: Eigentaste: A Constant Time Collaborative Filtering Algorithm. Inf. Retr. **4**(2), 133–151 (2001)
14. Griffith, J., O'Riordan, C., Sorensen, H.: Investigations into user rating information and predictive accuracy in a collaborative filtering domain. In: ACM Symposium on Applied Computing, pp. 937–942 (2012)
15. GroupLens: MovieLens datasets (2016). http://grouplens.org/datasets/movielens/
16. Herlocker, J.L., Konstan, J.A., Terveen, L.G., Riedl, J.T.: Evaluating collaborative filtering recommender systems. ACM Trans. Inf. Syst. **22**(1), 5–53 (2004)

17. Kanda, J., de Carvalho, A., Hruschka, E., Soares, C., Brazdil, P.: Meta-learning to select the best meta-heuristic for the Traveling Salesman Problem: A comparison of meta-features. Neurocomputing **205**, 393–406 (2016)
18. Kück, M., Crone, S.F., Freitag, M.: Meta-learning with neural networks and landmarking for forecasting model selection - an empirical evaluation of different feature sets applied to industry data meta-learning with neural networks and landmarking for forecasting model selection. In: International Joint Conference on Neural Networks. pp. 1499–1506 (2016)
19. Kuhn, M.: caret: Classification and Regression Training (2016). https://CRAN. R-project.org/package=caret, r package version 6.0-73
20. Ler, D., Koprinska, I., Chawla, S.: Utilizing regression-based landmarkers within a meta-learning framework for algorithm selection. School of Information Technologies University of Sydney, Technical report (2005)
21. Ler, D., Koprinska, I., Chawla, S.: Utilizing regression-based landmarkers within a meta-learning framework for algorithm selection. In: Proceedings of the ICML-2005 Workshop on Metalearning, pp. 44–51 (2005)
22. Lü, L., Medo, M., Yeung, C.H., Zhang, Y.C., Zhang, Z.K., Zhou, T.: Recommender systems. Phys. Rep. **519**(1), 1–49 (2012)
23. Matuszyk, P., Spiliopoulou, M.: Predicting the performance of collaborative filtering algorithms. In: International Conference on Web Intelligence, Mining and Semantics. pp. 38:1–38:6 (2014)
24. McAuley, J., Leskovec, J.: Hidden factors and hidden topics: understanding rating dimensions with review text. In: ACM Conference on Recommender Systems, pp. 165–172 (2013)
25. Pfahringer, B., Bensusan, H., Giraud-Carrier, C.: Meta-learning by landmarking various learning algorithms. In: International Conference on Machine Learning, pp. 743–750 (2000)
26. Pinto, F., Soares, C., Mendes-Moreira, J.: Towards automatic generation of Metafeatures. In: Pacific Asia Conference on Knowledge Discovery and Data Mining, pp. 215–226 (2016)
27. Prudêncio, R.B., Ludermir, T.B.: Meta-learning approaches to selecting time series models. Neurocomputing **61**, 121–137 (2004)
28. Rossi, A.L.D., de Carvalho, A.C.P.D.L.F., Soares, C., de Souza, B.F.: MetaStream: A meta-learning based method for periodic algorithm selection in time-changing data. Neurocomputing **127**, 52–64 (2014)
29. Serban, F., Vanschoren, J., Bernstein, A.: A survey of intelligent assistants for data analysis. ACM Comput. Surv. **47**(212), 1–35 (2013)
30. Vanschoren, J.: Understanding machine learning performance with experiment databases. Ph.D. thesis, Katholieke Universiteit Leuven (2010)
31. Wang, H., Lu, Y., Zhai, C.: Latent Aspect Rating Analysis Without Aspect Keyword Supervision. In: ACM SIGKDD, KDD 2011, pp. 618–626. ACM (2011)
32. Yahoo!: Webscope datasets (2016). https://webscope.sandbox.yahoo.com/
33. Yang, X., Guo, Y., Liu, Y., Steck, H.: A survey of collaborative filtering based social recommender systems. Comput. Commun. **41**, 1–10 (2014)
34. Yelp: Yelp dataset challenge (2016). https://www.yelp.com/dataset_challenge
35. Zafarani, R., Liu, H.: Social computing data repository at ASU (2009). http:// socialcomputing.asu.edu
36. Ziegler, C.N.C., McNee, S.M.S., Konstan, J.a.J., Lausen, G.: Improving recommendation lists through topic diversification. In: Proceedings of the 14th International Conference on World Wide Web WWW 2005, p. 22. ACM (2005)

Community Detection

Recursive Extraction of Modular Structure from Layered Neural Networks Using Variational Bayes Method

Chihiro Watanabe$^{(\boxtimes)}$, Kaoru Hiramatsu, and Kunio Kashino

NTT Communication Science Laboratories,
3-1, Morinosato Wakamiya, Atsugi-shi, Kanagawa Pref. 243-0198, Japan
{watanabe.chihiro,hiramatsu.kaoru,kashino.kunio}@lab.ntt.co.jp

Abstract. Deep neural networks have made a substantial contribution to the recognition and prediction of complex data in various fields, such as image processing, speech recognition and bioinformatics. However, it is very difficult to discover knowledge from the inference provided by a neural network, since its internal representation consists of many non-linear and hierarchical parameters. To solve this problem, an approach has been proposed that extracts a global and simplified structure for a neural network. Although it can successfully detect such a hidden modular structure, its convergence is not sufficiently stable and is vulnerable to the initial parameters. In this paper, we propose a new deep learning algorithm that consists of recursive back propagation, community detection using a variational Bayes, and pruning unnecessary connections. We show that the proposed method can appropriately detect a hidden inference structure and compress a neural network without increasing the generalization error.

Keywords: Layered neural networks · Network analysis · Community detection · Pruning · Variational Bayes method

1 Introduction

Layered neural networks have greatly improved the performance of tasks in various fields [1,9], including image processing [4,8], speech recognition [2,6] and bioinformatics [3,5]. Their deeply layered structures with many nonlinear parameters have made it possible to successfully perceive and recognize many complex real world data.

An open problem is that it is very difficult to understand or translate the inference provided by a layered neural network, since it consists of many nonlinear and complex parameters. Despite its powerful ability to represent or predict data, it is almost impossible to extract knowledge from the internal representation of a trained neural network in an interpretable way. This becomes an obstacle when introducing a neural network into applications that require guaranteed security and safety, such as driverless vehicles and medical use. In the

© Springer International Publishing AG 2017
A. Yamamoto et al. (Eds.): DS 2017, LNAI 10558, pp. 207–222, 2017.
DOI: 10.1007/978-3-319-67786-6_15

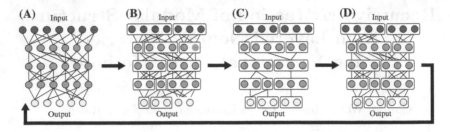

Fig. 1. The proposed deep learning algorithm, which consists of recursive (A) back propagation, (B) community and (C) modular structure detection using a variational Bayes, and (D) the pruning of unnecessary connections.

engineering sense, this causes a difficulty as regards optimizing the hyperparameters or initial parameters, since it requires experimental trials by hand.

Recently, to solve this problem, we proposed a method for extracting a modular representation, or a global and simplified structure for a layered neural network based on network analysis [14]. It detects communities of units that have similar patterns of connections, and then defines *bundled connections* between the communities. Although it can successfully detect the hidden modular structure of a neural network, its convergence is not sufficiently stable and is vulnerable to the initial parameters. This is partly because it employs an EM algorithm to find the optimal parameters and community assignments, with the result that an improved community detection method is required for more robust optimization.

In this paper, we propose a new algorithm for recursively detecting the communities of a layered neural network with a variational Bayes (or VB) method (Fig. 1). It is generally known that optimal parameters can be obtained with a VB method more robustly than with an EM algorithm. We also employed a method for compressing a neural network by pruning unimportant connections based on the resulting modular structure. By using this method, we can represent a trained neural network with essential parameters, without increasing the generalization error. The preciseness of the modular structure extraction can also be increased by pruning unnecessary connections.

We show that the proposed method can appropriately reveal the hidden modular structure of a trained neural network by applying it to both synthetic and real data in the experiment. By using synthetic data with a hidden embedded modular structure, we show that the generalization error remains low through the recursive modular structure extraction. In other words, the proposed method can properly select and maintain the important connections. The detection of the modular structure of real data provided various types of information about the inference given by the neural network. We discuss the details in Sect. 5.

2   Layered Neural Networks

The function and the training method used here are the same as those in [14]. Let $x \in \mathbb{R}^M$ and $y \in \mathbb{R}^N$ be input and output data, respectively. The probability density $q(x, y)$ is a function of x and y on $\mathbb{R}^M \times \mathbb{R}^N$. We assume that a training data set $\{(X_i, Y_i)\}_{i=1}^n$ with sample size n is generated independently from $q(x, y)$. In a layered neural network, a function $f(x, w)$ from $x \in \mathbb{R}^M$, $w \in \mathbb{R}^L$ to $\mathbb{R}^N$ is used to estimate the output y from input x and parameter w.

A layered neural network has two kinds of parameters, $w = \{\omega_{ij}^d, \theta_i^d\}$, where ω_{ij}^d is the weight of the connection between the i-th unit in the depth d layer and the j-th unit in the depth $d+1$ layer, and θ_i^d is the bias of the i-th unit in the depth d layer. Here, input and output layers correspond to the depth 1 and D layer, respectively. If a neural network consists of D layers, its function is given by

$$f_j(x, w) = \sigma(\sum_i \omega_{ij}^{D-1} o_i^{D-1} + \theta_j^{D-1}),$$

$$o_j^{D-1} = \sigma(\sum_i \omega_{ij}^{D-2} o_i^{D-2} + \theta_j^{D-2}), \quad \cdots, \quad o_j^2 = \sigma(\sum_i \omega_{ij}^1 x_i + \theta_j^1),$$

where $\sigma(x) = 1/(1 + \exp(-x))$ is a sigmoid function.

The training error $E(w)$ and the generalization error $G(w)$ are respectively defined by

$$E(w) = \frac{1}{n} \sum_{i=1}^n \|Y_i - f(X_i, w)\|^2, \quad G(w) = \int \|y - f(x, w)\|^2 q(x, y) dx dy,$$

where $\|\cdot\|$ is the Euclidean norm of $\mathbb{R}^N$. The generalization error is approximated by

$$G(w) \approx \frac{1}{m} \sum_{j=1}^m \|Y_j' - f(X_j', w)\|^2,$$

where $\{(X_j', Y_j')\}_{j=1}^m$ is a test data set that is independent of the training data set taken from $q(x, y)$.

By adopting the LASSO method [7,13], in which the minimized function is defined by $H(w)$ as given below, we can obtain a sparse neural network.

$$H(w) = \frac{n}{2} E(w) + \lambda \sum_{d,i,j} |\omega_{ij}^d|,$$

where λ is a hyperparameter. The parameters are trained by the stochastic steepest descent method. Let $H_i(w)$ be the training error computed from only the i-th sample (X_i, Y_i). The parameter update is given by

$$\Delta w = -\eta \nabla H_i(w) = -\eta \left(\frac{1}{2} \nabla \{\|Y_i - f(X_i, w)\|^2\} + \lambda \, \mathrm{sgn}(w) \right). \tag{1}$$

Here, η is defined for training time t such that $\eta(t) \propto \frac{1}{t}$, which is sufficient for the convergence of the stochastic steepest descent. Equation (1) is calculated numerically by the procedure shown in Algorithm 1 of [14], which is called error back propagation [12,15]. In this paper, we set the number of iterations per data a_1 in [14] at 2000. By using this algorithm with the LASSO method, we obtain a neural network whose redundant weight parameters are close to zero.

3   Recursive Extraction of Modular Structure from Layered Neural Networks Using Variational Bayes Method

3.1   Community Detection Using Variational Bayes Method

We propose a new algorithm for detecting communities from a layered neural network using the VB method. This method is an extension of the conventional method that detects communities using the EM algorithm [14], whose community detection method is based on the one proposed in [10].

Let π_c be the probability that a unit in the depth d layer belongs to the community c. The conditional probabilities of incoming and outgoing connection of the depth d layer are represented by the two parameters $\tau_{c,i}$ and $\tau'_{c,j}$, respectively. Here the parameter $\tau_{c,i}$ represents the conditional probability that a connection to a unit in the community c of the depth d layer comes from the i-th unit in the depth $d-1$ layer. The parameter $\tau'_{c,j}$ represents the conditional probability that a connection goes to the j-th unit in the depth $d+1$ layer. From the definition, the parameters $\pi = \{\pi_c\}$, $\tau = \{\tau_{c,i}\}$, $\tau' = \{\tau'_{c,j}\}$ satisfy the following normalization condition.

$$\sum_c \pi_c = 1. \quad \sum_i \tau_{c,i} = 1. \quad \sum_j \tau'_{c,j} = 1.$$

Let g_k be the community of the k-th unit in the depth d layer. We assume that the probability of a community assignment $g = \{g_k\}$ is defined as a function of the adjacency matrices A and B. Here, the adjacency matrix $A = \{A_{i,k}\}$ represents the connections between two layers of depths $d-1$ and d. The element $A_{i,k}$ is one if the absolute value of the connection weight between the i-th unit in the depth $d-1$ layer and the k-th unit in the depth d layer is larger than ξ, otherwise it is zero, where ξ is a weight removing hyperparameter. Similarly, each element $B_{k,j}$ of the adjacency matrix B is defined by the hyperparameter ξ and the connection weight between the k-th unit in the depth d layer and the j-th unit in the depth $d+1$ layer. The probability of the adjacency matrices A, B and the community assignment g for given parameters π, τ, τ' is given by

$$\Pr(A, B, g | \pi, \tau, \tau') = \Pr(A, B | g, \pi, \tau, \tau') \Pr(g | \pi, \tau, \tau'),$$

where

$$\Pr(A, B | g, \pi, \tau, \tau') = \prod_k \left\{ \prod_i \left(\tau_{g_k, i} \right)^{A_{i,k}} \right\} \left\{ \prod_j \left(\tau'_{g_k, j} \right)^{B_{k,j}} \right\}, \tag{2}$$

and

$$\Pr(g|\pi, \tau, \tau') = \prod_k \pi_{g_k}. \tag{3}$$

Let $y_{k,c} \equiv \delta(c, g_k)$. The variable $y_{k,c}$ takes a value of one if the k-th unit belongs to the community c, and it is zero otherwise. From Eqs. (2) and (3), the probabilities of A, B and y for given parameters are given by

$$\Pr(A, B|y, \pi, \tau, \tau') = \prod_c \prod_k \left\{ \prod_i (\tau_{c,i})^{A_{i,k} y_{k,c}} \right\} \left\{ \prod_j (\tau'_{c,j})^{B_{k,j} y_{k,c}} \right\}$$

$$= \prod_c \left\{ \prod_i \exp\left(\sum_k A_{i,k} y_{k,c} \log \tau_{c,i} \right) \right\} \left\{ \prod_j \exp\left(\sum_k B_{k,j} y_{k,c} \log \tau'_{c,j} \right) \right\}, \tag{4}$$

and

$$\Pr(y|\pi, \tau, \tau') = \prod_k \prod_c (\pi_c)^{y_{k,c}} = \prod_c \exp\left(\sum_k y_{k,c} \log \pi_c \right). \tag{5}$$

From Eqs. (4) and (5), we obtain the following equation.

$$\Pr(A, B, y|\pi, \tau, \tau') = \prod_c \left[\exp\left(\sum_k y_{k,c} \log \pi_c \right) \right.$$

$$\left. \left\{ \prod_i \exp\left(\sum_k A_{i,k} y_{k,c} \log \tau_{c,i} \right) \right\} \left\{ \prod_j \exp\left(\sum_k B_{k,j} y_{k,c} \log \tau'_{c,j} \right) \right\} \right]. \tag{6}$$

Let the Dirichlet distributions $\phi_1(\pi|\alpha)$, $\phi_2(\tau|\beta)$ and $\phi_3(\tau'|\gamma)$ be the prior probability distributions of the parameters π, τ and τ', respectively, where $\alpha = \{\alpha_c\}$, $\beta = \{\beta_{c,i}\}$ and $\gamma = \{\gamma_{c,j}\}$ are the hyperparameters. The probability distributions $\phi_1(\pi|\alpha)$, $\phi_2(\tau|\beta)$ and $\phi_3(\tau'|\gamma)$ are given by

$$\phi_1(\pi|\alpha) = \frac{1}{Z_1(\alpha)} \prod_c (\pi_c)^{\alpha_c - 1} = \frac{1}{Z_1(\alpha)} \prod_c \exp\left((\alpha_c - 1) \log \pi_c \right),$$

$$\phi_2(\tau|\beta) = \frac{1}{Z_2(\beta)} \prod_c \prod_i (\tau_{c,i})^{\beta_{c,i} - 1} = \frac{1}{Z_2(\beta)} \prod_c \prod_i \exp\left((\beta_{c,i} - 1) \log \tau_{c,i} \right),$$

$$\phi_3(\tau'|\gamma) = \frac{1}{Z_3(\gamma)} \prod_c \prod_j (\tau'_{c,j})^{\gamma_{c,j} - 1} = \frac{1}{Z_3(\gamma)} \prod_c \prod_j \exp\left((\gamma_{c,j} - 1) \log \tau'_{c,j} \right), \tag{7}$$

where

$$\log Z_1(\alpha) = \sum_c \log \Gamma(\alpha_c) - \log \Gamma\left(\sum_c \alpha_c \right),$$

$$\log Z_2(\beta) = \sum_c \left\{ \sum_i \log \Gamma(\beta_{c,i}) - \log \Gamma\left(\sum_i \beta_{c,i} \right) \right\},$$

$$\log Z_3(\gamma) = \sum_c \left\{ \sum_j \log \Gamma(\gamma_{c,j}) - \log \Gamma\left(\sum_j \gamma_{c,j} \right) \right\}.$$

Here, $\Gamma(x)$ is a gamma function. From Eq. (7), the joint probability of the adjacency matrices, the community assignment and the parameters is given by

$$\Pr(A, B, y, \pi, \tau, \tau') = \Pr(A, B, y | \pi, \tau, \tau') \; \phi_1(\pi | \alpha) \; \phi_2(\tau | \beta) \; \phi_3(\tau' | \gamma)$$

$$\propto \exp\Big(\sum_c \Big\{ \Big(\sum_k y_{k,c} + \alpha_c - 1\Big) \log \pi_c \Big\} + \sum_c \Big\{ \sum_i \Big(\Big(\sum_k A_{i,k} y_{k,c} + \beta_{c,i} - 1\Big)$$

$$\log \tau_{c,i}\Big) \Big\} + \sum_c \Big\{ \sum_j \Big(\Big(\sum_k B_{k,j} y_{k,c} + \gamma_{c,j} - 1\Big) \log \tau'_{c,j}\Big) \Big\} \Big). \tag{8}$$

Let $\Pr(y, \pi, \tau, \tau') = \Pr(y, \pi, \tau, \tau' | A, B, \alpha, \beta, \gamma)$ be the posterior distribution of (y, π, τ, τ') for a given $(A, B, \alpha, \beta, \gamma)$. From Eq. (8), it is given by

$$\Pr(y, \pi, \tau, \tau') \propto \exp\Big(\sum_c \Big\{ \Big(\sum_k y_{k,c} + \alpha_c - 1\Big) \log \pi_c \Big\}$$

$$+ \sum_c \Big\{ \sum_i \Big(\Big(\sum_k A_{i,k} y_{k,c} + \beta_{c,i} - 1\Big) \log \tau_{c,i}\Big) \Big\}$$

$$+ \sum_c \Big\{ \sum_j \Big(\Big(\sum_k B_{k,j} y_{k,c} + \gamma_{c,j} - 1\Big) \log \tau'_{c,j}\Big) \Big\} \Big). \tag{9}$$

We approximate the above posterior distribution with the product of two independent functions: $q(y)r(\pi, \tau, \tau')$. The self-consistent condition that minimizes the Kullback-Leibler divergence from $q(y)r(\pi, \tau, \tau')$ to $\Pr(y, \pi, \tau, \tau')$ is given by

$$q(y) = \mathbb{E}_{r(\pi,\tau,\tau')}[\Pr(y, \pi, \tau, \tau')], \quad r(\pi, \tau, \tau') = \mathbb{E}_{q(y)}[\Pr(y, \pi, \tau, \tau')], \tag{10}$$

which are equivalent to

$$q(y) \propto \exp\Big(\iiint r(\pi, \tau, \tau') \log \Pr(y, \pi, \tau, \tau') d\pi d\tau d\tau'\Big),$$

$$r(\pi, \tau, \tau') \propto \exp\Big(\int q(y) \log \Pr(y, \pi, \tau, \tau') dy\Big). \tag{11}$$

Here, variables $\hat{y}_{k,c}, \hat{\alpha}_c, \hat{\beta}_{c,i}, \hat{\gamma}_{c,j}$ exist such that Eq. (11) is rewritten as follows.

$$q(y) \propto \exp\Big(\sum_c \sum_k y_{k,c}(\log \hat{y}_{k,c} - C_0)\Big),$$

$$r(\pi, \tau, \tau') \propto \exp\Big(\sum_c (\hat{\alpha}_c - 1) \log \pi_c + \sum_c \sum_i (\hat{\beta}_{c,i} - 1) \log \tau_{c,i}$$

$$+ \sum_c \sum_j (\hat{\gamma}_{c,j} - 1) \log \tau'_{c,j}\Big), \tag{12}$$

where C_0 is a constant and $\sum_c \hat{y}_{k,c} = 1$. The optimal variables $\hat{y}_{k,c}, \hat{\alpha}_c, \hat{\beta}_{c,i}, \hat{\gamma}_{c,j}$ are found with a VB algorithm. The variables $\hat{\alpha}_c, \hat{\beta}_{c,i}, \hat{\gamma}_{c,j}$ with a given $\hat{y}_{k,c}$ are recursively optimized.

Theorem 1. *If variables $\hat{y}_{k,c}, \hat{\alpha}_c, \hat{\beta}_{c,i}, \hat{\gamma}_{c,j}$ minimize the Kullback-Leibler divergence from $q(y)r(\pi, \tau, \tau')$ to $\mathrm{Pr}(y, \pi, \tau, \tau')$, then they satisfy*

$$\hat{\alpha}_c = \alpha_c + \sum_k \hat{y}_{k,c}, \quad \hat{\beta}_{c,i} = \beta_{c,i} + \sum_k A_{i,k}\hat{y}_{k,c}, \quad \hat{\gamma}_{c,j} = \gamma_{c,j} + \sum_k B_{k,j}\hat{y}_{k,c}, \qquad (13)$$

and

$$\log \hat{y}_{k,c} = \psi(\hat{\alpha}_c) - \psi(\sum_c \hat{\alpha}_c) + \sum_i A_{i,k}(\psi(\hat{\beta}_{c,i}) - \psi(\sum_i \hat{\beta}_{c,i}))$$

$$+ \sum_j B_{k,j}(\psi(\hat{\gamma}_{c,j}) - \psi(\sum_j \hat{\gamma}_{c,j})) + C_0, \qquad (14)$$

where $\psi(x)$ is a digamma function, and the free energy F is given by

$$F = -\log Z_1(\hat{\alpha}) + \log Z_1(\alpha) - \log Z_2(\hat{\beta}) + \log Z_2(\beta) - \log Z_3(\hat{\gamma}) + \log Z_3(\gamma)$$

$$- \sum_k \sum_c \hat{y}_{k,c} \log \hat{y}_{k,c}. \qquad (15)$$

Proof. Equation (12) is equivalent to the following equations.

$$q(y) \propto \prod_c \prod_k (\hat{y}_{k,c})^{y_{k,c}},$$

$$r(\pi, \tau, \tau') \propto \prod_c (\pi_c)^{\hat{\alpha}_c - 1} \times \prod_c \prod_i (\tau_{c,i})^{\hat{\beta}_{c,i} - 1} \times \prod_c \prod_j (\tau'_{c,j})^{\hat{\gamma}_{c,j} - 1}. \qquad (16)$$

Equation (16) shows that $q(y)$ is a multinomial distribution and $r(\pi, \tau, \tau')$ is a Dirichlet distribution, so the expected value of each variable is given by

$$\mathbb{E}_{q(y)}[y_{k,c}] \propto \hat{y}_{k,c}, \quad \mathbb{E}_{r(\pi,\tau,\tau')}[\log \pi_c] = \psi(\hat{\alpha}_c) - \psi(\sum_c \hat{\alpha}_c),$$

$$\mathbb{E}_{r(\pi,\tau,\tau')}[\log \tau_{c,i}] = \psi(\hat{\beta}_{c,i}) - \psi(\sum_i \hat{\beta}_{c,i}),$$

$$\mathbb{E}_{r(\pi,\tau,\tau')}[\log \tau'_{c,j}] = \psi(\hat{\gamma}_{c,j}) - \psi(\sum_j \hat{\gamma}_{c,j}). \qquad (17)$$

From Eqs. (9) and (17), the following equations hold.

$$\mathbb{E}_{q(y)}[\log \mathrm{Pr}(y, \pi, \tau, \tau')] = \sum_c \left\{ (\sum_k \hat{y}_{k,c} + \alpha_c - 1) \log \pi_c \right\} + \sum_c \left\{ \sum_i ((\sum_k A_{i,k} \right.$$

$$\hat{y}_{k,c} + \beta_{c,i} - 1) \log \tau_{c,i}) \right\} + \sum_c \left\{ \sum_j ((\sum_k B_{k,j}\hat{y}_{k,c} + \gamma_{c,j} - 1) \log \tau'_{c,j}) \right\} + C_1,$$

$$\mathbb{E}_{r(\pi,\tau,\tau')}[\log \mathrm{Pr}(y, \pi, \tau, \tau')] = \sum_c \left\{ (\sum_k y_{k,c} + \alpha_c - 1)(\psi(\hat{\alpha}_c) - \psi(\sum_c \hat{\alpha}_c)) \right\}$$

$$+ \sum_c \left\{ \sum_i ((\sum_k A_{i,k}y_{k,c} + \beta_{c,i} - 1)(\psi(\hat{\beta}_{c,i}) - \psi(\sum_i \hat{\beta}_{c,i}))) \right\}$$

$$+ \sum_c \left\{ \sum_j ((\sum_k B_{k,j}y_{k,c} + \gamma_{c,j} - 1)(\psi(\hat{\gamma}_{c,j}) - \psi(\sum_j \hat{\gamma}_{c,j}))) \right\} + C_2, \qquad (18)$$

Algorithm 1. Community detection algorithm based on VB method

---

Input: $\alpha_c = \alpha$, $\beta_{c,i} = \beta$, $\gamma_{c,j} = \gamma$, where in this paper $\alpha = \beta = \gamma = 1$.
Output: probability of community assignment $\{\hat{y}_{k,c}\}$.

    for each layer, **do**

        Randomly choose variables $\{\hat{y}_{k,c}\}$. In this paper, we defined $\hat{y}_{k,c} = (10 + dy_{k,c})/\sum_c\left(10 + dy_{k,c}\right)$, where $dy_{k,c}$ was independently generated from a uniform distribution on $(0, 1)$.

        for $t = 1$ to 100 **do**

            Update $\hat{\alpha}_c, \hat{\beta}_{c,i}, \hat{\gamma}_{c,j}$ with Eq. (13).

            Update $\hat{y}_{k,c}$ with $\hat{y}_{k,c} = \exp(L_{k,c} - Max_k)/\sum_c \exp(L_{k,c} - Max_k)$,
            where $L_{k,c} = \psi(\hat{\alpha}_c) - \psi(\sum_c \hat{\alpha}_c) + \sum_i A_{i,k}\{\psi(\hat{\beta}_{c,i}) - \psi(\sum_i \hat{\beta}_{c,i})\} + \sum_j B_{k,j}\{\psi(\hat{\gamma}_{c,j}) - \psi(\sum_j \hat{\gamma}_{c,j})\}$, and $Max_k = \max_c L_{k,c}$.

        end for

    end for

---

where C_1 and C_2 are constants.

From Eqs. (10) and (12), the following equations hold.

$$\sum_c(\hat{\alpha}_c - 1)\log \pi_c + \sum_c\sum_i(\hat{\beta}_{c,i} - 1)\log \tau_{c,i} + \sum_c\sum_j(\hat{\gamma}_{c,j} - 1)\log \tau'_{c,j} + C_3$$

$$= \sum_c\left\{\left(\sum_k \hat{y}_{k,c} + \alpha_c - 1\right)\log \pi_c\right\} + \sum_c\left\{\sum_i((\sum_k A_{i,k}\hat{y}_{k,c} + \beta_{c,i} - 1)\log \tau_{c,i})\right\}$$

$$+ \sum_c\left\{\sum_j((\sum_k B_{k,j}\hat{y}_{k,c} + \gamma_{c,j} - 1)\log \tau'_{c,j})\right\} + C_1, \tag{19}$$

and

$$\sum_c\sum_k y_{k,c}\log \hat{y}_{k,c} + C_4 = \sum_c\left\{(\sum_k y_{k,c} + \alpha_c - 1)(\psi(\hat{\alpha}_c) - \psi(\sum_c \hat{\alpha}_c))\right\}$$

$$+ \sum_c\left\{\sum_i((\sum_k A_{i,k}y_{k,c} + \beta_{c,i} - 1)(\psi(\hat{\beta}_{c,i}) - \psi(\sum_i \hat{\beta}_{c,i})))\right\}$$

$$+ \sum_c\left\{\sum_j((\sum_k B_{k,j}y_{k,c} + \gamma_{c,j} - 1)(\psi(\hat{\gamma}_{c,j}) - \psi(\sum_j \hat{\gamma}_{c,j})))\right\} + C_2, \tag{20}$$

where C_3 and C_4 are constants. From Eq. (19), $C_1 = C_3$ and Eq. (13) hold, and from Eq. (20), Eq. (14) holds, where C_0 is a constant. ◻

In this paper, we define the constant C_0 so that $\sum_c \hat{y}_{k,c} = 1$. The whole algorithm of community detection based on the VB method is shown in Algorithm 1.

3.2   Modular Representation of Layered Neural Networks

With the previous method [14], we determined bundled connections that summarize multiple connections between pairs of communities, thus defining the modular representation of a layered neural network. In this paper, we used *Method 2*

Algorithm 2. A recursive algorithm of modular structure extraction and pruning

Input: a set of variables $\kappa = \{\kappa_{ij}^d\}$ is initially defined by $\kappa_{ij}^d = 0$ for all d, i and j.
Output: a set of parameters of a neural network $\{\omega_{ij}^d, \theta_i^d\}$.

  for $u = 1$ to 5 **do**

    (1) Train a layered neural network based on the steepest descent method (Sect. 2). Here, do not update the connection weight ω_{ij}^d iff $\kappa_{ij}^d = 1$ holds.

    (2) Define the adjacency matrices A and B from the connection weights ω.

    (3) Extract the modular structure of the trained neural network based on the adjacency matrices A and B (Algorithm 1).

    (4) Any connection weight ω_{ij}^d is set at zero iff the two communities that it connects do not have a bundled connection.

    (5) Define the variables $\{\kappa_{ij}^d\}$ by $\kappa_{ij}^d = \delta(\omega_{ij}^d, 0)$.

  end for

---

and 3 in [14]. We used these methods to define bundled connections based on the connection ratio between pairs of communities. In the experiment, we set the threshold ζ for defining bundled connections at 0.2.

3.3 Recursive Extraction of Modular Structure and Pruning

The modular structure obtained by the method shown in Sects. 3.1 and 3.2 serves as a clue for selecting important connections in a layered neural network. Here, we propose a new algorithm for compressing a layered neural network by recursively performing back propagation learning, modular structure extraction, and pruning redundant connections (Fig. 1). The modular structure of a layered neural network reveals the strength of connections between communities by bundled connections (Sect. 3.2). Therefore, we assume that the connections between two communities that have no bundled connection are relatively unimportant in terms of inference. Pruning, or setting the weight values at zero, of all of such unimportant connections would enable us to compress a neural network structure and thus make it simpler without any increase in the generalization error. After pruning the unimportant connections, only the weights of the remaining connections are trained again, using the current weights as the initial values for the second training. Thus, the connection weights and modular structure of a layered neural network can be optimized by recursively applying training, community detection and pruning (Algorithm 2).

4 Experiment

We applied the proposed method to synthetic (Sect. 4.1) and real (Sect. 4.2) data to discover the hidden modular structure of layered neural networks. In both experiments, the following process was performed:

1. The data normalization, the initial parameter settings and the visualization of the result were undertaken as in the experiment described in [14] (1), (2), (3), (6), and (7).

2. The connection matrix $A_{ij}^d = a \times 0.99$ if the absolute value of the connection weight between the i-th unit in the depth $d - 1$ layer and the j-th unit in the depth d layer is larger than a certain threshold ξ, otherwise $A_{ij}^d = a \times 0.01$. Note that $a \times 0.99$ and $a \times 0.01$ are used instead of a and 0 for stable computation. Similarly, B_{ij}^d is defined from the connection weight between the i-th unit in the depth d layer and the j-th unit in the depth $d+1$ layer. All units are removed that have no connections to other units. The hyperparameter a is used to increase the apparent number of data or connections in the neural network. By using this, the units are more likely to be decomposed into multiple different communities. We set ξ at 0.1, and a at one for Exp.1 (Sect. 4.1) and at five for Exp.2 (Sect. 4.2).

3. For each layer in a trained neural network, 100 community detection trials were performed. We defined the community detection result as one that achieved the largest expected log likelihood for an EM algorithm or the lowest free energy for a VB algorithm in the last of 100 iterations of the algorithm.

4. The hyperparameters were set as follows:
   - Number of training data sets n: 3000 (Exp.1) and 1044 (Exp.2).
   - Number of test data sets m: 3000 (Exp.1) and 0 (Exp.2).
   - Number of units in the input, hidden and output layers: $\{45, 45, 45\}$ (Exp.1) and $\{31, 20, 16\}$ (Exp.2).
   - Number of hidden layers: 1 (Exp.1) and 2 (Exp.2).
   - Hyperparameter of LASSO λ: 1.0×10^{-7} (Exp.1) and 3.0×10^{-7} (Exp.2).
   - Number of communities per layer: 5 (Exp.1) and 10 (Exp.2).
   - Method for defining bundled connections: Method 2 in [14] (Exp.1) and 3 (Exp.2).
   - Minimum and maximum values of normalized input data $\{x_{\min}, x_{\max}\}$: $\{-3, 3\}$ (Exp.1) and $\{-1, 1\}$ (Exp.2).

4.1   Discovery of a Hidden Structure in a Layered Neural Network

We show that the proposed method can properly discover the hidden modular structure of a neural network. We made synthetic data sets using a layered neural network with the underlying modular structure shown in Fig. 2, trained another layered neural network using such data, and applied the proposed method.

First, we defined a layered neural network based on Fig. 2. For a pair of communities with a bundled connection shown in Fig. 2, each connection weight followed $\omega_{i,j}^d \overset{\text{i.i.d.}}{\sim} \mathcal{N}(0, 2)$, and for one without a bundled connection, no connections existed. For the pairs of communities connected with dotted line, a connection exists with a probability of 0.5. The connection weights with absolute values of one or smaller were replaced by 0. The biases of each layer followed $\theta_j^d \overset{\text{i.i.d.}}{\sim} \mathcal{N}(0, 0.5)$. Then, input data with 45 dimensions were generated, and their values followed: $x_j^n \overset{\text{i.i.d.}}{\sim} \mathcal{N}(0, 3)$. The output data were generated by inputting these data into the above neural network, and then adding independent noise following $\mathcal{N}(0, 0.05)$. Finally, we used this data set to train another neural network, and applied the proposed method to the trained neural network.

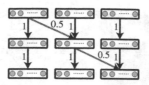

Fig. 2. Ground truth modular structure of a layered neural network trained by synthetic data. Each community contains 15 units.

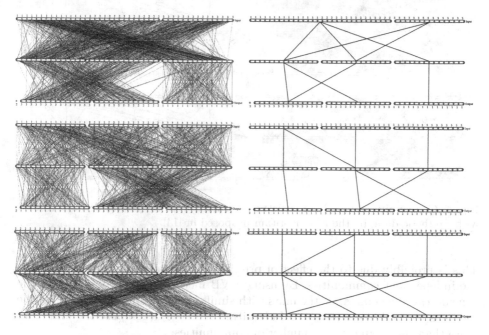

Fig. 3. Community and modular structure extracted from a neural network trained with synthetic data, by a VB method. The top, middle and bottom figures show the results of the first, second and third iterations of recursive extraction, respectively. Solid lines represent connections with positive weights, while dotted lines represent connections with negative weights.

The results of the community and modular structure of the trained neural network are shown in Figs. 3 and 4. For comparison, we also performed recursive modular structure detection based on the EM algorithm method [14], whose results are shown in the bottom three sets of figures. The top, middle and bottom figures show the results of the first, second and third iterations of recursive modular structure detection, respectively. The numbers above the input layer and below the output layer are the indices of the ground truth communities of the units. These results show that the proposed method can properly extract the modular structure underlying the trained neural network defined as in Fig. 2. The figures second from the top in Fig. 3 show that recursive compression enabled us to decompose two communities that were initially combined into one.

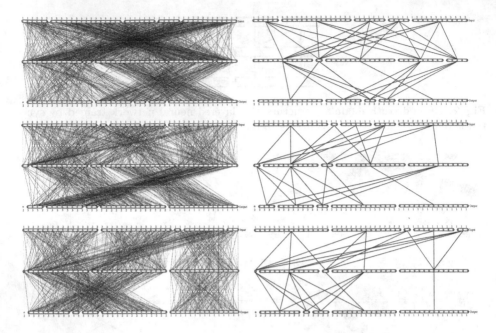

Fig. 4. Community and modular structure extracted from the neural network trained with synthetic data, by the EM algorithm proposed in [14].

This is probably due to the effect of pruning unnecessary connections between the independent communities. By using a VB method, the proposed technique can successfully summarize the units with similar connection patterns, while the previous method based on the EM algorithm tends to decompose the units into more than the ground truth number of communities.

Figure 5 shows the relationship between the training, generalization error and connection ratio of the neural network. Through recursive pruning, the number of connections decrease, thus performing an inference with fewer parameters. As shown by the gray lines in Fig. 5, an inappropriate community detection result causes inappropriate pruning, and leads to an increase in the generalization error in the subsequent training. It is shown that the proposed method can keep the generalization error low through five iterations, while compressing the number of connections to about 24% of total connections.

4.2 Analysis of STUDENT ALCOHOL CONSUMPTION Data Set

We analyzed the layered neural network trained with the STUDENT ALCO-HOL CONSUMPTION Data Set [11], using our proposed method. This dataset contains the characteristics of secondary school students, including basic information such as their sex and age, studying time, grade, and alcohol consumption. We set the input and output data as Table 1 (Inputs are I1 to 31 and outputs are O1 to 16). The data of I1 to 3, I5 to 7, I10 to 22, I24 to 29, O1 to 4, O7, and O8 are defined from the original data as shown in Table 1.

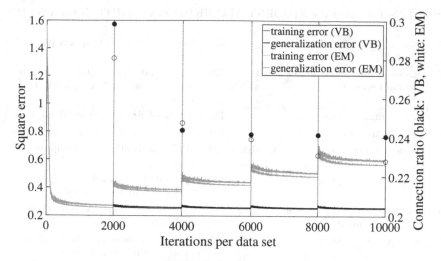

Fig. 5. Training and generalization errors (four lines) and connection ratio in a neural network (black and white circles) when using a VB algorithm and an EM algorithm.

The extracted modular structures are shown in Fig. 6. From these results, we can derive various types of information regarding the inference provided by a neural network.

1. Community A1 includes student's age, home address type, student's guardian and so on. This community A1 appears to provide most of the basic information about a student, and therefore most outputs were inferred using the inputs in A1.
2. Community A2 consists of three outputs: whether or not the students choose their school by school reputation (02), weekly study time (O5), and whether or not they do extra-curricular activities (O7), and this community seems to indicate whether or not they are diligent students.
3. With recursive modular structure detection, the resulting communities are different each time. However, there are some communities that always appeared in the resulting modular structure. For instance, community B1 contains family size, mother's education, and whether or not the parent's job is at home, and this community appeared in all three iterations. Therefore, it can be inferred that the inputs in B1 are closely related. Similarly, workday alcohol consumption and number of school absences seem to have a strong relationship (B2).
4. The bundled connections between communities also provide information about the relationship between the input, hidden and output units. For example, the communities C2 and C3 are used only in inferring the frequency of going out with friends (C4) from whether or not the father's job is healthcare related (C1) in the third iteration, and these four communities compose an independent neural network with units in other communities.

Table 1. Notations of the STUDENT ALCOHOL CONSUMPTION Data Set [11].

I1	Course (0: math, 1: Portuguese language)	I25	Family educational support (0: no, 1: yes)
I2	School (0: Mousinho da Silveira, 1: Gabriel Pereira)	I26	Extra paid classes within the course subject (math or Portuguese) (0: no, 1: yes)
I3	Sex (0: male, 1: female)	I27	Attended nursery school (0: no, 1: yes)
I4	Age (from 15 to 22)	I28	Internet access at home (0: no, 1: yes)
I5	Home address type (0: rural, 1: urban)	I29	With a romantic relationship (0: no, 1: yes)
I6	Family size (0: ≤ 3, 1: >3)	I30	Quality of family relationships (from 1 - very bad to 5 - excellent)
I7	Parent's cohabitation status (0: living together, 1: apart)	I31	Current health status (from 1 - very bad to 5 - very good)
I8	Mother's education[a]	O1	Reason for choosing this school (1: close to home, 0: otherwise)
I9	Father's education[a]	O2	Reason for choosing this school (1: school reputation, 0: otherwise)
I10	Mother's job (1: teacher, 0: otherwise)	O3	Reason for choosing this school (1: course preference, 0: otherwise)
I11	Mother's job (1: healthcare related, 0: otherwise)	O4	Reason for choosing this school (1: other, 0: otherwise)
I12	Mother's job (1: civil services (e.g. administrative or police), 0: otherwise)	O5	Weekly study time (1: $<2\,$h, 2: 2 to $5\,$h, 3: 5 to $10\,$h, or 4: $>10\,$h)
I13	Mother's job (1: at home, 0: otherwise)	O6	Number of past class failures (n if $1 \leq n < 3$, else 4)
I14	Mother's job (1: other, 0: otherwise)	O7	Extra-curricular activities (0: no, 1: yes)
I15	Father's job (1: teacher, 0: otherwise)	O8	Wants to take higher education (0: no, 1: yes)
I16	Father's job (1: healthcare related, 0: otherwise)	O9	Free time after school[b]
I17	Father's job (1: civil services (e.g. administrative or police), 0: otherwise)	O10	Going out with friends[b]
I18	Father's job (1: at home, 0: otherwise)	O11	Workday alcohol consumption[b]
I19	Father's job (1: other, 0: otherwise)	O12	Weekend alcohol consumption[b]
I20	Guardian (1: mother, 0: otherwise)	O13	Number of school absences (from 0 to 93)
I21	Guardian (1: father, 0: otherwise)	O14	First period grade (from 0 to 20)
I22	Guardian (1: other, 0: otherwise)	O15	Second period grade (from 0 to 20)
I23	Home to school travel time (1: $<15\,$min., 2: 15 to $30\,$min., 3: $30\,$min. to $1\,$h, or 4: $>1\,$h)	O16	Final grade (from 0 to 20)
I24	Extra educational support (0: no, 1: yes)		

[a] 0: none, 1: primary education (4th grade), 2: 5th to 9th grade, 3: secondary education or 4: higher education.
[b] from 1 - very low to 5 - very high.

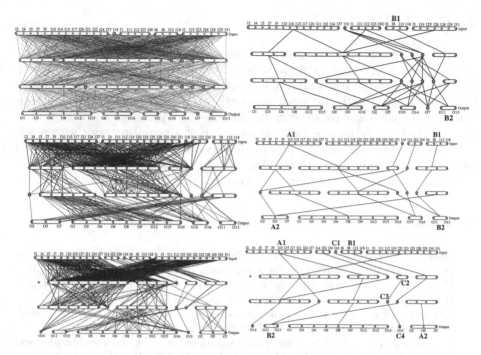

Fig. 6. Community and modular structure extracted from a neural network trained with the STUDENT ALCOHOL CONSUMPTION Data Set [11] by the proposed method.

5 Discussion

Here, we discuss the proposed method from two viewpoints: the understandability of modular structure detection and the appropriateness of neural network compression.

When using real data, the proposed VB algorithm tends to classify units into a small number of communities, without the constant multiplication of the adjacency matrix values. This is because the number of data or connections is insufficient to determine that there are multiple communities in a layer. This is undesirable in terms of understanding the inference provided by a neural network, so a method is needed to alleviate such 'over-summarization'.

From the perspective of generalization error through neural network compression, in the future we replace the pruning method with a more sophisticated approach. For example, jump of generalization error might be abated by employing a method to gradually make the connection weights close to zero between the communities without a bundled connection, instead of completely forgetting them.

6   Conclusion

Layered neural networks have greatly improved the performance when recognizing and predicting complex data in various fields. Despite its powerful ability to represent data in the real world, it has been very difficult to discover or interpret knowledge from the inference provided by a neural network, since it consists of many nonlinear parameters. In this paper, we proposed a new algorithm for recursively detecting the modular structure of a neural network with a VB method, and compressing it by pruning unnecessary connections. The proposed method can properly detect the hidden modular structure of a neural network, and retain important connections while maintaining a low generalization error.

References

1. Bengio, Y., Courville, A., Vincent, P.: Representation learning: a review and new perspectives. IEEE Trans. Pattern Anal. Mach. Intell. **35**(8), 1798–1828 (2013)
2. Hinton, G., et al.: Deep neural networks for acoustic modeling in speech recognition. IEEE Signal Process. Mag. **29**(6), 82–97 (2012)
3. Xiong, H., et al.: The human splicing code reveals new insights into the genetic determinants of disease. Science **347**(6218) (2015)
4. Tompson, J., et al.: Joint training of a convolutional network and a graphical model for human pose estimation. In: Advances in Neural Information Processing Systems (2014)
5. Leung, M., et al.: Deep learning of the tissue-regulated splicing code. Bioinformatics **30**(12), i121–i129 (2014)
6. Sainath, T., et al.: Deep convolutional neural networks for LVCSR. In International Conference on Acoustics, Speech and Signal Processing (2013)
7. Ishikawa, M.: A structural connectionist learning algorithm with forgetting. J. Jap. Soc. Artif. Intell. **5**(5), 595–603 (1990)
8. Krizhevsky, A., Sutskever, I., Hinton, G.: Imagenet classification with deep convolutional neural networks. In: Advances in Neural Information Processing Systems (2012)
9. LeCun, Y., Bengio, Y., Hinton, G.: Deep learning. Nature **521**, 436–444 (2015)
10. Newman, M., Leicht, E.: Mixture models and exploratory analysis in networks. Proc. Natl. Acad. Sci. **104**(23), 9564–9569 (2007)
11. Pagnotta, F., Amran, H.: Using data mining to predict secondary school student alcohol consumption (2016). https://www.researchgate.net/publication/296695247_USING_DATA_MINING_TO_PREDICT_SECONDARY_SCHOOL_STUDENT_ALCOHOL_CONSUMPTION
12. Rumelhart, D., Hinton, G., Williams, R.: Learning representations by back-propagating errors. Nature **323**, 533–536 (1986)
13. Tibshirani, R.: Regression shrinkage and selection via the lasso. J. Roy. Stat. Soc. Ser. B **58**, 267–288 (1994)
14. Watanabe, C., Hiramatsu, K., Kashino, K.: Modular representation of layered neural networks (2017). arXiv:1703.00168
15. Werbos, P.: Beyond regression : new tools for prediction and analysis in the behavioral sciences. Ph.D. thesis. Harvard University (1974)

Discovering Hidden Knowledge in Carbon Emissions Data: A Multilayer Network Approach

Kartikeya Bhardwaj[✉], HingOn Miu, and Radu Marculescu

Carnegie Mellon University, Pittsburgh, PA 15213, USA
kbhardwa@andrew.cmu.edu, miuhingon@gmail.com, radum@cmu.edu

Abstract. In this paper, we construct the first human carbon emissions network which connects more than a thousand geographical locations based on their daily carbon emissions. We use this network to enable a data-driven analysis for a myriad of scientific knowledge discovery tasks. Specifically, we demonstrate that our carbon emissions network is strongly correlated with oil prices and socio-economic events like regional wars and financial crises. Further, we propose the first multilayer network approach that couples carbon emissions with climate (temperature) anomalies and identifies climate anomaly outlier locations across 60 years of documented carbon emissions data; these outlier locations, despite having different emission trends, experience similar temperature anomalies. Overall, we demonstrate how using network science as a key data analysis technique can reveal a treasure trove of knowledge hidden beneath the carbon emissions data.

Keywords: Scientific knowledge discovery · Multilayer networks · Carbon emissions networks · Network science · Community detection

1 Introduction

Network science has emerged as an important data mining tool for knowledge discovery in many scientific and engineering domains [27]. Applications of network science range from social networks [20,29], World Wide Web, biological networks [12], all the way to urban systems [19] and climate networks [16]. Such single layer networks reveal an enormous amount of knowledge hidden behind these complex systems. However, as no system can evolve in isolation, a *multilayer network* approach coupling several systems is often more desirable [10,14]. For instance, to accurately capture the climate change dynamics, studying climate networks alone is *not* sufficient. Since climate change is mainly driven by human-made carbon emissions, the study of climate networks must be coupled with carbon emissions in a multilayer network framework.

To discover latent knowledge in climate systems, prior work uses massive timeseries meteorological data to construct climate anomaly networks[1] (*e.g.*,

---

[1] In this work, we restrict the climate anomalies to temperature anomalies (*i.e.*, deviation of observed temperature data from long-term means).

© Springer International Publishing AG 2017
A. Yamamoto et al. (Eds.): DS 2017, LNAI 10558, pp. 223–238, 2017.
DOI: 10.1007/978-3-319-67786-6_16

temperature anomaly networks [2,6,11,16,30].) Further, few papers have also constructed climate networks based on multiple climate variables such as pressure, humidity, wind, precipitation, *etc.* [23]. This prior art can help discover a large number of long-range links (or *teleconnections*) between distant locations in the climate system; these teleconnections also represent the transport of energy across the planet and, therefore, are very important to understand the underlying climate dynamics. This prior climate networks research, however, does *not* address one of the main driving forces behind climate change, *i.e.*, the anthropogenic carbon emissions. Hence, coupling climate networks with carbon emissions is imperative for capturing the climate change more accurately.

In this paper, therefore, we are concerned with scientific knowledge discovery using complex networks in a completely new problem space, namely, the anthropogenic carbon emissions. Specifically, we answer the following **key questions**:

- Can we use principles of network science to construct a global carbon emissions network? Further, can we construct multilayer networks coupling global climate system with human-made carbon emissions?
- Can we uncover the key socio-economic and political drivers behind the anthropogenic carbon emissions using network science?

To address these questions, we first construct a global carbon emissions network using daily timeseries carbon emissions data. Then, we use the properties of this network to infer various socio-economic factors that drive the carbon emissions across the world. We also couple carbon emissions with temperature anomalies in a multilayer network framework. We then analyze the resulting network using multilayer community detection tool to reveal its latent multiscale community structure [17]. Overall, we make the following **key contributions**:

1. We construct the *first* global carbon emissions networks connecting more than 1500 locations using 60 years of daily carbon emissions data.
2. We *discover* that our carbon emissions network demonstrates strong correlations with many socio-economic factors such as regional wars, financial crises, oil prices, trade, rate of development, GDP growth, *etc.*
3. We propose the *first* Carbon Emissions - Temperature Anomaly Multilayer Network (CETA-MLN) to understand the coupled human-climate dynamics.
4. We analyze our proposed CETA-MLN for about 60 years using multilayer community detection and identify communities at multiple scales, from within counties to beyond countries and even across continents. These communities reveal certain *climate anomaly outlier locations* which have different carbon emission trends while experiencing similar temperature anomalies.

Taken together, our contributions demonstrate how our *data-driven* approach can reveal a treasure trove of knowledge hidden behind the carbon emissions data. Such latent insights can be used to understand the impact of political and socio-economic events on the global carbon emissions.

2   Data Collection and Preparation

We primarily use two sources of data: (i) Daily temperature anomalies from HadGHCND dataset [4], and (ii) Monthly carbon emissions from Carbon Dioxide Information Analysis Center (CDIAC) [1], for the years 1951–2010. Both temperature anomaly and carbon emissions span all land-surface areas at different spatial grids − 2.5° × 3.75° (1° × 1°) latitude-longitude resolution for temperature anomaly (carbon emissions) data. Next, we convert the temporal scale of carbon emissions data from monthly to daily using weekly scale factors [18] which have been created specifically to improve the temporal resolutions of CDIAC datasets. We further preprocess the carbon emissions data to make the resolution of spatial grids comparable for both variables. Therefore, both carbon emissions and temperature anomaly data are now at the same spatial and temporal scales.

Temperature anomaly data in HadGHCND dataset is missing for some parts of Australia and most of Southern Africa and South America. Therefore, after taking missing data into account, we have temperature anomaly data for 932 locations, and carbon emissions data for 1533 locations around the world. Figure 1(a) shows locations for which either temperature anomaly or carbon emissions data is available. Of note, these locations serve as network nodes in our approach.

3   Proposed Multilayer Network

In this section, we first describe our multilayer network construction approach in detail. We then explain the multilayer community detection used in the paper.

3.1   Building a Multilayer Network

Single-layer networks are represented by an adjacency matrix A, where $A_{ij} = W$, and W is the link weight between nodes i and j. Multilayer networks consist of two or more single layer networks in which same set of nodes have different connectivity patterns in different layers. Figure 1(b) shows a multilayer network with layers r ($\{A_{ijr}\}$) and s ($\{A_{ijs}\}$), coupled by a coefficient, ω. Similarly, Fig. 1(c) illustrates our Carbon Emissions-Temperature Anomaly Multilayer Network (CETA-MLN). Next, we construct this multilayer network.

We start with 60 years (1951–2010) of daily timeseries data for temperature anomaly and carbon emissions. Let $\{T_1, T_2, T_3, \ldots, T_p\}$ denote the *temperature anomaly network nodes* (*i.e.*, the locations) for which temperature anomaly data is available. Similarly, let $\{C_1, C_2, C_3, \ldots, C_q\}$ denote the *carbon emissions network nodes* (*i.e.*, the locations) at which carbon emissions data is available. Here, $p = 932$ and $q = 1533$ are the number of temperature anomaly and carbon emissions nodes, respectively. Each layer of CETA-MLN is generated using a cross-covariance based approach [11]. This prior approach constructs a network from timeseries data, as summarized below for the temperature anomaly layer:

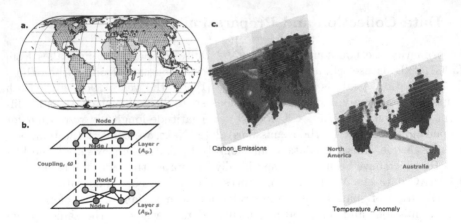

Fig. 1. (a) Locations for which either temperature anomaly or carbon emissions data is available. (b) Schematic of a multilayer network where same set of nodes have different connectivity patterns in different layers. For instance, nodes i and j in layers r and s are coupled by a coupling coefficient, ω. (c) Example of CETA-MLN for the year 2008. (Color figure online)

1. For a given year y and each temperature anomaly node T_i, $i = 1, 2, 3, \ldots, p$, compute the link weight between nodes T_i and T_j ($W^y_{T_i, T_j}$) as follows:

$$W^y_{T_i, T_j} = \frac{\text{MAX } C^y_{T_i, T_j} - \text{MEAN } C^y_{T_i, T_j}}{\text{STD } C^y_{T_i, T_j}}, \tag{1}$$

where, $C^y_{T_i, T_j}$ is a cross-covariance function between nodes T_i and T_j. This cross-covariance function is computed using daily data for the years $y, y + 1, y + 2$ and the lag parameter (τ) used by cross-covariance is ± 72 days [11]. We use this lag since Guez *et al.* have demonstrated in [11] that the lag of ± 72 days is sufficient to capture interesting climate phenomena and is long enough so that the weight values ($W^y_{T_i, T_j}$) are *not* sensitive to the choice of lag. Moreover, -72 to $+72$ days is equivalent to almost 5 months which is sufficient to capture the seasonal behavior of temperature and carbon emission timeseries. We further conduct an experiment to verify that doubling this lag does not have a significant impact on the most significant links of the network. Finally, since $W^y_{T_i, T_j} = W^y_{T_j, T_i}$, this approach results in an undirected network. Repeating this process for all possible combinations of T_i and T_j, we get a fully connected network. We call this network the *real network of temperature anomalies* as it is based on real (observed) data. We obtain such networks for every year $y \in \{1951, 1952, \ldots, 2008\}$.

2. Next, links that are present only due to statistical properties of the data (*e.g.*, distribution of values, autocorrelation, *etc.*) must be removed. Such links do not represent physical (real) dependencies among nodes. Towards this, we construct a surrogate dataset [11] by preserving the order of data within each year, while randomly shuffling the order of years for each node.

3. We now construct a *surrogate network* by repeating step 1 for this new shuffled dataset. Surrogate network links occur only due to statistical properties. Figure 2(a) shows the probability density of the link weights obtained for *all years* 1951–2008, both for real and surrogate temperature anomaly networks. Clearly, the link weights in the surrogate network (green) do not take values more than $W = 4$. Hence, links with weights ≤ 4 in the real network occur purely due to statistical properties and not due to physical dependencies; we remove these spurious links by setting a threshold of 4.

4. Removing spurious links yields a set of adjacency matrices $\{A_T^y\}$, $y \in \{1951, \ldots, 2008\}$ which represent the final temperature anomaly networks.

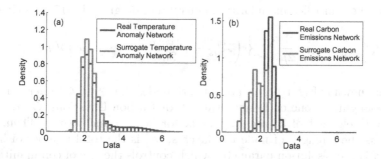

Fig. 2. Link weight histograms for (a) temperature anomaly, and (b) carbon emissions real and surrogate networks. Link weight threshold for temperature anomaly (carbon emissions) networks is 4 (3). Two sample Kolmogorov-Smirnov (KS) Test for temperature anomaly and carbon emissions link weight distributions resulted in KS statistics, $k = 0.1286$ and $k = 0.6187$ (Asymptotic p-value, $p = 0$ for both cases), respectively. Surrogate and the real link weights, hence, come from different distributions. (Color figure online)

Next, we repeat the above steps to generate the Carbon Emissions (CE) network. Figure 2(b) shows the probability density of link weights for real (blue) and surrogate (green) carbon emissions network. The threshold in this case is $W = 3$. A sequence of adjacency matrices $\{A_C^y\}$, $y \in \{1951, 1952, \ldots, 2008\}$ representing the final carbon emissions networks is thus obtained.

Intuitively, temperature anomaly networks represent a network of *co-occurring temperature anomalies* across the world, thereby indicating how different regions are impacted by similar climate change. In contrast, carbon emissions networks represent a proxy for the *rate of development* in different parts of the world and how trends for carbon emissions are similar in different regions.

Both temperature anomaly and carbon emissions networks have many common nodes as they both cover land-surface areas. Therefore, for each year, we can represent this system as a *multilayer network* [10] containing the temperature anomaly layer and the carbon emissions layer for that year. We will call this Carbon Emissions and Temperature Anomaly Multilayer Network as CETA-MLN throughout the paper. Earlier, we visualized the CETA-MLN for the year 2008 in Fig. 1(c) using MuxViz software [5]. It is evident that both network layers are very dense in terms of number of links and are characterized by the teleconnections (*i.e.*, the long-range links) that exist even across continents (*e.g.*, see the links connecting Canada and Australia in the temperature anomaly layer).

3.2 Community Detection on Multilayer Networks

Community structure on a multilayer network can be detected by maximizing a multislice modularity quality function. A completely general framework for multilayer community detection is derived in [17]. For the problem at hand, we have only two network layers and both layers are changing every year. Therefore, we adopt the multislice modularity quality function given in [17] as follows:

$$Q^y{}_{multislice} = \frac{1}{2\mu} \sum_{ijsr} \left\{ \left(A^y_{ijs} - \gamma \frac{k_{is}k_{js}}{2m_s} \right) \delta_{sr} + \delta_{ij}\omega \right\} \delta(g_{is}, g_{jr}) \qquad (2)$$

where, as shown in Fig. 1(b), i, j are network nodes, $s, r \in \{T, C\}$ refer to network layers belonging to one of Temperature, T or Carbon Emissions, C layers, and A^y_{ijs} is ij^{th} element of adjacency matrix for layer s and year y. Then, $k_{is} = \sum_j A^y_{ijs}$ is the strength of node i in layer s, m_s is the total number of links in layer s, γ is the resolution parameter which controls the size of communities and all the δ's are Kronecker Delta (see [17] for more details). Essentially, the terms associated with δ_{sr} and δ_{ij} account for *intra-layer* and *inter-layer* contribution to community structure, respectively. The last Kronecker Delta ensures that the summation takes place only if the nodes i in layer s and j in layer r are in the same group in the multilayer community structure.

The modularity (Eq. 2) measures the *stability of communities* in the multilayer network [15]. Consequently, maximizing the modularity gives the optimal community structure of the network. We run this multilayer community detection on CETA-MLN for different values of resolution, γ and coupling, ω parameters. Intuitively, increasing the γ, increases the number of communities in the network (*i.e.*, the resolution of communities increases). Moreover, increasing the ω makes communities in both layers more and more similar (since the layers get more tightly coupled). Of note, we have used Generalized Louvain algorithm [13] in MATLAB which essentially detects the community structure using the approach explained in [17]. Also, we have implemented all of our codes in MATLAB and visualized the global maps using MATLAB's mapping toolbox.

4 Results and Discussion

In this section, we present the main results and an in-depth analysis of our proposed carbon emissions network and CETA-MLN.

4.1   Knowledge Discovery in Carbon Emissions Networks

We analyze the characteristics of our carbon emissions network in Fig. 3. Specifically, we plot the crude oil prices in Fig. 3(a), while Fig. 3(b) shows the dynamics of total number of links in the carbon emissions network with time. As evident, the number of links in carbon emissions network closely follows the trends for crude oil prices. For instance, around the year 1980, the oil prices as well as number of links see a sharp hike, followed by a gradual decrease and an increase around 2008 again. As with the oil prices, these trends correlate to socio-political-economic factors like Iran-Iraq war that led to oil crisis in 1980, or the global financial crisis in 2007–2008, both of which increased the oil prices. In fact, it has been explicitly documented that a sharp rise in carbon emissions was observed after the 2007–2008 global financial crisis [22]. Since our carbon emissions network captures such events, it can, therefore, improve our understanding of latent factors behind rising emissions due to financial issues.

In Fig. 3(c), we demonstrate that there is a strong correlation (=0.65) between oil prices (blue) and number of links in carbon emissions network (red) lagged by 5 years ($p < 0.05$). We show results only for lag $= 5$ years since the correlation gets maximized at this lag. This shows that there is indeed a statistically significant correlation between oil prices and carbon emissions network links. Furthermore, our carbon emissions network captures the same global physical processes such as wars and financial crises that govern the oil prices. Therefore, the carbon emissions network properties such as teleconnections, number of links, *etc.*, can improve our understanding of how fossil fuels are being used globally and how their usage is affected by such socio-economic factors.

Next, we identify the most significant links in our network by finding the top 0.1% links with highest weights or the top 0.1% links that cover longest distances (*i.e.*, the longest teleconnections). Of note, these most significant links dynamically change across the years as the network structure changes. In Fig. 3(d), we show top 0.1% *highest weighted links* for the year 1974. We observe that a set of *strong links* (shown in red) connect two countries which do not seem to have anything in common – Sweden in Europe and Suriname in South America. In fact, analyzing the raw Gross Domestic Product (GDP) data for both countries does *not* yield any direct relationship between Sweden and Suriname.

We next analyze the normalized GDP for Sweden and Suriname for the 1960–1980 period (see Fig. 3(e)). Note that, we used this period since we observed the Sweden-Suriname strong links for the year 1974 which lies within 1960–1980 (and, hence, we did not choose, say, the 1981–2010 period). Moreover, since GDP data is at yearly timescale, 20 years of data yields 20 data points which allows a more reliable calculation of statistics such as correlations than, say, 5 years of data. Figure 3(e) clearly demonstrates that there was indeed a very strong correlation of 0.985 ($p < 0.05$) between the GDP of Sweden and Suriname during the 1960–1980 period [24]. This shows that our carbon emissions network can capture strong economic correlations among different regions, which can possibly be explained by their rate of development (GDP). Furthermore, this demonstrates that our Carbon Emissions network can reveal hidden insights

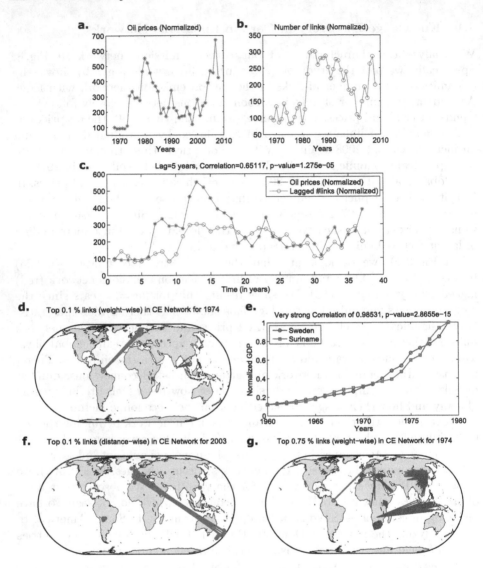

Fig. 3. Characteristics of carbon emissions network. (a) Crude Oil prices from 1968–2008 (Data Source: [8]). (b) Number of links in the carbon emissions network with time. (c) Carbon emissions network shows strong correlation with Crude Oil Prices. (d) The top 0.1% weighted links reveal strong links (shown in red) between Suriname and Sweden for the year 1974. (e) Normalized Gross Domestic Product (GDP) of Suriname closely follows that of Sweden, which shows that the two countries exhibit a very high correlation of 0.98 ($p < 0.05$) during 1960–1980 (Data Source: [24]). (f) Top 0.1% *longest* links of carbon emissions network persistently connect New Zealand with Europe and U.K. for all the years which could represent their long-term trade relations. (g) Finally, doubling the lag τ does not impact the most significant links (*e.g.*, Sweden-Suriname links shown in red). Carbon emissions network, hence, spans several dimensions from fossil fuel usage, oil prices, wars, GDP, to international trade. (Color figure online)

which are not so obvious initially (*e.g.*, we can answer questions like why two very different countries – Sweden and Suriname – could be connected in Carbon Emissions network).

Let us now examine the top 0.1% longest teleconnection links (*i.e.*, the links with *longest distances*) in our carbon emissions network. As shown in Fig. 3(f), there are links (shown in red) connecting New Zealand with U.K. and the rest of Europe. Although, Fig. 3(f) shows these links only for 2003, we found that such links exist persistently throughout all years. The presence of these links can be explained by long-term trade relations between New Zealand, U.K. and Europe. In fact, Europe is the second largest trading partner of New Zealand, only after Australia[2] [9]. Moreover, previous literature also shows that carbon emissions can get transferred across countries as a result of international trade [21]. Of note, even though Europe is the second largest trading partner of New Zealand, New Zealand is not one of the largest trading partners of Europe. Yet, these links show up in the *undirected* network because directionality cannot be accounted for in the current cross covariance-based method (since link weights are derived using Eq. 1). Nonetheless, directionality is an important issue and will make an excellent future research work (*e.g.*, similar to recent work reported in [30]).

Finally, we explore the sensitivity of cross-covariance-based link weights to lag parameter τ by doubling it from ±72 days to ±144 days. The lag of -72 to $+72$ days almost amounts to 5 months which should take care of the seasonality in carbon emissions. For instance, in the Sweden-Suriname case (Fig. 3(d)), Sweden is expected to have more emissions during the winter. Since the timeseries is lagged by up to 5 months, our approach should automatically account for this seasonality in emission timeseries data. However, even if we were to consider that 5 months is not fully sufficient to capture this seasonality, a total lag of ±144 days (which amounts to almost 10 months) will definitely account for this seasonality. Figure 3(g) shows that for the ±144 days lag, the Sweden-Suriname links are still strong (occurring within the top 0.75% weights); in fact, we were able to reproduce other results (*e.g.*, Figs. 2 and 3(f)) for the ±144 days lag. Therefore, we conclude that the lag, $\tau = \pm72$ days is sufficient, and is also long enough such that the link weights are not sensitive to our choice of lag.

All these latent insights (Fig. 3(a)–(g)) show that our proposed carbon emissions network spans several dimensions from global fossil fuel usage to oil prices and wars, to GDP and financial crises, all the way to international trade. As a result, our carbon emissions network can play a fundamental role in the scientific discovery of how global fossil fuel usage and their resulting carbon emissions get affected by different political and socio-economic factors.

Correlation vs. Causation. In many scientific problems, correlations are often important prerequisites for a detailed physics, economics and/or machine learning based research which can confirm those correlations and in turn improve the overall understanding of the phenomena. In the present work, therefore, our results yield correlations and *not* causation. Much like the existing literature

---

[2] Since we are plotting the 0.1% longest links (distance-wise), the links between Australia and New Zealand do not appear due to their close proximity.

on climate networks which establishes correlations among climate anomalies in distant parts of world and attempts to explain them using relevant physical phenomena, we identify the presence of the latent correlations in the carbon emissions network and give *possible* explanations as to why those links might be present. To this end, we demonstrate that these links can be possibly explained by underlying *socio-economic drivers* of emissions such as wars, trade, financial crises, rates of development in different regions, *etc.* These correlations can then be used to guide research in more specialized domains to improve the understanding of latent factors that drive the global anthropogenic emissions and to establish causal mechanisms between these factors and their resulting emissions.

Indeed, many important studies use correlations to establish a necessary first step in a new direction. For example, in a completely different domain of cancer etiology, correlations have recently been used to hypothesize a new source of cancer − random mutations during DNA replication which yield a strong correlation of 0.81 between cancer risk and number of stem cell divisions in a given tissue [25,26]. Another, much more relevant example is [7], where the authors establish correlation between phone call communication networks and socio-economic well-being of communities. The authors of that work clearly acknowledge that their work cannot establish causality and yet *"social network diversity seems to be at the very least a strong structural signature for the economic development of a community"*. Similar in spirit, our work establishes such correlations between the aforementioned socio-economic drivers and anthropogenic carbon emissions or vice versa. Clearly, understanding how these underlying hidden factors affect global emissions can have important implications for climate change.

4.2 Community Detection on CETA-MLN

Figure 4 shows the results of multilayer community detection on the CETA-MLN for three years between 1951 and 2008. Each color represents a community and locations with same color belong to the same community (for simplicity and better visualization, we have plotted only the top 20 communities in MATLAB). Further, we observe that the communities for carbon emissions and temperature anomaly layers are identical for most of South America and Africa. This is due to the missing temperature anomaly data for these locations which makes them disconnected in the temperature anomaly layer, while still being connected in the carbon emissions layer. Therefore, the multilayer community structure is entirely governed by the links in the carbon emissions layer for these locations. As a result, we do not draw any conclusions about South America and Africa.

To emphasize the community structure details, we zoomed the indicated portions of Fig. 4(a) as shown in Fig. 4(b, c), where the top panel shows the carbon emission communities, while the bottom panel shows the temperature anomaly communities. Figure 4(a) shows that certain communities extend across continents, *e.g.*, the yellow carbon emissions community shown in Fig. 4(a)–(c) is spread across Alaska, contiguous USA, France, India *etc.*, thereby connecting North America, Europe and Asia; many other similar examples exist across the years. Recall that links in our CETA-MLN use cross-covariance between

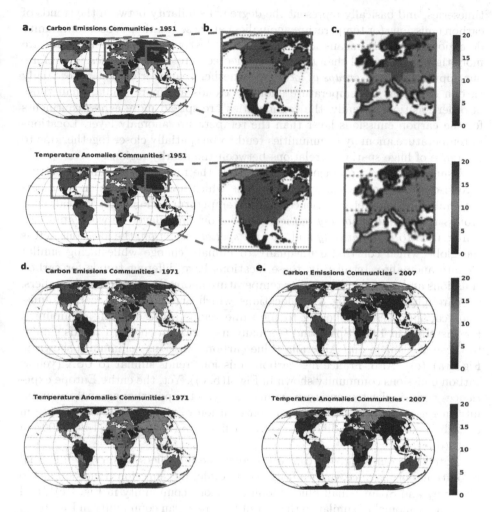

Fig. 4. Multilayer community detection on CETA-MLN reveals climate anomaly outliers. Each color represents a community and locations with same color belong to the same community. The colorbar on extreme right represents community labels. For instance, in the top panel of (a) and (b), USA (Canada) is shown in yellow (blue) color which means that its community label is 12 (6). Community structure for the year (a) 1951. Zoomed community structure of (a) is shown in (b) for North America, and (c) for Europe (top-panel: carbon emissions communities, and bottom panel: temperature anomaly communities). (b, c) clearly show the climate anomaly outliers: USA and France have similar emissions (yellow emissions community). Germany, France and Spain contributed unequally to climate change, and yet the entire Europe still experienced very similar anomalies (cyan anomaly community). Other smaller outliers include central Italy which has emission trends similar to Germany (blue) but anomaly trends similar to the entire Europe (cyan). Similar outliers are shown in (d) Green region north of India in 1971, and (e) Eastern Russia in 2007. (Color figure online)

timeseries, and basically represent the degree of similarity between the trends of carbon emissions (or temperature anomalies) at these locations. Also, community detection intuitively means locations (*i.e.*, nodes) within the communities are more tightly connected than they are connected across communities. Therefore, our approach reveals *groups of locations*, whether nearby or distant, whose daily carbon emissions or temperature anomaly trends are more similar than those of other locations. Finally, the communities are spread more across continents for the carbon emissions layer than the temperature anomaly layer. Locations in temperature anomaly communities tend to be spatially closer together due to existence of huge spatial correlations between the nearby regions.

Using the above arguments, we note that the multilayer community structure also reveals certain groups of locations which have different trends for carbon emissions and yet experience similar temperature anomalies. We call such groups the *climate anomaly outliers*. Hence, our multilayer network approach can be used to *automatically* find, at *high spatial resolution*, the locations across the globe which contribute unequally to climate change while facing similar climate anomalies. Intuitively, these locations have different trends for carbon emissions and yet face very similar temperature anomalies. To find such outliers, we carefully examine Fig. 4 for locations which are in different carbon emissions communities, but belong to the same temperature anomaly community. For instance, in 1951, Spain and Germany have carbon emission trends similar to those of Canada and Australia (blue carbon emissions community shown in Fig. 4(a)–(c)). Also, France has carbon emission trends similar to USA (yellow carbon emissions community shown in Fig. 4(b, c)). Yet, the entire Europe experiences very similar temperature anomalies (cyan temperature anomaly community in Fig. 4(c)), even though their contribution to climate change in the form of carbon emissions was different. More outliers are shown for 1971 and 2007 in Fig. 4(d, e).

Note that, our approach can also detect these outliers at very high spatial resolution. For instance, central Italy (near Rome) has emission trends similar to Germany and Spain (small blue carbon emissions community in Fig. 4(c)), and yet it faces anomalies similar to the rest of Europe (cyan community in Fig. 4(c)). This makes our approach truly *multiscale* since the community structure thus revealed, ranges from small counties to beyond countries and even across continents. Therefore, our approach precisely *reveals* the locations or groups of countries which experience same degree of climate change despite not equally contributing to the climate change. Of note, the teleconnection links and outliers detected using multilayer community structure are *latent* insights which obviously cannot be obtained without using network science.

4.3   Impact of Varying Coupling and Resolution Parameters

In Fig. 5, we show the impact of increasing the resolution parameter, γ on multilayer community structure. As expected, by increasing γ, we are able to detect communities and, hence, the climate anomaly outliers at higher spatial resolutions. For instance, we can clearly see that for Australia in Fig. 5(c) and for

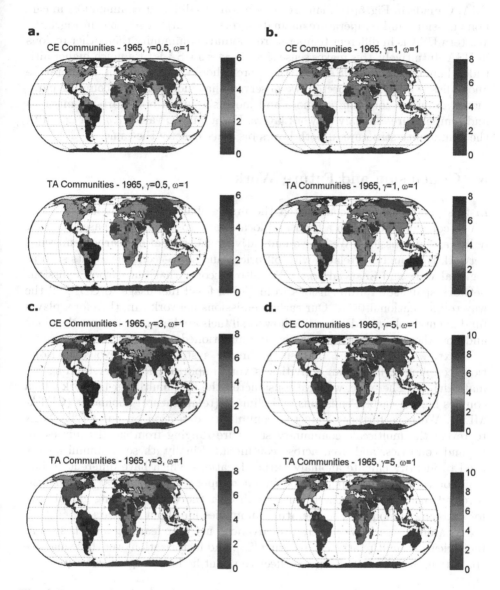

Fig. 5. Impact of varying resolution parameter γ on multilayer community structure: (a) $\gamma = 0.5$, (b) $\gamma = 1$, (c) $\gamma = 3$, and (d) $\gamma = 5$. We have plotted $log_2(community\ id)$ for better visualization since, as expected, number of communities increases as γ increases. Using higher values of γ, it is possible to detect community structure and, hence, the climate anomaly outliers, at higher spatial resolutions. For instance, we can clearly see that for Australia in (c) and for USA, Canada in (d), many regions belong to different communities in CE and TA network layers. Similar considerations hold for Europe and rest of the world. (Color figure online)

USA, Canada in Fig. 5(d), many regions belong to different communities in carbon emissions and temperature anomaly layers. We further see that in Fig. 5(d), Western USA is in different temperature anomaly community than other regions in USA. Intuitively, this makes sense since western USA is at a much higher altitude than the rest of USA and, therefore, should face different temperature anomalies. Finally, as expected, we observed that increasing the coupling parameter, ω makes communities more and more similar in both carbon emissions and temperature anomaly network layers. For a large ω value (*e.g.*, for $\omega = 10$), the community structure in both layers becomes exactly the same.

5   Conclusion and Future Work

In this paper, we have constructed the first global carbon emissions network which connects more than a thousand locations across the world based on their daily carbon emissions. We have quantitatively shown that our carbon emissions network is strongly correlated with oil prices and socio-economic events such as regional wars and financial crises. We also discover that our carbon emissions network spans several dimensions from global fossil fuel usage to GDP, all the way to international trade. Our carbon emissions network can, therefore, play a fundamental role in understanding how fossil fuels are being used across the globe and how they can be affected by the aforementioned socio-economic factors.

Next, we have argued that since climate system cannot evolve in isolation, coupling it to human activities is very important for understanding climate change. Therefore, we have constructed the first multilayer network which couples human carbon emissions and climate (temperature) anomalies (CETA-MLN). We have further conducted community detection on our CETA-MLN to reveal the multiscale community structure ranging from small counties, to beyond countries, and even across continents. Finally, these communities are used to identify climate anomaly outlier locations which have different carbon emission trends while experiencing similar temperature anomalies. Our analysis revealed such climate anomaly outliers all over the world by using global data for carbon emissions and temperature anomalies for about 60 years.

As a future work, we plan to use advanced machine learning models such as multiview [3], multirelational clustering [28], and factorization over these massive spatiotemporal datasets for more effective scientific knowledge discovery.

Acknowledgments. This work was supported in part by the US National Science Foundation (NSF) under CyberSEES Grant CCF-1331804. The authors also acknowledge useful discussions and constructive feedback received from Dr. Da-Cheng Juan from Google Research in the early stages of the manuscript.

References

1. Andres, R., Boden, T., Marland, G.: Monthly Fossil-Fuel CO_2 Emissions: Mass of Emissions Gridded by One Degree Latitude by One Degree Longitude. CDIAC, Oak Ridge National Laboratory, U.S. Department of Energy, USA (2013)
2. Berezin, Y., Gozolchiani, A., Guez, O., Havlin, S.: Stability of climate networks with time. Natu. Sci. Rep. **2**, 1–8 (2012)
3. Bickel, S., Scheffer, T.: Multi-view clustering. In: ICDM, vol. 4, pp. 19–26 (2004)
4. Caesar, J., Alexander, L., Vose, R.: Large-scale changes in observed daily maximum and minimum temperatures: creation and analysis of a new gridded data set. J. Geophys. Res. **111**, 1–10 (2006)
5. Domenico, M.D., Porter, M.A., Arenas, A.: MuxViz: a tool for multilayer analysis and visualization of networks. J. Complex Netw. **3**(2), 159–176 (2014)
6. Donges, J.F., Zou, Y., Marwan, N., Kurths, J.: The backbone of the climate networks. Europhys. Lett. **87**(4), 1–6 (2009)
7. Eagle, N., Macy, M., Claxton, R.: Network diversity and economic development. Science **328**(5981), 1029–1031 (2010)
8. EIA: Historical Crude Oil Prices. Energy Information Administration (1968–2008). http://www.eia.gov/finance/markets/crudeoil/spot_prices.cfm
9. EU: European Commission - Trade (2016). http://ec.europa.eu/trade/policy/countries-and-regions/countries/new-zealand/
10. Gao, J., Li, D., Havlin, S.: From a single network to a network-of-networks. Natl. Sci. Rev. **1**, 346–356 (2014)
11. Guez, O., et al.: Global climate network evolves with North Atlantic Oscillation phases: coupling to Southern Pacific Ocean. Europhys. Lett. **103**, 1–5 (2013)
12. Jeong, H., Tombor, B., Albert, R., Oltvai, Z.N., Barabási, A.L.: The large-scale organization of metabolic networks. Nature **407**(6804), 651–654 (2000)
13. Jutla, I.S., Jeub, L.G.S., Mucha, P.J.: A generalized Louvain method for community detection implemented in MATLAB (2011–2014). http://netwiki.amath.unc.edu/GenLouvain
14. Kivela, M., Arenas, A., et al.: Multilayer networks. J. Complex Netw. **2**, 203–271 (2014)
15. Lambiotte, R., Delvenne, J.C., Barahona, M.: Laplacian dynamics and multiscale modular structure in networks. arXiv preprint arXiv:0812.1770 (2008)
16. Ludescher, J., Gozolchiani, A., Bogachev, M.I., Bunde, A., Havlin, S., Schellnhuber, H.J.: Improved El Nino forecasting by cooperativity detection. Proc. Natl. Acad. Sci. (PNAS) **110**(29), 11742–11745 (2013)
17. Mucha, P.J., et al.: Community structure in time-dependent, multiscale, and multiplex networks. Science **328**, 876–878 (2010)
18. Nassar, R., Napier-Linton, L., Gurney, K., et al.: Improving the temporal and spatial distribution of CO_2 emissions from global fossil fuel emission datasets. J. Geophys. Res. **118**, 917–933 (2013)
19. Ohara, K., Saito, K., Kimura, M., Motoda, H.: Accelerating computation of distance based centrality measures for spatial networks. In: Calders, T., Ceci, M., Malerba, D. (eds.) DS 2016. LNCS, vol. 9956, pp. 376–391. Springer, Cham (2016). doi:10.1007/978-3-319-46307-0_24
20. Pereira, F.S.F., de Amo, S., Gama, J.: On using temporal networks to analyze user preferences dynamics. In: Calders, T., Ceci, M., Malerba, D. (eds.) DS 2016. LNCS, vol. 9956, pp. 408–423. Springer, Cham (2016). doi:10.1007/978-3-319-46307-0_26

21. Peters, G.P., et al.: Growth in emission transfers via international trade from 1990 to 2008. Proc. Natl. Acad. Sci. **108**(21), 8903–8908 (2011)
22. Peters, G.P., et al.: Rapid growth in CO_2 emissions after the 2008–2009 global financial crisis. Nat. Clim. Change **2**, 2–4 (2012)
23. Steinhaeuser, K., et al.: Multivariate and multiscale dependence in the global climate system revealed through complex networks. Clim. Dyn. **39**, 889–895 (2012)
24. The World Bank: GDP Data for Suriname and Sweden. World Development Indicators (1960–1980). http://data.worldbank.org
25. Tomasetti, C., Li, L., Vogelstein, B.: Stem cell divisions, somatic mutations, cancer etiology, and cancer prevention. Science **355**(6331), 1330–1334 (2017)
26. Tomasetti, C., Vogelstein, B.: Variation in cancer risk among tissues can be explained by the number of stem cell divisions. Science **347**(6217), 78–81 (2015)
27. Wu, X., Zhu, X., Wu, G.Q., Ding, W.: Data mining with big data. IEEE Trans. Knowl. Data Eng. **26**(1), 97–107 (2014)
28. Yin, X., Han, J., Philip, S.Y.: Crossclus: user-guided multi-relational clustering. Data Min. Knowl. Disc. **15**(3), 321–348 (2007)
29. Zhang, Y., et al.: COSNET: connecting heterogeneous social networks with local and global consistency. In: Proceedings of the 21th ACM SIGKDD International Conference on Knowledge Discovery and Data Mining, pp. 1485–1494. ACM (2015)
30. Zhou, D., Gozolchiani, A., Ashkenazy, Y., Havlin, S.: Teleconnection paths via climate network direct link detection. Phys. Rev. Lett. **115**, 1–5 (2016)

Topic Extraction on Twitter Considering Author's Role Based on Bipartite Networks

Takako Hashimoto[1]([✉]), Tetsuji Kuboyama[2], Hiroshi Okamoto[3],
and Kilho Shin[4]

[1] Chiba University of Commerce, Ichikawa, Japan
takako@cuc.ac.jp
[2] Gakushuin University, Tokyo, Japan
kuboyama@tk.cc.gakushuin.ac.jp
[3] RIKEN Brain Science Institute, Saitama, Japan
hiroshi.okamoto299792458@gmail.com
[4] University of Hyogo, Kobe, Japan
kilhoshin314@gmail.com

Abstract. This paper proposes a quality topic extraction on Twitter based on author's role on bipartite networks. We suppose that author's role which means who were in what group, affects the quality of extracted topics. Our proposed method expresses relations between authors and words as bipartite networks, explores author's role by forming clusters using our original community detection technique, and finds quality topics considering the semantic accuracy of words and author's role.

Keywords: Topic extraction · Social media analysis · Twitter analysis · Bipartite network · Data mining · Community detection

1 Introduction

This paper proposes a quality topic extraction on Twitter based on author's role on bipartite networks. Basically, topic structures on Twitter are not clear and same words sometimes belong to different topics. Therefore, quality topic extraction tends to be difficult. We suppose that considering author's role which means who were in what group, makes the quality of extracted topics higher. If there are two topics that have similar words but have different authors, they may sometimes be considered different topics. We already proposed topic extraction method by community detection on bipartite networks [1,2]. Our proposed method expresses relations between authors and words as bipartite networks and form clusters by a random walk technique. In addition to that, this paper explores author's role by forming clusters using our original technique, and finds quality topics considering the semantic accuracy of words and author's role. The paper identifies topics from millions of tweets after the Great East Japan Earthquake as well. We compute the coherence value [3] that can evaluate the semantic accuracy, and jaccard coefficient [4] that shows author groups' similarity, and present that topics with high coherence value and low jaccard coefficient could be considered as quality topics.

© Springer International Publishing AG 2017
A. Yamamoto et al. (Eds.): DS 2017, LNAI 10558, pp. 239–247, 2017.
DOI: 10.1007/978-3-319-67786-6_17

2 Related Work

Most conventional social media analysis methods follow this basic template.

1. Creation of Author-Word Count Matrices: Form high dimensional matrices (or bipartite graphs) of connections between authors and words over time
2. Clustering: For each matrix, adopt a topic model technique such as LSA [5] and LDA [6] to form clusters as topics
3. Feature Selection: For each cluster, define important keywords to explain the contents (LDA also produces keyword importance scores)
4. Topic Detection: Analyze each cluster contents by extracted keywords (feature words) to identify topics and compute time series similarity between neighboring clusters to detect changes over time

   This conventional methods have problems. Existing data mining techniques such as graph based methods, LSA and LDA basically can not form quality clusters. To extract important keywords from clusters, existing techniques generally use word scoring methods such as TF-IDF [7] or term-score [8]. However, such scoring methods are based on word occurrence, and high-frequency words tend to be extracted. Word scoring methods cannot always explain each cluster with high precision. Sometimes these methods identify false similarities between clusters over time.

   Bhattacharya et al. [9] tried to improve the topic quality of LDA by improved Query Classifier. However, for their classifier, the corpus based on topics should be prepared in advance. Endoh et al. [10] proposed emerging topic extraction method based on Non-negative Matrix Factorization (NMF). They tried to efficiently extract quality topics related to a specific theme, however user should input specified query words in advance. Fujino et al. [11] and Wang et al. [12] analyzed tweets over time based on LDA. Zhao et al. [13] analyzed twitter and news article using LDA. These conventional topic extraction methods are based on LDA and require topic categories or corpus in advance to improve topic extraction quality. They are still facing the accuracy problem.

3 Community Detection Technique from Bipartite Networks

This section briefly introduces the community detection technique from bipartite networks that was proposed by [1, 2].

   In network science, a "communities" refers to a group of nodes that are densely connected each other and are more sparsely connected with nodes outside the group [14]. Detecting communities in networks is of essential importance for finding functional modules of complex systems described by networks. To achieve soft clustering of words appearing in tweets, therefore, the present study has conducted community detection on bipartite networks of users and words. Community detection in the present study has adopted a technique which allows for defining overlapping between communities. This technique exploits a random walk in the network from which we wish to detect communities, and is also applicable to bipartite networks such as those examined in the present study.

The following briefly surveys this technique. Let $p_t(n)$ be the probability that a random walker (say, Mr. X) moving in the network is observed at node n. The $p(n)$ is given as the steady-state solution of the recursive equation.

$$p_t(n) = \Sigma_{m=1}^{N} T_{nm} p_{t-1}(m) \tag{1}$$

where $T_{mn} \equiv A_{nm}/\Sigma_{n'=1} A_{n'm}$ is the transition probability from node to node n; N is the total number of nodes comprising the network, and A_{nm} denotes the weight of the link from node m to node n. Suppose that $p(n)$ is decomposed as

$$p(n) = \Sigma_{k=1}^{K} \pi_k p(n \mid k) \tag{2}$$

where $p(n \mid k)$ is the conditional probability that Mr. X is observed at node n provided that he is staying in community k ; π_k is the probability of community k and satisfies.

$$\Sigma_{k=1}^{K} \pi_k = 1 \tag{3}$$

Individual communities are characterized by the conditional probabilities because $p(n \mid k)$ can be viewed as the relative importance of node n in community k. Therefore, community detection is accomplished once $p(n \mid k)$ and π_k in decomposition (2) are known. Indeed, this decomposition can be solved by EM algorithm. The $p(n \mid k)$ is generally positive for different k's; this means that node n belongs to more than one community. One can thus define overlapping between communities. The algorithm has only one parameter, α. The magnitude of this parameter controls the resolution of community detection; the smaller its magnitude, the larger the number of detected communities. In addition, the number of communities k that should be detected is provided by users.

4   Proposed Method

4.1   Step 1: Creation of Author-Word Count Matrices

First, following conventional methods, we group the tweets by a certain period (e.g. hour) during which they were sent. We then create the sequence of author-word count matrices, $\langle A_0, A_1 \ldots, A_t, \ldots, A_T \rangle$ that summarizes the words used in tweets by each author during each time slot. These time series matrices, $A_0, \ldots, A_T$, are obviously sparse. We assume that any significant event does not happen in the first time period $t = 0$, and let A_0 be the initial matrix representing an ordinary state.

4.2   Step 2: Clustering and Step 3: Feature Selection

We leverage our technique that is explained in Sect. 3, because Author-Word Count Matrices can be considered as bipartite networks that consist of authors and words (See Fig. 1). From these bipartite networks, our method forms clusters and selects feature words. Based on the number of communities (clusters) that we want to form, our method does soft-clustering and makes clusters C_{kt}. Where k is the number of clusters and t identifies the corresponding time slot. Each cluster has the set of words W_{kt} and the set of authors A_{kt}. For each word and for each author id, fuzzy belongingness are calculated respectively. Words and authors with high fuzzy belongingness should be recognized as feature words and feature authors in each cluster.

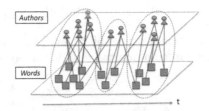

Fig. 1. Bipartite networks by authors and words

4.3   Step 4: Topic Detection

To analyze the semantic accuracy of topics, we utilize the topic coherence measure [3] that shows the meaning semantic coherence. The coherence is a human judged quality that depends on the semantics of the words. Mimno et al. [3] proposed an automated evaluation metric for identifying topics that does not rely on human annotators or reference collections outside the training data. We also compute jaccard coefficient J of authors in every topic to evaluate author's group similarity between topics. In our experiment, J is computed for every pair of topics and the average jaccard coefficient should be calculated for each topics. If J is small, that means the topic is not similar with other topics. Basically our method extracted number of topics (according to the number of topics k), but it is not easy to detect actual quality topics from them. Therefore we propose the way to detect quality topics by coherence value and jaccard coefficient. We propose that topics with high coherence value and low jaccard coefficient is quality. We use the scattering diagram with the coherence value and the jaccard coefficient to evaluate topic quality.

5   Experimental Result

5.1   Target Data

Our target data is over 200 million tweets sent around the time of the Great East Japan Earthquake that happened at 14:47 on March 11, 2011. The social media monitoring company Hottolink [15] tracked users who used one of 43 hashtags (for example, #jishin, #nhk, and #prayforjapan) or one of 21 keywords related to the disaster. Later, they captured all tweets sent by all of these users between March 9th (2 days prior to the earthquake) and March 29th. This resulted in an archive of around 200 million tweets, sent by around 1 million users. An average of about 8 million tweets were posted by around 200 thousand authors per day. The average data size per day was around 8 GB, and the total data size was over 150 GB. This dataset offers a significant document of users' responses to a crisis, but its size presents a challenge.

5.2   Step 1: Creation of Author-Word Count Matrices

We began by creating author-word count matrices for our dataset. To segment tweets that may not have used spaces to delineate word boundaries, we employed the fast and customizable Japanese morphological analyzer, MeCab [16]. Then we created author-word count matrices for every 30 min.

5.3   Step 2: Clustering and Step 3: Feature Selection

We set $\alpha = 0.001$ and $k = 100$ and run the program developed based on our technique that was explained in Sect. 3. We also set the number of EM iteration as 600. Our method formed clusters (≤ 100) until they converges. As Step 3, feature authors and words were also extracted by our method.

5.4   Step 4: Topic Detection

We made the scattering diagram with the coherence value and the jaccard coefficient. Figure 2(a), (b) and (c) show the scatter diagrams for Mar. 11 15:00–15:30, 15:30–16:00 and 16:00–16:30 topics. Each dot shows one topic.

Generally, topics with high coherence values are considered as quality topics. On the other hand, the topics shown in Table 1 do not have high coherence values, but have low jaccard coefficient values. Actually topic #29 shows the topic about Disneyland situation. There was an unique topic about the Disney area on Twitter after the earthquake, because the Disney area was very affected by the quake and guests were stranded there. It was quite unique topic. Topic #11 is about the Hanshin earthquake that happened in 1995. The Hansihn earthquake was another big earthquake and people talked about the shortage of water, body warmer and so on and the bath problem as our previous experiment. It was also unique topic. Topic #64 is about the play cancellation at the Imperial Theater due to the earthquake. It was also very specific topic discussed by specific people.

In Table 2, Topic #22 shows the topic about Disney area situation too. There was still the specific situation there after the earthquake, and people posted tweets continuously. Topic #6 is about how to treat injured people with towel and so on. We suppose this kind of topics happened, because people were aware

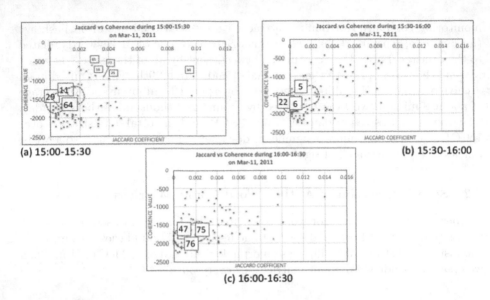

Fig. 2. Coherence Value vs Jaccard Coefficient for topics in Mar. 11 15:00–16:30

Table 1. Topics with low Jaccard Coefficient value in Mar. 11 15:00–15:30

Topic id	Coherence	Jaccard	Feature words (top 30)
29	-1688.53	0.00056	Room, Bookshelf, Condition, Panic, Danger, Ariel, Catastrophe, Sho, Smiling, Disney, Audience, Wire, Performance, Hard, Leaving, Generation, Applause, Mermaid LagoonTheater, Seat, Chaos, Shower, Contents, earthquake, stand, mistake, total destruction, lazy metal, peeling, changing clothes, e-mail magazine
11	-1421.55	0.00132	place, securing, gas, Hanshin earthquake, door, main plug, bath, confirmation, evacuation, experience, food, wasteful, collecting, case, family, evacuation route, escape path, blanket, route, bathtub, secondary disaster, body warmer, Earthquake, Telephone, Water Supply, Okay, Escape Route, Drinking Water, Aftershock, Pipe
64	-1707.03	0.00095	Canceled, DOCOMO, Today, Performance, Graduation, Sorry, Imperial theater, Lobby, After, Stage, Opposition, Cotton, Laughter, Range, Imperial Palace, Ticket, Interluption, Works, Dance, Earthquake, Points, Courage, Cancer, Metropolitan Police Department, last year, announcement, huge earthquake, okay

of injured people surrounding them. Topic #5 is about family, communication tools. Most of public transportations stopped and people could not return home. Some people started a discussion to communicate with their family. In Table 3, Topic #75 shows the topic about the liquefaction phenomenon in Minato-ku area. The liquefaction phenomenon happened in some places, especially in Tokyo. Besides the earthquake damage, some people discussed the special status in Tokyo. Topic #47 is still the topic about Disney area situation. Topic #76 is

Table 2. Topics with low Jaccard Coefficient value in Mar. 11 15:30–16:00

Topic id	Coherence	Jaccard	Feature words (top 30)
22	-1633.70	0	occurrence, panic, ariel, state, DisneySea, smile, danger, show, audience, performance, wire, hard, bean idling theater, shake, exit, audience, applause, clapping, place, earthquake, Miyagi oki earthquake, leakage, evacuation direction, within, not possible, Rokkasho village, nice, various places
6	-1732.08	0.00055	Account, Hemostasis, Heart, Bleeding, Compression, Radio, Person, Injury, Direct, Towel, Consciousness, Self-Standing, Location Information, Partial, Floor, Address, Handkerchief, Massage, Earthquake, Allowance, Antenna, Part, one hand, cut, congestion, close, backwards
5	-1426.32	0.00109	family, communication, means, report, service, entrance, load, condition, emergency, via, clothing, distance, search, easy, guard, operation, utilization, head, excess, entrance, preparation, answering machine, food, Bulletin board, backpack, message, loud voice, preparation, exchange, carrying

Table 3. Topics with low Jaccard Coefficient value in Mar. 11 16:00–16:30

Topic id	Coherence	Jaccard	Feature words (top 30)
75	-1739.20	0.00089	moving, outing, feet, liquefaction phenomenon, Minato-ku, chome, metropolitan, sufficient, you, posting, happy, applicable, point, welcome, list, Tokyo Tower, crisis management, house, case, Chang, Qiwad, Concrete, Nice, then, fashion, squid, earthquake, part, tomorrow, also spread
47	-1321.52	0.00071	shake, danger, state, panic, Ariel, DisneySea, smile, outbreak, show, audience, hard, performance, wire, beim idagun Theater, cluster, exit, audience, applause, Himeji, Uno, Yuki, Chugoku, Tempozan, Shodoshima, Kansai, Kobe, earthquake, Wakayama, Ariel, Shuya
76	-1909.79	0.00107	Abandoned, Impact, Today, Schedule, Shinjuku, Bus, Future, Live, Announcement, Sorry, Sure, Opening, Performance, Street, Customer, University, Nuisance, Immediately, Event, Camera, Postponement, Display, station front, acknowledgment, intention, paralysis, wait

about a specific event cancellation. The organizer announced their event cancellation. Some specific people discussed the topic. We basically evaluate topics with high coherence values. But, some topics that don't have high coherence values but have low jaccard values of author groups that show low similarity with other author groups can be considered as quality topics as well. We propose to consider the low similarity of author groups to evaluate the quality of topics.

6   Conclusion

This paper proposed the method considering author's role to extract accurate topics from tweets. By our method, quality topics can be extracted from tweets. For our future work, we intend to improve our methods considering forming the general model on topic life cycle and apply the method to time series data. We plan to compare our method with conventional methods, too.

Acknowledgment. This paper was supported by the Grant-in-Aid for Scientific Research (KAKENHI Grant Numbers 26280090, 15K00314, and 17H00762) from the Japan Society for the Promotion of Science.

References

1. Qiu, X., Inagi, A.S., Nukui, S., Murata, T., Okamoto, H.: Random walk based community detection from bipartite networks. In: Proceedings of The 30th Annual Conference of the Japanese Society for Artificial Intelligence (2016)
2. Hashimoto, T., Kuboyama, T., Okamoto, H., Shin, K.: Topic extraction from millions of tweets based on community detection in bipartite networks. In: Proceedings of 27th International Conference on Information Modelling and Knowledge Bases, pp. 409–424 (2017)
3. Mimno, D., Wallach, H.M., Talley, E., Leenders, M., McCallum, A.: Optimizing semantic coherence in topic models. In: Proceedings of the Conference on Empirical Methods in Natural Language Processing, pp. 262–272. Association for Computational Linguistics (2011)
4. Jaccard, P.: The distribution of the flora in the alpine zone. New Phytol. **11**(2), 37–50 (1912)
5. Sayyadi, H., Raschid, L.: A graph analytical approach for topic detection. ACM Trans. Internet Technol. **13**(2), 4:1–4:23 (2013). Article No. 4
6. Blei, D.M., Ng, A.Y., Jordan, M.I.: Latent Dirichlet allocation. J. Mach. Learn. Res. **3**(4–5), 993–1022 (2003). doi:10.1162/jmlr.2003.3.4-5.993
7. Wu, H.C., Luk, R.W.P., Wong, K.F., Kwok, K.L.: Interpreting TF-IDF term weights as making relevance decisions. ACM Trans. Inf. Syst. **26**(3), 13:1–13:37 (2008). doi:10.1145/1361684.1361686. Article No. 13
8. Blei, D.M., Lafferty, J.D.: Topic models. In: Srivastava, A., Sahami, M. (eds.) Text Mining: Theory and Applications. Taylor and Francis, UK (2009)
9. Bhattacharya, I., Sil, J.: Query classification using LDA topic model and sparse representation based classifier. In: Proceedings of the 3rd IKDD Conference on Data Science 2016, p. 24. ACM (2014)
10. Endo, Y., Toda, H., Koike, Y.: What's hot in the theme: query dependent emerging topic extraction from social streams. In: Proceedings of the 24th International Conference on World Wide Web, pp. 31–32. ACM (2015)
11. Fujino, I., Hoshino, Y.: A method for identifying topics in twitter and its application for analyzing the transition of topics. In: Proceedings of DEIM Forum 2014. C4-2 (2014)
12. Wang, Y., Agichtein, E., Benzi, M.: TM-LDA: efficient online modeling of latent topic transitions in social media. In: Proceedings of the 18th ACM SIGKDD International Conference on Knowledge Discovery and Data Mining, pp. 123–131. ACM (2012)
13. Zhao, W.X., Jiang, J., Weng, J., He, J., Lim, E.-P., Yan, H., Li, X.: Comparing twitter and traditional media using topic models. In: Clough, P., Foley, C., Gurrin, C., Jones, G.J.F., Kraaij, W., Lee, H., Mudoch, V. (eds.) ECIR 2011. LNCS, vol. 6611, pp. 338–349. Springer, Heidelberg (2011). doi:10.1007/978-3-642-20161-5_34

14. Newman, M.E.J.: Communities, modules and large-scale structure in networks. Nature Phys. **8**(1), 25–31 (2012)
15. Hottolink Inc. http://www.hottolink.co.jp/english
16. MeCab: Yet Another Part-of-Speech and Morphological Analyzer. http://taku910. github.io/mecab/

Pattern Mining

Mining Strongly Closed Itemsets
from Data Streams

Daniel Trabold[1(✉)] and Tamás Horváth[1,2]

[1] Fraunhofer IAIS, Schloss Birlinghoven, 53754 St. Augustin, Germany
{daniel.trabold,tamas.horvath}@iais.fraunhofer.de
[2] Department of Computer Science, University of Bonn, Bonn, Germany

Abstract. We consider the problem of mining strongly closed itemsets from transactional data streams. Compactness and stability against changes in the input are two characteristic features of this kind of itemsets that make them appealing for different applications. Utilizing their algebraic and algorithmic properties, we propose an algorithm based on reservoir sampling for approximating this type of itemsets in the landmark streaming setting, prove its correctness, and show empirically that it yields a considerable speed-up over a straightforward naive algorithm without any significant loss in precision and recall. As a motivating application, we experimentally demonstrate the suitability of strongly closed itemsets to concept drift detection in transactional data streams.

1 Introduction

It is a well-known fact that *closed* frequent itemsets provide a compact representation of frequent itemsets [11]. The concept of closedness has been generalized in [2]: An itemset is *strongly* or more precisely, Δ-*closed* for some $\Delta > 0$ integer, if all of its extensions result in a drop of at least Δ transactions in its support set. Clearly, Δ-closed itemsets are ordinary closed (i.e., 1-closed) for any $\Delta \geq 1$. With increasing Δ, the number of Δ-closed itemsets becomes usually much smaller than that of ordinary closed itemsets [2]. Despite the fact that strongly closed itemsets provide only lossy representations of frequent itemsets, this typically *tiny* subset of ordinary closed itemsets is still able to capture some *essential* information about the data at hand that is *stable* against changes [2].

Compactness and stability make strongly closed itemsets attractive, among others, for streaming applications. As a motivating example, we consider the *transactional* data stream scenario in which the objects arriving continuously are subsets of some ground set (set of all items) and are generated by some *unknown* distribution that may change over time. Such changes are referred to as *concept drifts*. Concept drift detection is essential for most algorithms building some model from data streams with changing distributions. We experimentally demonstrate that changes in the family of strongly closed itemsets are good indicators for detecting concept drifts in transactional data streams. For example, we could reliably detect concept drifts by monitoring the changes in around 250 Δ-closed itemsets out of more than 45,000 ordinary closed itemsets.

© Springer International Publishing AG 2017
A. Yamamoto et al. (Eds.): DS 2017, LNAI 10558, pp. 251–266, 2017.
DOI: 10.1007/978-3-319-67786-6_18

Motivated by this and other practical applications, we present an efficient algorithm for mining strongly closed itemsets from transactional data streams in the *landmark* model.[1] To make the algorithm practically feasible for *massive* transactional data streams, we consider a random subset of the data stream generated by *reservoir sampling* [13] and approximate the family of strongly closed itemsets in the data stream by that in the sample. The size of the sample is chosen in a way that with high probability, it preserves the relative frequencies of itemsets within some small error. Our algorithm calculates the family of strongly closed itemsets from this sample upon request or after a certain number of new transactions have arrived since the last update.

Reservoir sampling allows us to record the changes from the last update not in the sample, but in two separate databases. As the replacement of a transaction in the sample is equivalent to removing the old transaction and inserting the new one, the two databases correspond to the sets of transactions to be *deleted* from and those to be *added* to the sample. The motivation behind splitting replacement into deletion and insertion is that in contrast to the method in [2], strong closedness of an itemset can be decided much faster when the support set of the itemset is empty in at least one of the two databases. With increasing stream length this situation becomes more and more typical as the number of changes in the sample and accordingly, the size of the two databases decreases.

Our algorithm is based on the fact that strongly closed itemsets of the sample form a *closure system* [2]. We make use of this property and calculate the update by traversing the old strongly closed itemsets with a divide and conquer algorithm. It is based on a folklore algorithm (see, e.g., [4]) that lists all closed sets of a set system. This algorithm has a number of advantageous algorithmic properties [1,4] utilized by our algorithm as well.

We empirically evaluated the *speed-up* and *quality* (in terms of precision and recall) of our algorithm on artificial and real-world benchmark datasets. To measure the speed-up, we compared the batch algorithm generating strongly closed itemsets from the new sample from scratch with our incremental algorithm for different number of changes in the sample. For small changes, which is the case for long data streams, we obtained a speed-up of up to two orders of magnitude. Regarding the quality, in most cases we achieved very high precision and recall values (close to 1). Thus, as the empirical results demonstrate, our algorithm is much faster than the algorithm computing from scratch and still calculates a close approximation of the family of strongly closed itemsets.

Outline. The rest of the paper is organized as follows. We briefly discuss related work in Sect. 2, define the necessary concepts and the problem setting in Sect. 3, and describe our algorithm in Sect. 4. In Sect. 5 we first present some experimental results demonstrating the suitability of strongly closed itemsets for concept drift detection and then empirically evaluate the speed and approximation quality of our algorithm on benchmark and real-world datasets. For space limitations, we focus on the speed and quality results only in this short version. We finally mention some interesting directions for future work in Sect. 6.

---

[1] Due to space limitations we omit frequency constraints in this short version.

2 Related Work

Mining ordinary closed itemsets is the special case of mining Δ-closed itemsets for $\Delta = 1$. Out of the different streaming models (e.g., sliding window, time-fading, landmark) studied for mining closed frequent itemsets in data streams, we discuss only the *landmark* model, corresponding to our problem setting. It considers all transactions from a landmark starting time in the past to the current time point. For space limitations, we only discuss the two most relevant algorithms FP-CDS [9] and LC-CloStream [6], which mine ordinary closed frequent itemsets under the landmark model. We note that we consider a *more general* problem that has not been studied before to the best of our knowledge.

FP-CDS [9] processes the transactions in batches. It first constructs a local tree structure for the current batch of transactions and then merges this tree with a global one built for the entire data stream from the landmark starting time up to the previous batch. Closed patterns are generated from this global tree from scratch. Using the idea in [10], the algorithm considers all patterns of frequency at least ϵ for some appropriately chosen $\epsilon < \theta$, where θ is the frequency threshold. Our approach is, however, fundamentally different, as we process the transactions with reservoir sampling. Furthermore, while our algorithm incrementally updates the family of closed itemsets, FP-CDS computes it from scratch.

LC-CloStream [6] combines the main features of the algorithms LossyCounting [10] mining frequent itemsets and CloStream [14] mining closed frequent itemsets from data streams in the sliding window model. Similarly to Lossy-Counting, LC-CloStream computes an ϵ-approximation of ordinary closed frequent itemsets and similarly to CloStream, it calculates ordinary closed itemsets from intersections of transactions. LC-CloStream returns all closed itemsets which are estimated to be frequent. Its output is incomplete, in contrast to ours, which is complete with respect to the transactions in the reservoir.

Finally we note that there is a vast literature on concept drift detection in data streams (see, e.g., [3] for a survey). We omit the discussion of the related literature, as concept drift detection is only one of the potential applications of our *general-purpose* algorithm mining Δ-closed itemsets in transactional databases.

3 The Problem Setting

In this section we define the problem setting for this work. We first provide the necessary notions and fix the notation. For all $m \in \mathbb{N}$, $[m]$ denotes the set $\{1, \ldots, m\}$. Given some finite ground set E (*items*), the concepts of *itemsets* and *transactions* (i.e., subsets of E), and *transaction databases* over E (i.e., multisets of transactions) are used in the standard way. A *transactional data stream* over E (in what follows, simply a *data stream*) is a sequence $\mathcal{S}_t = \langle T_1, T_2, \ldots, T_t \rangle$, where the T_i's are non-empty transactions, i.e., $\emptyset \neq T_i \subseteq E$ for all $i \in [t]$. To calculate the family of strongly closed itemsets for $\mathcal{S}_t$, the order of the transactions in $\mathcal{S}_t$ does not matter. For this operation $\mathcal{S}_t$ can therefore be regarded as a transaction database (i.e., multi-set) $\mathcal{D}_t$ over E. We make use of this property and formulate most definitions below for transaction databases.

Let $\mathcal{D}$ be a transaction database over E. The *support set* of an itemset $X \subseteq E$ in $\mathcal{D}$, denoted $\mathcal{D}[X]$, is defined by the multi-set $\{T \in \mathcal{D} : X \subseteq T\}$; the *support count* of X by the cardinality $|\mathcal{D}[X]|$ of $\mathcal{D}[X]$. For a threshold $\tilde{\Delta} \in [0,1]$, an itemset $X \subseteq E$ is *relatively $\tilde{\Delta}$-closed* in $\mathcal{D}$ if $|\mathcal{D}[X]|/|\mathcal{D}| - |\mathcal{D}[Y]|/|\mathcal{D}| \geq \tilde{\Delta}$ holds for all Y with $X \subsetneq Y \subseteq E$. That is, any proper extension of X decreases its relative frequency by at least $\tilde{\Delta}$. Thus, $\tilde{\Delta}$ indicates the *strength* of the closure. If it is clear from the context, we omit the adverb "relatively". Motivated by different real-world applications (e.g., concept drift detection, computer aided product configuration), we consider the following mining problem:

$\tilde{\Delta}$-**Closed Set Listing Problem:** *Given* a single pass over a data stream $\mathcal{S}_t = \langle T_1, T_2, \ldots, T_t \rangle$ over a set E of items, a threshold $\tilde{\Delta} \in [0,1]$, and an integer $t' \in [t]$, *list* all itemsets $X \subseteq E$ that are $\tilde{\Delta}$-closed in $\mathcal{S}_{t'} = \langle T_1, T_2, \ldots, T_{t'} \rangle$.

Note that the definition of *relative $\tilde{\Delta}$-closedness* for $\mathcal{S}_t$ above can equivalently be reformulated by that of *absolute Δ-closedness* [2] as follows: X is relatively $\tilde{\Delta}$-*closed* in $\mathcal{S}_t$ if and only if it is absolutely Δ-*closed* in $\mathcal{S}_t$ for $\Delta = \lceil t\tilde{\Delta} \rceil$, that is, $|\mathcal{S}_t[X]| - |\mathcal{S}_t[Y]| \geq \Delta$ for all Y with $X \subsetneq Y \subseteq E$. The adverb "absolutely" will be omitted when it is clear from the context. Clearly, ordinary closed itemsets are 1-closed itemsets. For this reason, Δ-closed itemsets will also be referred to as *strongly* closed itemsets [2] when there is no emphasis on Δ. In general, the family of Δ-closed itemsets in a transaction database $\mathcal{D}$ is denoted by $\mathcal{C}_{\Delta,\mathcal{D}}$. In particular, the family of Δ-closed itemsets in $\mathcal{S}_t$ above is denoted by $\mathcal{C}_{\Delta,\mathcal{S}_t}$. The relevance of absolute Δ-closed itemsets to our work is that we approximate the family of relative $\tilde{\Delta}$-closed itemsets in $\mathcal{S}_t$ by that in a random *sample* of $\mathcal{S}_t$ for some *fixed* size s. In this way, we can make use of some advantageous algebraic and algorithmic properties of *absolute Δ-closed* itemsets [2] and work with them for $\Delta = \lceil s\tilde{\Delta} \rceil$.

We recall some basic algebraic and algorithmic properties of Δ-closed itemsets from [2]. We start with the definition of closure operators. Let E be some finite set and $\sigma : 2^E \to 2^E$ be a function, where 2^E denotes the power set of E. Then σ is *extensive* if $X \subseteq \sigma(X)$, *monotone* if $X \subseteq Y$ implies $\sigma(X) \subseteq \sigma(Y)$, and *idempotent* if $\sigma(X) = \sigma(\sigma(X))$ for all $X, Y \subseteq E$. If σ is extensive and monotone then it is a *preclosure*; if, in addition, it is idempotent then it is a *closure* operator on E. A set $X \subseteq E$ is *closed* if it is a fixed point of σ (i.e., $X = \sigma(X)$).

For a transaction database $\mathcal{D}$ over E and integer $\Delta > 0$, let $\hat{\sigma}_{\Delta,\mathcal{D}} : 2^E \to 2^E$ be defined by $\hat{\sigma}_{\Delta,\mathcal{D}}(X) = X \cup \{e \in E \setminus X : |\mathcal{D}[X]| - |\mathcal{D}[X \cup \{e\}]| < \Delta\}$ for all $X \subseteq E$. It holds that $\hat{\sigma}_{\Delta,\mathcal{D}}$ is a preclosure on E that is not idempotent [2]. For an itemset $X \subseteq E$, consider the sequence $\hat{\sigma}^0_{\Delta,\mathcal{D}}(X), \hat{\sigma}^1_{\Delta,\mathcal{D}}(X), \hat{\sigma}^2_{\Delta,\mathcal{D}}(X), \ldots$ with $\hat{\sigma}^0_{\Delta,\mathcal{D}}(X) = X$, $\hat{\sigma}^1_{\Delta,\mathcal{D}}(X) = \hat{\sigma}_{\Delta,\mathcal{D}}(X)$, and $\hat{\sigma}^{l+1}_{\Delta,\mathcal{D}}(X) = \hat{\sigma}_{\Delta,\mathcal{D}}(\hat{\sigma}^l_{\Delta,\mathcal{D}}(X))$ for all $l \geq 1$. This sequence has a smallest fixed point, giving rise to the following definition: For all $X \subseteq E$, let $\sigma_{\Delta,\mathcal{D}} : 2^E \to 2^E$ be defined by $\sigma_{\Delta,\mathcal{D}}(X) = \hat{\sigma}^k_{\Delta,\mathcal{D}}(X)$ with $k = \min\{l \in \mathbb{N} : \hat{\sigma}^l_{\Delta,\mathcal{D}}(X) = \hat{\sigma}^{l+1}_{\Delta,\mathcal{D}}(X)\}$. The proof of the claims in the theorem below can be found in [2].

Theorem 1. *Let $\mathcal{D}$ be a transaction database over some finite ground set E and $\Delta > 0$ an integer. Then (i) for all $X \subseteq E$, X is Δ-closed in $\mathcal{D}$ if and only if*

Algorithm 1. CLOSURE [2]

---

input: $X \subseteq E$ and integer $\Delta > 0$
require: dataset $\mathcal{D}$ over E and $\mathcal{D}' = \mathcal{D}[X]$
output: $\sigma_{\Delta, \mathcal{D}}(X)$

1: $C \leftarrow X$
2: **repeat**
3:     **for all** $e \in E \setminus C$ **do**
4:         **if** $|\mathcal{D}'| - |\mathcal{D}'[e]| < \Delta$ **then** $C \leftarrow C \cup \{e\}$; $\mathcal{D}' \leftarrow \mathcal{D}'[e]$
5: **until** $\mathcal{D}'$ has not been changed in Loop 3–4
6: **return** C

$X = \sigma_{\Delta, \mathcal{D}}(X)$, (ii) $\sigma_{\Delta, \mathcal{D}}$ is a closure operator over E, and (iii) for all $X \subseteq E$, the closure $\sigma_{\Delta, \mathcal{D}}(X)$ of X can be computed by Algorithm 1.

Using the fact that $\sigma_{\Delta, \mathcal{D}}$ is a closure operator, the family of all Δ-closed itemsets of a dataset $\mathcal{D}$ can be enumerated by the following divide and conquer folklore algorithm (see, e.g., [4]): Generate first recursively all Δ-closed supersets of a set that contain a certain item $e \in E$, and then all that do not. This algorithm lists all Δ-closed itemsets non-redundantly, with polynomial delay and in polynomial space (see [1] for some further properties of this algorithm).

4   The Mining Algorithm

In this section we present our algorithm for the $\tilde{\Delta}$-Closed Set Listing problem defined in Sect. 3. To tackle massive data streams in feasible time, we approximate the $\tilde{\Delta}$-closed sets for a data stream $\mathcal{S}_t = \langle T_1, \ldots, T_t \rangle$ at time t from a random sample $\mathcal{D}_t$ generated from $\mathcal{S}_t$ without replacement. Since the order of the elements in the sample does not matter, $\mathcal{D}_t$ is regarded as a transaction database. The size s of $\mathcal{D}_t$ is chosen in a way that for all $X \subseteq E$, the discrepancy between the relative frequency of X in $\mathcal{S}_t$ and that in $\mathcal{D}_t$ is at most ϵ with probability at least $1 - \delta$. The parameters ϵ (*error*) and δ (*confidence*) are specified by the user. Our extensive experiments in Sect. 5.2 show that a very close approximation of the true family of $\tilde{\Delta}$-closed itemsets can be obtained in this way.

Our algorithm recalculates the family of $\tilde{\Delta}$-closed itemsets *not* after each new transaction, but either upon request or after b new transactions have been received since the last update, where b, the *buffer size*, is specified by the user. Given $\mathcal{S}_t = \langle T_1, \ldots, T_t \rangle$ and $\mathcal{S}_{t'} = \langle T_1, \ldots, T_t, T_{t+1}, \ldots, T_{t'} \rangle$ with $t' - t \leq b$, the new sample $\mathcal{D}_{t'}$ of $\mathcal{S}_{t'}$ is computed from the old sample $\mathcal{D}_t$ by $\mathcal{D}_{t'} = \mathcal{D}_t \ominus \mathcal{D}_{\mathrm{del}} \oplus \mathcal{D}_{\mathrm{ins}}$, where $\mathcal{D}_{\mathrm{del}}$ (resp. $\mathcal{D}_{\mathrm{ins}}$) is the multiset of transactions to be removed from (resp. added to) $\mathcal{D}_t$, and $\ominus$ and $\oplus$ denote the set difference and the union operations on multisets.

We sketch the sampling algorithm in Sect. 4.1 and describe the algorithm updating the family of $\tilde{\Delta}$-closed itemsets from $\mathcal{D}_t$ to $\mathcal{D}_{t'}$ in Sect. 4.2.

4.1 Sampling

We use *reservoir sampling* [7,13] for generating a random sample $\mathcal{D}_t$ of size s for a data stream $\mathcal{S}_t = \langle T_1, \ldots, T_t \rangle$, as this method does not require the stream length to be known in advance. The general scheme of reservoir algorithms is that they first add $T_1, \ldots, T_s$ to a "reservoir" and then throw a biased coin with probability s/k of head for all $k = s + 1, \ldots, t$. If the outcome is head they replace one of the elements selected from the reservoir uniformly at random with T_k. This naive version of reservoir sampling, attributed to A.G. Waterman by D. Knuth in [7], generates a random sample $\mathcal{D}_t$ of $\mathcal{S}_t$ without replacement uniformly at random. That is, all elements of $\mathcal{S}_t$ have probability s/t of being part of the sample after $\mathcal{S}_t$ has been processed. We have implemented Vitter's more sophisticated version, called Algorithm Z in [13].

Given a sample $\mathcal{D}_t$ of a data stream $\mathcal{S}_t = \langle T_1, \ldots, T_t \rangle$, the sample $\mathcal{D}_{t'}$ for $\mathcal{S}_{t'} = \langle T_1, \ldots, T_t, T_{t+1}, \ldots, T_{t'} \rangle$ is computed from $\mathcal{D}_t$ by repeatedly applying Algorithm Z to $\mathcal{D}_t$ and the elements in $\langle T_{t+1}, \ldots, T_{t'} \rangle$. Recall that $t' - t \leq b$, where b is the buffer size. If a transaction in the sample is replaced by a new transaction $T \in \{T_{t+1}, \ldots, T_{t'}\}$, we appropriately update a database $\mathcal{D}_{\text{del}}$ containing the transactions to be removed from $\mathcal{D}_t$ and a database $\mathcal{D}_{\text{ins}}$ containing the transactions to be added to $\mathcal{D}_t$. Clearly, $|\mathcal{D}_{\text{del}}| = |\mathcal{D}_{\text{ins}}|$. Furthermore

$$\mathbb{E}[|\mathcal{D}_{\text{del}}|] = \mathbb{E}[|\mathcal{D}_{\text{ins}}|] \leq \frac{bs}{t'} . \tag{1}$$

This follows directly from the linearity of the expectation and from $\mathbb{E}[X_k] = s/t'$, where X_k is the indicator random variable for the event that T_k is selected for $\mathcal{S}_{t'}$. The RHS of (1) approaches 0 as t' approaches infinity. For example, it is only 15 for $b = 10\,\text{k}$, $t' = 100\,\text{M}$, $\epsilon = 0.005$, $\delta = 0.001$, and $s = 150\,\text{k}$, where the sample size $s = s(\epsilon, \delta)$ is calculated by Hoeffding's inequality[2], i.e., $s = \lceil \frac{1}{2\epsilon^2} \ln \frac{2}{\delta} \rceil$.

4.2 Incremental Update

For any transaction database $\mathcal{D}$ of size s and $\tilde{\Delta} \in [0, 1]$, the family of relatively $\tilde{\Delta}$-closed itemsets of $\mathcal{D}$ is equal to the family $\mathcal{C}_{\Delta, \mathcal{D}}$ of absolutely Δ-closed itemsets for $\Delta = \lceil s\tilde{\Delta} \rceil$. Thus, we consider the following equivalent mining problem:

Δ**-Closed Set Listing Problem:** *Given* $\mathcal{D}_t$, $\mathcal{D}_{\text{del}}$, $\mathcal{D}_{\text{ins}}$ *for* $\mathcal{S}_t$ *and* $\mathcal{S}_{t'}$ *as* defined in Section 4.1, an integer $\Delta > 0$, and the family $\mathcal{C}_{\Delta, \mathcal{D}_t}$ of Δ-closed itemsets of $\mathcal{D}_t$, *generate* all elements of $\mathcal{C}_{\Delta, \mathcal{D}_{t'}}$ for $\mathcal{D}_{t'} = \mathcal{D}_t \ominus \mathcal{D}_{\text{del}} \oplus \mathcal{D}_{\text{ins}}$.

Instead of generating $\mathcal{C}_{\Delta, \mathcal{D}_{t'}}$ from scratch, our goal is to minimize the number of evaluations of the closure operator $\sigma_{\Delta, \mathcal{D}_{t'}}$ for the new sample $\mathcal{D}_{t'}$, as this step is the most expensive part of the algorithm. We make use of the fact that the expected number of changes in $\mathcal{D}_{t'}$ w.r.t. $\mathcal{D}_t$ becomes smaller and smaller as t' increases (cf. Eq. (1)). Accordingly, our focus in the design of the updating

---

[2] We note that Hoeffding's inequality applies to samples without replacement as well [5]. A tighter bound can be derived from Serfling's inequality [12]. The improvement becomes however marginal with increasing data stream length.

Algorithm 2. UPDATE $\mathcal{C}_{\Delta,\mathcal{D}_t}$

---

input: datasets $\mathcal{D}_{\text{del}}, \mathcal{D}_{\text{ins}}$ over E and $\Delta \in \mathbb{N}$
require: totally ordered set $(E, \leq)$, dataset $\mathcal{D}_t$ over E, and $\mathcal{C}_{\Delta,\mathcal{D}_t}$
output: $\mathcal{C}_{\Delta,\mathcal{D}_{t'}}$ for $\mathcal{D}_{t'} = \mathcal{D}_t \ominus \mathcal{D}_{\text{del}} \oplus \mathcal{D}_{\text{ins}}$

MAIN:

1: $\mathcal{C}_{\Delta,\mathcal{D}_{t'}} \leftarrow \{\emptyset\}$
2: LISTCLOSED$(\emptyset, \emptyset, \min E)$

LISTCLOSED(C, N, i):

1: $X \leftarrow \{k \in E \setminus C : k \geq i\}$
2: **if** $X \neq \emptyset$ **then**
3:    $e \leftarrow \min X$; $C_e \leftarrow C \cup \{e\}$
4:    **if** $\mathcal{D}_{\text{del}}[C_e] = \emptyset \wedge \mathcal{D}_{\text{ins}}[C_e] = \emptyset$ **then** $C' \leftarrow$ CLOSURE_$\alpha(C, e, \mathcal{C}_{\Delta,\mathcal{D}_t})$ $\triangleright$ Case (α)
5:    **else if** $\mathcal{D}_{\text{ins}}[C_e] = \emptyset$ **then** $C' \leftarrow$ CLOSURE_$\beta(C, e, \mathcal{D}_{\text{del}}[C_e], \mathcal{C}_{\Delta,\mathcal{D}_t})$ $\triangleright$ Case (β)
6:    **else if** $\mathcal{D}_{\text{del}}[C_e] = \emptyset$ **then** $C' \leftarrow$ CLOSURE_$\gamma(C, e, \mathcal{D}_{\text{ins}}[C_e], \mathcal{C}_{\Delta,\mathcal{D}_t})$ $\triangleright$ Case (γ)
7:    **else** $C' \leftarrow \sigma_{\Delta,\mathcal{D}_{t'}}(C_e)$ $\triangleright$ Case (δ)
8:    **if** $C' \cap N = \emptyset$ **then**
9:       add $(C, e, N, C', \uparrow)$ to $\mathcal{C}_{\Delta,\mathcal{D}_{t'}}$
10:       LISTCLOSED$(C', N, e+1)$
11:    **else**
12:       add $(C, e, N, C', \downarrow)$ to $\mathcal{C}_{\Delta,\mathcal{D}_{t'}}$
13: $Y \leftarrow \{k \in E \setminus C : k > e\}$
14: **if** $Y \neq \emptyset$ **then**
15:    $e' \leftarrow \min Y$
16:    LISTCLOSED$(C, N \cup \{e\}, e')$

---

algorithm is on *quickly* deciding whether an element $C' \in \mathcal{C}_{\Delta,\mathcal{D}_t}$ remains Δ-closed in $\mathcal{D}_{t'}$, where C' is obtained by $C' = \sigma_{\Delta,\mathcal{D}_t}(C \cup \{e\})$ for some $C \in \mathcal{C}_{\Delta,\mathcal{D}_t}$ and $e \in E$. Below we show that in all of the cases when at least one of the support sets $\mathcal{D}_{\text{del}}[C \cup \{e\}]$, $\mathcal{D}_{\text{ins}}[C \cup \{e\}]$ is empty, the problem above can be decided much faster than with the naive way of using Algorithm 1. As we empirically demonstrate in Sect. 5, a considerable speed-up over the naive algorithm can be achieved in this way.

We first briefly sketch the algorithm computing $\mathcal{C}_{\Delta,\mathcal{D}_{t'}}$ (see Algorithm 2). It is a divide and conquer algorithm that recursively calls LISTCLOSED with some Δ-closed set $C \in \mathcal{C}_{\Delta,\mathcal{D}_{t'}}$, forbidden set $N \subseteq E$, and minimum candidate generator element i. It first determines the next smallest generator element e (Line 3) and calculates the closure $C' = \sigma_{\Delta,\mathcal{D}_{t'}}(C \cup \{e\})$ in Lines 4–7; these steps are discussed in detail below. We store C', together with some auxiliary information (Lines 8 and 12) and the algorithm calls LISTCLOSED recursively for generating further Δ-closed supersets of C'. In particular, if C' does not contain any forbidden item from N then the last element of the quintuple stored for C' is $\uparrow$ (Line 9); o/w it is $\downarrow$ (Line 12). After all elements of $\mathcal{C}_{\Delta,\mathcal{D}_{t'}}$ have been generated that are supersets of C, contain $\{e\}$, but do not contain any element in N, the algorithm

---

Algorithm 3. CLOSURE_α

input: $C \in \mathcal{C}_{\Delta,\mathcal{D}_{t'}}$ with $\mathcal{D}_{t'} = \mathcal{D}_t \ominus \mathcal{D}_{\mathrm{del}} \oplus \mathcal{D}_{\mathrm{ins}}$, $e \in E$, and $\mathcal{C}_{\Delta,\mathcal{D}_t}$
require: $\mathcal{D}_t$
output: $\sigma_{\Delta,\mathcal{D}_{t'}}(C \cup \{e\})$

1: **if** $\exists (C, e, N, C', q) \in \mathcal{C}_{\Delta,\mathcal{D}_t}$ for some N, C', and q **then return** C'
2: **else return** $\sigma_{\Delta,\mathcal{D}_t}(C \cup \{e\})$ $\triangleright$ $\sigma_{\Delta,\mathcal{D}_t}(C \cup \{e\}) = \sigma_{\Delta,\mathcal{D}_{t'}}(C \cup \{e\})$ for this case

---

generates all closed sets in $\mathcal{C}_{\Delta,\mathcal{D}_{t'}}$ that are supersets of C and do not contain any element from $N \cup \{e\}$ (Lines 13–16).

Theorem 2. *Algorithm 2 generates all elements of $\mathcal{C}_{\Delta,\mathcal{D}_{t'}}$ correctly, irredundantly, with polynomial delay, and in polynomial space.*

Proof. The proof of all of the four properties are given in [1,4]; we only need to show that C' computed in Lines 4–7 satisfies $C' = \sigma_{\Delta,\mathcal{D}_{t'}}(C \cup \{e\})$. The correctness of CLOSURE_α (Algorithm 3), CLOSURE_β (Algorithm 4), and CLOSURE_γ (Algorithm 5) are shown below in Lemmas 1, 2, and 3, respectively.   □

In the rest of this section we give the algorithms for the cases distinguished in Lines 4–6 (case (δ) is trivial) and prove their correctness.

Case (α). We first consider the case that the set $C \cup \{e\}$ with $C \in \mathcal{C}_{\Delta,\mathcal{D}_{t'}}$ and $e \in E$ to be extended for further Δ-closed sets satisfies

$$\mathcal{D}_{\mathrm{del}}[C \cup \{e\}] = \emptyset \text{ and } \mathcal{D}_{\mathrm{ins}}[C \cup \{e\}] = \emptyset \tag{2}$$

(Line 4 of Algorithm 2). The closure $\sigma_{\Delta,\mathcal{D}_{t'}}(C \cup \{e\})$ for this case can be computed by Algorithm 3; the correctness of Algorithm 3 is stated in Lemma 1 below.

Lemma 1. *Algorithm 3 is correct, i.e., for all $C \in \mathcal{C}_{\Delta,\mathcal{D}_{t'}}$ and for all $e \in E$, the output of the algorithm is $\sigma_{\Delta,\mathcal{D}_{t'}}(C \cup \{e\})$.*

Proof. Condition (2) implies that $\mathcal{D}_t[C \cup \{e\}] = \mathcal{D}_{t'}[C \cup \{e\}]$, where $\mathcal{D}_{t'} = \mathcal{D}_t \ominus \mathcal{D}_{\mathrm{del}} \oplus \mathcal{D}_{\mathrm{ins}}$. Hence, $\sigma_{\Delta,\mathcal{D}_{t'}}(C \cup \{e\}) = \sigma_{\Delta,\mathcal{D}_t}(C \cup \{e\})$ and $\sigma_{\Delta,\mathcal{D}_t}(C \cup \{e\}) \in \mathcal{C}_{\Delta,\mathcal{D}_t}$, from which the proof is immediate for both cases considered in Lines 1–2.   □

Case (β). We now turn to the case that $C \in \mathcal{C}_{\Delta,\mathcal{D}_{t'}}$ and $e \in E$ fulfill

$$\mathcal{D}_{\mathrm{del}}[C \cup \{e\}] \neq \emptyset \text{ and } \mathcal{D}_{\mathrm{ins}}[C \cup \{e\}] = \emptyset \tag{3}$$

(Line 5 of Algorithm 2). In Proposition 1 below we first prove some monotonicity results that will be used also for case (γ).

Proposition 1. *Let $\mathcal{D}_1$ and $\mathcal{D}_2$ be transaction databases over E. If $\mathcal{D}_1 \subseteq \mathcal{D}_2$ then for all $\Delta \in \mathbb{N}$,*

$$\mathcal{C}_{\Delta,\mathcal{D}_1} \subseteq \mathcal{C}_{\Delta,\mathcal{D}_2}. \tag{4}$$

Furthermore, for all $\Delta \in \mathbb{N}$ and for all $X \subseteq E$,

$$\sigma_{\Delta,\mathcal{D}_1}(X) \supseteq \sigma_{\Delta,\mathcal{D}_2}(X). \tag{5}$$

Algorithm 4. CLOSURE_β

input: $C \in \mathcal{C}_{\Delta,\mathcal{D}_{t'}}$ with $\mathcal{D}_{t'} = \mathcal{D}_t \ominus \mathcal{D}_{\mathrm{del}} \oplus \mathcal{D}_{\mathrm{ins}}$, $e \in E$, $\mathcal{D}_{\mathrm{del}}[C \cup \{e\}]$, and $\mathcal{C}_{\Delta,\mathcal{D}_t}$
require: $\mathcal{D}_t$
output: $\sigma_{\Delta,\mathcal{D}_{t'}}(C \cup \{e\})$

1: $b \leftarrow$ FALSE
2: **if** there exists (C, e, N, C', q) in $\mathcal{C}_{\Delta,\mathcal{D}_t}$ for some N, C', and q **then**
3: $\quad$ C'.count $\leftarrow$ C'.count $- |\mathcal{D}_{\mathrm{del}}[C']|$
4: $\quad$ **for all** $i \in E \setminus C'$ **do**
5: $\quad\quad$ **if** C'.count $- C'.\Delta_i + |\mathcal{D}_{\mathrm{del}}[C' \cup \{i\}]| < \Delta$ **then**
6: $\quad\quad\quad$ $b \leftarrow$ TRUE; **break**
7: $\quad\quad$ **else** $C'.\Delta_i \leftarrow C'.\Delta_i - |\mathcal{D}_{\mathrm{del}}[C' \cup \{i\}]|$
8: **else** $b \leftarrow$ TRUE
9: **if** $b =$ TRUE **then**
10: $\quad$ $C' \leftarrow \sigma_{\Delta,\mathcal{D}_{t'}}(C \cup \{e\})$; C'.count $\leftarrow |\mathcal{D}_{t'}[C']|$
11: $\quad$ **for all** $i \in E \setminus C'$ **do** $C'.\Delta_i \leftarrow |\mathcal{D}_{t'}[C' \cup \{i\}]|$
12: **return** C'

Proof. Let $C \in \mathcal{C}_{\Delta,\mathcal{D}_1}$ for some $\Delta \in \mathbb{N}$ and let $\mathcal{D}' = \mathcal{D}_2 \ominus \mathcal{D}_1$. Then, for any $e \in E \setminus C$, we have

$$\begin{aligned}
|\mathcal{D}_2[C \cup \{e\}]| &= |\mathcal{D}_1[C \cup \{e\}]| + |\mathcal{D}'[C \cup \{e\}]| \\
&\leq |\mathcal{D}_1[C]| - \Delta + |\mathcal{D}'[C]| \\
&= |\mathcal{D}_2[C]| - \Delta ,
\end{aligned}$$

where the inequality follows from $C \in \mathcal{C}_{\Delta,\mathcal{D}_1}$ and from the anti-monotonicity of support sets. Hence $C \in \mathcal{C}_{\Delta,\mathcal{D}_2}$ completing the proof of (4).

To show (5), suppose that during the calculation of $\sigma_{\Delta,\mathcal{D}_2}(X)$, the items in $\sigma_{\Delta,\mathcal{D}_2}(X) \setminus X$ have been added to X in the order $e_1, \ldots, e_k$. Let $X_0 = X$ and $X_i = X \cup \{e_1, \ldots, e_{i-1}, e_i\}$ for all $i \in [k]$. Then $|\mathcal{D}_2[X_{i-1}]| - |\mathcal{D}_2[X_i]| < \Delta$ for all $i \in [k]$ (see Algorithm 1). Since $\mathcal{D}_2[X_{i-1}] \supseteq \mathcal{D}_2[X_i]$ and $\mathcal{D}_1 \subseteq \mathcal{D}_2$, we have $|\mathcal{D}_1[X_{i-1}]| - |\mathcal{D}_1[X_i]| < \Delta$ for all i. Thus, as Algorithm 1 is Church-Rosser, all e_i will be added to $\sigma_{\Delta,\mathcal{D}_1}(X)$ as well, implying (5). $\qquad\square$

Using Proposition 1, we have the following result for Algorithm 4 concerning case (β):

Lemma 2. *Algorithm 4 is correct, i.e., for all $C \in \mathcal{C}_{\Delta,\mathcal{D}_{t'}}$ and for all $e \in E$, the output of the algorithm is $\sigma_{\Delta,\mathcal{D}_{t'}}(C \cup \{e\})$.*

Proof. By Condition (3), $\mathcal{D}_{t'}[C \cup \{e\}] \subseteq \mathcal{D}_t[C \cup \{e\}]$ and hence Proposition 1 implies that there is no $Y \in \mathcal{C}_{\Delta,\mathcal{D}_{t'}}$ with $C \cup \{e\} \subsetneq Y \subsetneq \sigma_{\Delta,\mathcal{D}_t}(C \cup \{e\})$. Furthermore, if $\sigma_{\Delta,\mathcal{D}_t}(C \cup \{e\}) \notin \mathcal{C}_{\Delta,\mathcal{D}_{t'}}$ then $\sigma_{\Delta,\mathcal{D}_t}(C \cup \{e\}) \subsetneq \sigma_{\Delta,\mathcal{D}_{t'}}(C \cup \{e\})$. Thus, to check whether $C' = \sigma_{\Delta,\mathcal{D}_t}(C \cup \{e\})$ remains closed in $\mathcal{D}_{t'}$, it suffices to test whether

$$|\mathcal{D}_{t'}[C']| - |\mathcal{D}_{t'}[C' \cup \{i\}]| \geq \Delta \qquad (6)$$

Algorithm 5. CLOSURE_γ

input: $C \in \mathcal{C}_{\Delta,\mathcal{D}_{t'}}$ with $\mathcal{D}_{t'} = \mathcal{D}_t \ominus \mathcal{D}_{\mathrm{del}} \oplus \mathcal{D}_{\mathrm{ins}}$, $e \in E$, $\mathcal{D}_{\mathrm{ins}}[C \cup \{e\}]$, and $\mathcal{C}_{\Delta,\mathcal{D}_t}$
require: $\mathcal{D}_t$
output: $\sigma_{\Delta,\mathcal{D}_{t'}}(C \cup \{e\})$

1: **if** there exists (C, e, N, C', q) in $\mathcal{C}_{\Delta,\mathcal{D}_t}$ for some N, C', and q **then**
2:      $C'' \leftarrow C \cup \{e\}$; $\mathcal{D}' \leftarrow (\mathcal{D}_t \oplus \mathcal{D}_{\mathrm{ins}})[C'']$
3:      **repeat**
4:          **for all** $i \in C' \setminus C''$ **do**
5:              **if** $|\mathcal{D}'| - |\mathcal{D}'[i]| < \Delta$ **then**
6:                  $C'' \leftarrow C'' \cup \{i\}$; $\mathcal{D}' \leftarrow \mathcal{D}'[i]$
7:      **until** $\mathcal{D}'$ has not been changed in Loop 4–6
8:      **return** C''
9: **else**
10:      **return** $\sigma_{\Delta,\mathcal{D}_{t'}}(C \cup \{e\})$

---

further holds for all items $i \in E \setminus C'$ (Lines 3–7 of Algorithm 4). By storing $|\mathcal{D}_t[C' \cup \{i\}]$ in $C'.\Delta_i$ for all $i \in E \setminus C'$, (6) can be decided from $\mathcal{D}_{\mathrm{del}}$ without any access to $\mathcal{D}_t$ (Line 5), implying the correctness of Algorithm 4 for the case that $C' \in \mathcal{C}_{\Delta,\mathcal{D}_{t'}}$ (i.e., $b = $ FALSE after loop 4–7); the claim is trivial for the other case (Lines 9–11). $\square$

The values of $C'.\Delta_i$ for further updates ($i \in E \setminus C'$) are (re)calculated in Line 7 if C' remains Δ-closed; o/w in Line 11. We note that with increasing stream length, the number of elements to be deleted from $\mathcal{C}_{\Delta,\mathcal{D}_t}$ becomes smaller and typically, most of the elements of $\mathcal{C}_{\Delta,\mathcal{D}_{t'}}$ are calculated by avoiding Lines 10–11.

Case (γ). Finally we discuss the case that $C \in \mathcal{C}_{\Delta,\mathcal{D}_{t'}}$ and $e \in E$ satisfy the condition

$$\mathcal{D}_{\mathrm{del}}[C \cup \{e\}] = \emptyset \text{ and } \mathcal{D}_{\mathrm{ins}}[C \cup \{e\}] \neq \emptyset \tag{7}$$

(see Line 6 of Algorithm 2). The proof for this case is shown also by using Proposition 1.

Lemma 3. *Algorithm 5 is correct, i.e., for all $C \in \mathcal{C}_{\Delta,\mathcal{D}_{t'}}$ and for all $e \in E$, the output of the algorithm is $\sigma_{\Delta,\mathcal{D}_{t'}}(C \cup \{e\})$.*

Proof. The proof is automatic for the case that the condition in Line 1 of Algorithm 5 is false. Consider the case that it is true. Proposition 1 with Condition (7) implies that $\mathcal{C}_{\Delta,\mathcal{D}_t} \subseteq \mathcal{C}_{\Delta,\mathcal{D}_{t'}}$ (i.e., all Δ-closed itemsets in $\mathcal{C}_{\Delta,\mathcal{D}_t}$ are preserved) and that $\sigma_{\Delta,\mathcal{D}_{t'}}(C \cup \{e\}) \subseteq \sigma_{\Delta,\mathcal{D}_t}(C \cup \{e\})$. Thus, when calculating $\sigma_{\Delta,\mathcal{D}_{t'}}(C \cup \{e\})$ in Loop 3–7, it suffices to consider only the elements in $\sigma_{\Delta,\mathcal{D}_t}(C \cup \{e\}) \setminus (C \cup \{e\})$, from which the claim is immediate for this case. $\square$

Compared to case (β), we need to calculate support counts in the entire sample $\mathcal{D}_{t'}$ for this case. However, the inner loop (Lines 4–6) iterates over a typically much smaller set than the general closure algorithm (cf. Lines 2–5 of Algorithm 1). Analogously to case (β), the number of new Δ-closed itemsets to be added to $\mathcal{C}_{\Delta,\mathcal{D}_{t'}}$ becomes smaller with increasing stream length, and hence, most of the elements of $\mathcal{C}_{\Delta,\mathcal{D}_{t'}}$ are calculated in the "then" part (Line 2–8) of the "if" statement.

5 Empirical Evaluation

In this section we empirically evaluate our algorithm on artificial and real-world datasets. In particular, we experimentally demonstrate that it results in a considerable speed-up (Sect. 5.1) and has high approximation quality (Sect. 5.2). The data streams, all consisting of 5 M transactions, were generated from benchmark datasets from the UCI Machine Learning [8] and from the Frequent Itemset Mining Dataset[3] repositories (see Table 1). For each dataset $\mathcal{D}$, Table 1 contains the cardinality of the ground set ($|E|$), the number of transactions ($|\mathcal{D}|$), and the density defined by $\sum_{T \in \mathcal{D}} |T|/(|E||\mathcal{D}|)$.

Table 1. Benchmark datasets used in the experiments

Name	Kosarak	Mushroom	Poker-hand	Retail	T10I4D100K	T40I10D100K
$|E|$	41,270	119	95	16,470	870	942
$|\mathcal{D}|$	990,002	8,124	1,025,010	88,162	100,000	100,000
Density	0.000196	0.193277	0.115789	0.000626	0.011612	0.042044

Before presenting our results, we first motivate our work by demonstrating the suitability of strongly closed itemsets for *concept drift detection* in transactional data streams.[4] The data streams with concept drifts were generated from the datasets in Table 1 by repeatedly drawing transactions from the dataset. We investigate the dimensions (i) *pace*, i.e., the time required to completely replace the old distribution by the new one, and (ii) *commonality*, i.e., the overlap of the two distributions. For (i) we consider both *swift* and *gradual* replacements of distributions. Gradual replacements are generated as follows: We insert ℓ transactions between two data streams S_1 and S_2, where transaction i is taken from S_1 at random with probability $1 - i/\ell$ and from S_2 with probability i/ℓ. For (ii) we consider *separated* and *intersected* distributions. To obtain separated distributions, each item was replaced by a new symbol. For intersected distributions, items were removed from the transactions independently and uniformly at random. Combining (i) and (ii), we thus have four cases (swift–separated, swift–intersected, gradual–separated, gradual–intersected). For each data stream, we generated three concept drifts after 2 M, 4 M, and 6 M transactions respectively.

To detect concept drifts in the data streams generated, we started a new instance of our mining algorithm every 100k transactions, with parameter values

---

[3] http://fimi.ua.ac.be/data/.

[4] We are going to present further practical applications (e.g., computer aided product configuration) in the long version of this paper.

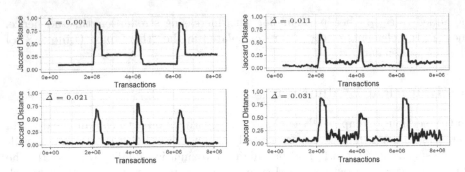

Fig. 1. Concept drift detection results for Poker-hand with swift–intersected concept drifts for $\tilde{\Delta} \in \{0.001, 0.011, 0.021, 0.031\}$ comparing current with two ago.

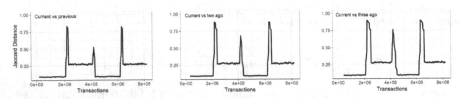

Fig. 2. Concept drift detection results for Poker-hand with swift–intersected concept drifts for varying detection delay at $\tilde{\Delta} = 0.001$.

$\epsilon = 0.01$, $\delta = 0.02$, and $b = 25\,\mathrm{k}$. As indicator for concept drifts, we used the Jaccard distance between the families of strongly closed sets returned by the algorithms started after each other. In Figs. 1 and 2 we present our results, for space limitations for Poker-hand and for swift–intersected concept drifts only, using probability 0.5 for the intersected distribution.

In particular, in Fig. 1 we investigate the influence of $\tilde{\Delta}$ ranging from 0.001 to 0.031, corresponding to $\Delta = 23$ and $\Delta = 714$, respectively. The upper limit 0.031 is chosen based on the values in Table 2. In case of Poker-hand for instance, it gives around 250 strongly closed itemsets out of 46,000 ordinary closed itemsets (i.e., around 0.5%). For all values of $\tilde{\Delta} \in \{0.001, 0.011, 0.021, 0.031\}$ the drifts are clearly visible. While they are smoother and more indicative for lower values of $\tilde{\Delta}$ (i.e., for larger subsets of ordinary closed itemsets), even as few as 250 strongly closed itemsets ($\tilde{\Delta} = 0.031$) suffice to detect the drifts.

Figure 2 shows the influence of the delay between two miners for which we calculate the Jaccard distance. On the left we compare the current miner with the previously started one, in the middle the current one with the one started two intervals before, and on the right the current one with the one started three intervals before. The three drifts are clearly visible in all graphs in both figures.

Finally we note that we obtained consistently very similar results for all other three cases (i.e., swift–separated, gradual–intersected and gradual–separated) and for all other datasets as well, indicating the suitability of strongly closed itemsets for concept drift detection in data streams.

5.1 Speedup

In this section we empirically study the speed-up obtained by our algorithm. To do so, we first sample $100\,\mathrm{k}$ random transactions, replace then $10\,\mathrm{k}$, $1\,\mathrm{k}$, 100, 10, and 1 transaction in the sample, and (i) run our algorithm as well as (ii) update the sample and run the algorithm that corresponds to Algorithm 2 with $\mathcal{C}_{\Delta,\mathcal{D}_t} = \emptyset$. Henceforth we refer to the algorithm for (ii) as the *batch* algorithm. Figure 3 shows the average runtime fraction of our algorithm in comparison to the batch algorithm as a function of the number of changed transactions for all datasets from Table 1. The runtime results are reported in detail for space limitations only for one dataset in Fig. 4 by noting that we observed a similar speedup for all other datasets. As the number of changes decreases, our streaming algorithm needs to evaluate considerably less database queries, implying that the smaller the change in the sample is, the more the runtime of the two settings differs. In Table 2 we present the number of strongly closed itemsets ($|\mathcal{C}_{\Delta,\mathcal{D}_t}|$) and the speed-up ($S$) of our algorithm for various values of $\tilde{\Delta}$ for experiments when only a single transaction has been changed. In most of the cases our algorithm is faster by at least one order of magnitude. Interestingly, the more $\tilde{\Delta}$-closed itemsets are calculated, the higher the speed-up.

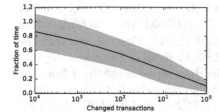

#Changes	Stream time	Batch time
10,000	6.0	6.0
1,000	4.7	6.0
100	4.2	6.0
10	1.1	6.0
1	0.3	6.0

Fig. 3. Fraction of the runtime of our streaming and the batch algorithm as a function of the number of changes (log scale): black: mean, gray: SD.

Fig. 4. Runtime in seconds of our streaming and the batch algorithm obtained for T10I4D100k for different number of changes and for $\tilde{\Delta} = 0.006$.

5.2 Approximation Quality

In this section we present empirical results demonstrating the high approximation quality of our algorithm measured in terms of precision and recall. For these experiments, we use data streams of length $5\,\mathrm{M}$ obtained by random enlargement of the benchmark datasets listed in Table 1, as well as two artificial data streams (T10I4D5M and T40I10D5M), each of length $5\,\mathrm{M}$, generated with the IBM Quest data generator. For the two artificial data streams we used the same parameters (except for the size) as for T10I4D100K and T40I10D100K. We run the experiments for the values $\tilde{\Delta} = 0.001 + 0.005i$ for $i = 0,1,\ldots,9$ and $b = 25\,\mathrm{k}$. For

Table 2. Number of $\tilde{\Delta}$-closed sets and speed-up (S) for changing a single transaction.

$\tilde{\Delta}$	Kosarak		Mushroom		Poker-hand		Retail		T10I4D100K		T40I10D100K	
	$\|\mathcal{C}_{\Delta,\mathcal{D}_t}\|$	S	$\|\mathcal{C}_{\Delta,\mathcal{D}_t}\|$	S	$\|\mathcal{C}_{\Delta,\mathcal{D}_t}\|$	S	$\|\mathcal{C}_{\Delta,\mathcal{D}_t}\|$	S	$\|\mathcal{C}_{\Delta,\mathcal{D}_t}\|$	S	$\|\mathcal{C}_{\Delta,\mathcal{D}_t}\|$	S
0.046	8	3.09	154	10.90	62	8.26	8	5.69	6	7.69	191	16.57
0.041	8	5.81	186	13.74	62	11.55	10	2.89	11	8.00	222	16.52
0.036	9	5.48	245	5.89	127	11.33	11	8.05	19	11.48	267	18.73
0.031	10	7.20	385	7.24	247	14.74	12	3.67	29	9.89	330	14.37
0.026	14	7.36	547	18.04	353	13.17	13	7.69	44	26.15	433	22.79
0.021	15	7.31	1105	19.44	578	19.45	13	9.28	83	24.57	649	23.40
0.016	24	11.13	2012	23.44	738	36.43	18	11.00	138	16.77	1146	43.30
0.011	38	14.87	4367	34.73	739	26.43	26	12.58	219	21.77	2780	96.39
0.006	86	23.61	11 k	33.96	4238	39.35	67	24.63	391	50.05	11 k	235.86
0.001	1148	99.79	82 k	37.41	46 k	59.66	1638	106.19	2574	148.09	346 k	1893

all datasets, we use $\Delta = \lceil \tilde{\Delta}t \rceil$ for the batch and $\Delta = \lceil \tilde{\Delta}s \rceil$ for our streaming algorithm, where s is the sample size. In particular, for $\epsilon = 0.005$ and $\delta = 0.001$ we have $s = 150\,\text{k}$ (see Sect. 4.1), corresponding to around 3% of the 5 M stream length. The output of the batch algorithm serves as a gold standard and will be compared to the results obtained by our algorithm. The results are reported in Table 3 in terms of precision (P) and recall (R), together with the number of $\tilde{\Delta}$-closed sets ($\|\mathcal{C}_{\Delta,\mathcal{D}_t}\|$). We note that for T40I10D5M, the batch algorithm was unable to compute the result for $\tilde{\Delta} = 0.001$ in 24 h. One can see that the precision and recall values are never below 0.80; in most of the cases they are actually close or equal to 1. The results on the data streams obtained from the benchmark datasets might be favorable for our algorithm due to the repetition of transactions. The two artificial data streams T10I4D5M and T40I10D5M do not have such a bias. Still, we obtained very good results for these data streams as well. Thus the repetition of transactions does not improve the results in favour of our algorithm. We have carried out experiments on several other artificial data streams generated by the IBM Quest data generator using other parame-

Table 3. Number of $\tilde{\Delta}$-closed sets ($\|\mathcal{C}_{\Delta,\mathcal{D}_t}\|$), precision ($P$) and recall ($R$) after processing 5 M transactions for various datasets and different values of $\tilde{\Delta}$.

$\tilde{\Delta}$	Kosarak			Mushroom			Poker-hand			Retail			T10I4D5M			T40I10D5M		
	$\|\mathcal{C}_{\Delta,\mathcal{D}_t}\|$	P	R	$\|\mathcal{C}_{\Delta,\mathcal{D}_t}\|$	P	R	$\|\mathcal{C}_{\Delta,\mathcal{D}_t}\|$	P	R	$\|\mathcal{C}_{\Delta,\mathcal{D}_t}\|$	P	R	$\|\mathcal{C}_{\Delta,\mathcal{D}_t}\|$	P	R	$\|\mathcal{C}_{\Delta,\mathcal{D}_t}\|$	P	R
0.046	8	1	1	155	0.99	1	6	1	1	8	1	1	6	1	1	190	1	0.99
0.041	8	1	1	190	1	0.99	62	1	1	10	1	1	11	0.92	1	223	0.98	0.99
0.036	9	1	1	225	0.98	0.98	127	1	1	11	1	1	19	1	1	267	0.99	0.99
0.031	10	1	1	382	0.99	1	248	1	1	12	1	1	29	1	0.93	330	0.99	0.99
0.026	14	1	1	676	0.97	0.98	353	1	1	13	1	1	44	0.96	0.98	433	1	0.98
0.021	16	1	0.94	1112	0.99	1	578	1	1	13	0.93	1	82	0.99	0.98	650	1	0.99
0.016	24	1	1	1934	0.97	1	738	1	1	18	1	1	140	1	0.99	1137	0.98	0.98
0.011	40	0.98	1	4361	0.84	0.84	739	1	1	27	1	0.96	218	0.98	0.99	2785	0.98	0.98
0.006	86	0.98	0.97	9469	0.80	0.93	4343	0.96	0.94	66	0.98	0.98	390	0.99	1	11 k	0.97	0.97
0.001	1153	0.93	0.96	76 k	0.93	0.98	47 k	1	1	1653	0.93	0.96	2591	0.96	0.94	—		

ters selected systematically (except for the size 5 M). For all datasets considered, the results were very similar to those obtained for T10I4D5M and T40I10D5M. For space limitations, we do not report these results in this short version.

6 Concluding Remarks

We have presented a general purpose algorithm for mining strongly closed itemsets from transactional data streams under the landmark model. The speed and approximation results of the previous section indicate the suitability of our algorithm for mining strongly closed itemsets from *massive* transactional data streams as well. Our empirical results give also evidence that strongly closed itemsets are of high practical relevance to concept drift detection in transactional data streams. Other practical applications, e.g., computer aided product configuration, will be discussed in the long version of this paper.

The experimental results motivate us to develop an algorithm *specific* to concept drift detection that is based on mining and monitoring the changes in strongly closed itemsets. Besides the landmark model, we are going to consider the problem of mining strongly closed itemsets under the *sliding window* model as well. This problem requires, however, an entirely different algorithmic approach.

The speed-up results reported in Sect. 5 can further be improved by utilizing that $|\mathcal{C}_{\Delta,\mathcal{D}_t}|$ is typically (much) smaller than the sample size s calculated by Hoeffding's inequality (see [2] for a detailed discussion on the size of $\mathcal{C}_{\Delta,\mathcal{D}_t}$). For such cases, the closure $\sigma_{\Delta,\mathcal{D}_{t'}}(C \cup \{e\})$ can be computed from $\mathcal{C}_{\Delta,\mathcal{D}_t}$ without any database access to $\mathcal{D}_{t'}$, even when the closure of $C \cup \{e\}$ has not been calculated for $\mathcal{D}_t$. For example, instead of computing $\sigma_{\Delta,\mathcal{D}_t}(C \cup \{e\})$ in Line 2 of Algorithm 3, we can return $\bigcap \{Y \in \mathcal{C}_{\Delta,\mathcal{D}_t} : C \cup \{e\} \subseteq Y\}$, as $\mathcal{C}_{\Delta,\mathcal{D}_t}$ is a closure system.

References

1. Boley, M., Horváth, T., Poigné, A., Wrobel, S.: Listing closed sets of strongly accessible set systems with applications to data mining. Theoret. Comput. Sci. **411**(3), 691–700 (2010)
2. Boley, M., Horváth, T., Wrobel, S.: Efficient discovery of interesting patterns based on strong closedness. Stat. Anal. Data Mining **2**(5–6), 346–360 (2009)
3. Gama, J., Žliobaitė, I., Bifet, A., Pechenizkiy, M., Bouchachia, A.: A survey on concept drift adaptation. ACM Comput. Surv. **46**(4), 44:1–44:37 (2014)
4. Gély, A.: A generic algorithm for generating closed sets of a binary relation. In: Ganter, B., Godin, R. (eds.) ICFCA 2005. LNCS, vol. 3403, pp. 223–234. Springer, Heidelberg (2005). doi:10.1007/978-3-540-32262-7_15
5. Hoeffding, W.: Probability inequalities for sums of bounded random variables. J. Am. Stat. Assoc. **58**(301), 13–30 (1963)
6. Iwanuma, K., Yamamoto, Y., Fukuda, S.: An on-line approximation algorithm for mining frequent closed itemsets based on incremental intersection. In: Proceedings of the 19th International Conference on Extending Database Technology, pp. 704–705 (2016)

7. Knuth, D.E.: The Art of Computer Programming. Seminumerical Algorithms, vol. 2. Addison-Wesley, Reading (1997)
8. Lichman, M.: UCI machine learning repository (2013)
9. Liu, X., Guan, J., Hu, P.: Mining frequent closed itemsets from a landmark window over online data streams. Comput. Math. Appl. **57**(6), 927–936 (2009)
10. Manku, G.S., Motwani, R.: Approximate frequency counts over data streams. In: Proceedings of the 28th International Conference on Very Large Data Bases (VLDB), pp. 346–357. VLDB Endowment (2002)
11. Pasquier, N., Bastide, Y., Taouil, R., Lakhal, L.: Efficient mining of association rules using closed itemset lattices. Inf. Syst. **24**(1), 25–46 (1999)
12. Serfling, R.J.: Probability inequalities for the sum in sampling without replacement. Ann. Statist. **2**(1), 39–48 (1974)
13. Vitter, J.S.: Random sampling with a reservoir. ACM Trans. Math. Softw. **11**(1), 37–57 (1985)
14. Yen, S.J., Wu, C.W., Lee, Y.S., Tseng, V.S., Hsieh, C.H.: A fast algorithm for mining frequent closed itemsets over stream sliding window. In: IEEE International Conference on Fuzzy Systems (FUZZ-IEEE), pp. 996–1002 (2011)

Extracting Mutually Dependent Multisets

Natsuki Kiyota[1], Sho Shimamura[2], and Kouichi Hirata[1(✉)]

[1] Department of Artificial Intelligence, Kyushu Institute of Technology,
Kawazu 680-4, Iizuka 820-8502, Japan
{kiyota,hirata}@dumbo.ai.kyutech.ac.jp
[2] Graduate School of Computer Science and Systems Engineering,
Kyushu Institute of Technology, Kawazu 680-4, Iizuka 820-8502, Japan
shimamura@dumbo.ai.kyutech.ac.jp

Abstract. In this paper, we extend mutually dependent patterns as itemsets introduced by Ma and Hellerstein (2001) to *mutually dependent multisets* allowing two or more occurrences of the same items. Then, by improving the algorithm to extract all of the mutually dependent patterns based on APRIORI with maintaining itemsets and their supports, we design the algorithm to extract all of the mutually dependent multisets based on APRIORITID with traversing a database just once and maintaining both multisets and their *tail occurrences* but without computing overall multiplicity of items in multisets. Finally, we give experimental results to apply the algorithm to both real data as antibiograms consisting of a date, a patient id, a detected bacterium, and so on and artificial data obtained by repeating items in transaction data.

1 Introduction

An *association rule mining* [9], which extracts association rules from a transaction database, is one of the most famous research areas in data mining. In the association rule mining, we first extract all of the sets of items (*itemsets*, for short) that occur frequently in a database under the *minimum support*. We call such itemsets *frequent itemsets*. Then, by dividing every frequent itemset into two parts, that is, a premise and a consequence, we extract all of the implications from the premise to the consequence, which we call *association rules*, that occur accurately in the database under the *minimum confidence* [1,9].

Here, an association rule claims that the probability that the consequence occurs in a database when the premise occurs is greater than the minimum confidence. Hence, we can regard that an association rule represents the *dependency* in the database between two itemsets of the premise and the consequence.

As a generalization of the dependency between itemsets, Ma and Hellerstein [5] have introduced a *mutually dependent pattern* that is an itemset such that every subset in the itemset occurs in a database together with other

This work is partially supported by Grant-in-Aid for Scientific Research 17H00762, 16H02870, 16H01743 and 15K12102 from the Ministry of Education, Culture, Sports, Science and Technology, Japan.

A. Yamamoto et al. (Eds.): DS 2017, LNAI 10558, pp. 267–280, 2017.
DOI: 10.1007/978-3-319-67786-6_19

subsets dependently. Here, the dependency is evaluated as the *conditional probability* of two subsets under the threshold called the *minimum dependency*. On the other hand, Xiong *et al.* [8] have introduced a *hyperclique pattern* that is an itemset such that, if every item in the itemset occurs in a database then other items also occur under the *minimum h-confidence*. They have shown that the hyperclique pattern is equivalent to the itemset under the minimum *all-confidence* [7]. Furthermore, it is also equivalent to the mutually dependent pattern.

In their researches [5,8], the patterns are itemsets and they do not contain two or more occurrences of the same items. On the other hand, it is natural to extend the patterns from itemsets (sets of items) to *multisets* of items allowing two or more occurrences of the same items in several transaction databases. For example, in this paper, we deal with an *antibiogram* consisting of a date, a patient id, a detected bacterium, and so on, as experiments. Then, in the antibiogram, the multiple dates, patient id's and bacteria are necessary to obtain the rules representing the causes of hospital acquired infection. Hence, we formulate a *mutually dependent multiset* as an extension of a mutually dependent pattern.

Both Ma and Hellerstein [5] and Xiong *et al.* [8] have designed the algorithm to extract all of the mutually dependent patterns based on the famous algorithm APRIORI [1]. In contrast, in this paper, we design the algorithm to extract all of the mutually dependent multisets based on APRIORITID [1] to traverse a transaction database just once.

Whereas the previous algorithms [5,8] are sufficient to maintain itemsets and their supports, just the supports are insufficient to represent the multiple occurrences of items in multisets. This is because, when determining a pattern as a multiset is included in a transaction as a multiset, it is necessary to compute multiplicity of every item in both the pattern and the transaction. Furthermore, since our algorithm traverses a transaction database just once, it is necessary to store the information to determine such an inclusion after traversing.

On the other hand, our algorithm constructs the candidates by adding an item to the tails of already constructed mutually dependent multisets same as APRIORITID [1]. According to this strategy, in this paper, we introduce a *tail occurrence* of a multiset in order to determine such an inclusion without computing overall multiplicity of items in multisets.

This paper is organized as follows. In Sect. 2, we formulate mutually dependent multisets. In Sect. 3, we introduce a tail occurrence of a multiset and present its properties. In Sect. 4, we design the algorithm to extract all of the mutually dependent multisets. In Sect. 5, we implement our algorithm, apply it to both real data as antibiograms consisting of a date, a patient id, a detected bacterium, and so on, and artificial data obtained by repeating items in FIMI repository [2], and evaluate the extracted mutually dependent multisets.

2   Mutually Dependent Multisets

Let $\mathcal{X}$ and $\mathcal{T}$ be finite sets. We call an element in $\mathcal{X}$ an *item* and an element of $\mathcal{T}$ a *transaction id* (*tid*, for short). Also we call a set $X \subseteq \mathcal{X}$ of items an *itemset*.

In this paper, we deal with a multiset of items on $\mathcal{X}$ as a set allowing the multiple occurrences of items explicitly. Formally, let $\mathbf{N}$ be natural numbers. Then, a *multiset* on $\mathcal{X}$ is a mapping $X : \mathcal{X} \to \mathbf{N}$. For a multiset X on $\mathcal{X}$, we say that $x \in \mathcal{X}$ is an *element* of X if $X(x) > 0$ and denote it by $x \in X$ (like as a standard set). When $X(x) = k$ (> 0) for $x \in \mathcal{X}$, we call k the *multiplicity* of x in X. The *cardinality* of X, denoted by $|X|$, is defined as $\sum_{x \in \mathcal{X}} X(x)$.

Let X and Y be multisets on $\mathcal{X}$. We say that X is a *sub-multiset of* Y, denoted by $X \sqsubseteq Y$, if it holds that $X(x) \leq Y(x)$ for every $x \in \mathcal{X}$, and $X = Y$ if $X \sqsubseteq Y$ and $Y \sqsubseteq X$. Also we call the multiset Z such that $Z(x) = \max\{X(x), Y(x)\}$ for every $x \in \mathcal{X}$ the *union* of X and Y and denote it by $X \sqcup Y$.

We assume a lexicographic order $\preceq$ on $\mathcal{X}$. Here, $x \prec y$ if $x \preceq y$ and $x \neq y$. Then, we represent a multiset X on $\mathcal{X}$ as a string $x_1^{k_1} \cdots x_n^{k_n}$ on $\mathcal{X}$ such that $x_i \in \mathcal{X}$, $k_i = X(x_i) > 0$ and $x_i \prec x_j$ for every $1 \leq i < j \leq n$. If $k_i = 1$, then we sometimes omit k_i and denote x_i simply, instead of x_i^1. Let $X = x_1^{k_1} \cdots x_n^{k_n}$ and $Y = y_1^{l_1} \cdots y_m^{l_m}$ be multisets on $\mathcal{X}$ in the above form. Then, it holds that $X \sqsubseteq Y$ if, for every i ($1 \leq i \leq n$) such that $x_i \in X$, there exists a j ($1 \leq j \leq m$) such that $x_i = y_j \in Y$ and $k_i \leq l_j$, and $X = Y$ if $n = m$, $x_i = y_i$ and $k_i = l_i$ for every i ($1 \leq i \leq n$).

Let X be a multiset $x_1^{k_1} \cdots x_n^{k_n}$ and x an item in $\mathcal{X}$ such that $x_n \preceq x$. Then, we define a *concatenation* Xx as $x_1^{k_1} \cdots x_n^{k_n} x$ if $x_n \prec x$ or $x_1^{k_1} \cdots x_n^{k_n+1}$ if $x_n = x$. Furthermore, for two items $x, y \in \mathcal{X}$ such that $x \preceq y$, we define Xxy as $(Xx)y$.

We call a pair $\langle t, W \rangle$ of a tid $t \in \mathcal{T}$ and a multiset W on $\mathcal{X}$ a *transaction*. Here, we assume that every tid has just one multiset as a transaction. Also we call the set of transactions a *transaction database*. For a transaction database $\mathcal{D}$ and a multiset X, we define a *tidset* $\tau_{\mathcal{D}}(X)$ of X in $\mathcal{D}$ as follows.

$$\tau_{\mathcal{D}}(X) = \{t \in \mathcal{T} \mid \langle t, W \rangle \in \mathcal{D}, X \sqsubseteq W\}.$$

Then, we define the *frequency* $freq_{\mathcal{D}}(X)$ and the *support* $supp_{\mathcal{D}}(X)$ of X in $\mathcal{D}$ as follows.

$$freq_{\mathcal{D}}(X) = |\tau_{\mathcal{D}}(X)|, \ \ supp_{\mathcal{D}}(X) = \frac{freq_{\mathcal{D}}(X)}{|\mathcal{D}|}.$$

Definition 1 (Frequent multiset). Let $\mathcal{D}$ be a transaction database, X a multiset and σ ($0 \leq \sigma \leq 1$) the *minimum support*. Then, we say that X is *frequent* in $\mathcal{D}$ if $supp_{\mathcal{D}}(X) \geq \sigma$.

Theorem 1 (*cf., [1]*). *If a multiset X is frequent in $\mathcal{D}$, then so is every submultiset $X' \sqsubseteq X$. In other words, $supp_{\mathcal{D}}(X') \geq supp_{\mathcal{D}}(X)$ for every $X' \sqsubseteq X$.*

Proof. Since it holds that $X' \sqsubseteq W$ whenever $X \sqsubseteq W$, it holds that $t \in \tau_{\mathcal{D}}(X')$ whenever $t \in \tau_{\mathcal{D}}(X)$, which implies that $|\tau_{\mathcal{D}}(X')| \geq |\tau_{\mathcal{D}}(X)|$. $\qquad\square$

For multisets X_1 and X_2, we define a *conditional probability* $P_{\mathcal{D}}(X_1 | X_2)$ in $\mathcal{D}$ that the occurrence of X_2 in a transaction in $\mathcal{D}$ implies that of X_1 as follows.

$$P_{\mathcal{D}}(X_1 | X_2) = \frac{supp_{\mathcal{D}}(X_1 \sqcup X_2)}{supp_{\mathcal{D}}(X_2)} = \frac{freq_{\mathcal{D}}(X_1 \sqcup X_2)}{freq_{\mathcal{D}}(X_2)}.$$

Hence, the occurrence of X_1 is more dependent on the occurrence of X_2 in $\mathcal{D}$ if $P_{\mathcal{D}}(X_1|X_2)$ is larger.

Let π $(0 \leq \pi \leq 1)$ be the threshold of conditional probability, which we call the *minimum dependency*. Then, we say that non-empty multisets X_1 and X_2 are *mutually dependent* in $\mathcal{D}$ if $P_{\mathcal{D}}(X_1|X_2) \geq \pi$ and $P_{\mathcal{D}}(X_2|X_1) \geq \pi$.

Definition 2 (Mutually dependent multiset). Let $\mathcal{D}$ be a transaction database, X a multiset and π the minimum dependency. Then, we say that X is *mutually dependent* in $\mathcal{D}$ if $P_{\mathcal{D}}(X_1|X_2) \geq \pi$ for every $X_1, X_2 \sqsubseteq X$ such that $X_1, X_2 \neq \emptyset$.

We call the minimum value of conditional probabilities for every pair of non-empty sub-multisets of a mutually dependent multiset X the *dependency* of X, which we will use to evaluate the mutually dependent patterns in Sect. 5.

A mutually dependent multiset without multiple occurrences of items (i.e., that is a set) coincides with a *mutually dependent pattern* [5]. The following properties also hold for mutually dependent multisets.

Theorem 2 *(cf., [5]).* *For a multiset X and the minimum dependency π, X is mutually dependent in $\mathcal{D}$ iff $P_{\mathcal{D}}(X|x) \geq \pi$ for every $x \in X$.*

Proof. The only-if direction is obvious by letting $X_1 = X$ and $X_2 = x$ for every $X_1, X_2 \sqsubseteq X$ in Definition 2. To show the if-direction, suppose that $P_{\mathcal{D}}(X|x) \geq \pi$ for every $x \in X$. Suppose that X_1 and X_2 are non-empty sub-multisets of X and $x \in X_2$. Since $X_1 \sqcup X_2 \sqsubseteq X$, $x \sqsubseteq X_2$ and by Theorem 1, it holds that $supp_{\mathcal{D}}(X_1 \sqcup X_2) \geq supp_{\mathcal{D}}(X)$ and $supp_{\mathcal{D}}(x) \geq supp_{\mathcal{D}}(X_2)$, which implies that $P_{\mathcal{D}}(X_1|X_2) = supp_{\mathcal{D}}(X_1 \sqcup X_2)/supp_{\mathcal{D}}(X_2) \geq supp_{\mathcal{D}}(X)/supp_{\mathcal{D}}(x) = P_{\mathcal{D}}(X|x) \geq \pi$. □

Theorem 3 *(cf., [5]).* *If a multiset X is mutually dependent in $\mathcal{D}$, then so is every sub-multiset $X' \sqsubseteq X$.*

Proof. Let X_1 and X_2 be non-empty multisets such that $X_1 \sqsubseteq X'$ and $X_2 \sqsubseteq X'$. Since $X' \sqsubseteq X$, it is obvious that $X_1 \sqsubseteq X$ and $X_2 \sqsubseteq X$. Since X is mutually dependent in $\mathcal{D}$, it holds that $P_{\mathcal{D}}(X_1|X_2) \geq \pi$, which implies that X' is also mutually dependent in $\mathcal{D}$. □

3   Tail Occurrence of Multisets

In the following, for a multiset X on $\mathcal{X}$, in order to strengthen the multiplicity of $x \in X$ and to avoid the confusion, we denote $X(x)$ by $m_X(x)$.

In order to extract mutually dependent multisets from a transaction database $\mathcal{D}$, it is necessary for every algorithm to determine whether or not $X \sqsubseteq W$ for a multiset X and a transaction $\langle t, W \rangle \in \mathcal{D}$, that is, to check $m_X(x) \leq m_W(x)$ for every $x \in X$. Furthermore, since we will design the algorithm based on APRIORITID instead of APRIORI, our algorithm traverses a transaction database $\mathcal{D}$ just once and it is necessary to store the information to determine whether or not $X \sqsubseteq W$ after traversing $\mathcal{D}$. Hence, as a key idea of our algorithm, in this paper, we introduce the following *tail occurrence*.

Definition 3 (Tail item). Let X be a multiset on $\mathcal{X}$. Then, a *tail item* of X, denoted by $tl(X)$, is an item $x \in X$ such that $y \preceq x$ for every $y \in X$.

Definition 4 (Tail occurrence). Let X and W be multisets on $\mathcal{X}$. Then, the *tail occurrence* $tloc(X, W)$ of X in W is the following value.

$$tloc(X, W) = \max\{m_W(tl(X)) - m_X(tl(X)) + 1, 0\}.$$

Example 1. For $X_1 = a^5 b^3$ and $W = a^3 b^4 c$, since $tl(X_1) = b$, it holds that $m_{X_1}(b) = 3$ and $m_W(b) = 4$, which implies that $tloc(X_1, W) = 4 - 3 + 1 = 2$. Similarly, for $X_2 = a^2 b^4$ and $X_3 = a^2 b^5$, it holds that $tloc(X_2, W) = 1$ and $tloc(X_3, W) = 0$.

Theorem 4. *For multisets X and W, if $tloc(X, W) = 0$, then it holds that $X \not\sqsubseteq W$. Furthermore, if X is of the form $Y x^k$ such that $Y \sqsubseteq W$, $x = tl(X)$ and $1 \leq k \leq m_X(x)$, then it holds that $tloc(X, W) = 0$ iff $X \not\sqsubseteq W$.*

Proof. If $tloc(X, W) = 0$, then it holds that $m_W(tl(X)) - m_X(tl(X)) + 1 \leq 0$, that is, $m_W(tl(X)) < m_X(tl(X))$, which implies that $X \not\sqsubseteq W$. Also suppose that $X = Y x^k$, $Y \sqsubseteq W$ and $x = tl(X)$. If $X \not\sqsubseteq W$, then it is necessary that $m_X(x) > m_W(x)$. Since $x = tl(X)$, it holds that $m_W(tl(X)) - m_X(tl(X)) + 1 \leq 0$, which implies that $tloc(X, W) = 0$.   □

Example 2. Consider X_2, X_3 and W in Example 1 again. Suppose that $Y = a^2 b^3$. Then, it holds that $b = tl(X_2) = tl(X_3)$, $X_2 = Yb$, $X_3 = Yb^2$ and $Y \sqsubseteq W$. Also, it holds that $tloc(X_2, W) = 1$ and $tloc(X_3, W) = 0$ by Example 1 and $X_2 \sqsubseteq W$ and $X_3 \not\sqsubseteq W$, which Theorem 4 claims.

Theorem 5. *Let W and X be multisets. Also let x and y be items such that $tl(X) \preceq x \preceq y$. Suppose that $Xx, Xy, Xxy \sqsubseteq W$. If $x \prec y$, then it holds that $tloc(Xxy, W) = tloc(Xy, W) = m_W(y)$. Otherwise, that is, if $x = y$, then it holds that $tloc(Xxx, W) = tloc(Xx, W) - 1$.*

Proof. Suppose that $x \prec y$. Then, it holds that $tl(X) \prec y$, which implies that $m_{Xy}(y) = m_{Xxy}(y) = 1$. Hence, it holds that $tloc(Xxy, W) = m_W(y) - m_{Xxy}(y) + 1 = m_W(y)$ and $tloc(Xy, W) = m_W(y) - m_{Xy}(y) + 1 = m_W(y)$.

Suppose that $x = y$. Then, it holds that $tloc(Xx, W) = m_W(x) - m_{Xx}(x) + 1$ and $tloc(Xxx, W) = m_W(x) - m_{Xxx}(x) + 1$. Since $m_{Xxx}(x) = m_{Xx}(x) + 1$, it holds that $tloc(Xxx, W) = tloc(Xx, W) - 1$.   □

Example 3. Let $W_1 = a^2 bc$, $W_2 = b^2 c^3$ and $X = \emptyset$.

Consider the case that $x = b$ and $y = c$. Then, it holds that $tloc(Xx, W_1) = tloc(b, a^2 bc) = 1$, $tloc(Xx, W_2) = tloc(b, b^2 c^3) = 2$, $tloc(Xy, W_1) = tloc(c, a^2 bc) = 1$ and $tloc(Xy, W_2) = tloc(c, b^2 c^3) = 3$. Hence, it holds that $tloc(Xxy, W_1) = tloc(bc, a^2 bc) = 1 = tloc(Xy, W_1)$ and $tloc(Xxy, W_2) = tloc(bc, b^2 c^3) = 3 = tloc(Xy, W_2)$.

Consider the case that $x = y = c$. Then, it holds that $tloc(Xx, W_1) = tloc(c, a^2 bc) = 1$ and $tloc(Xx, W_2) = tloc(c, b^2 c^3) = 3$. Hence, it holds that $tloc(Xxy, W_1) = tloc(c^2, a^2 bc) = 0 = tloc(Xx, W_1) - 1$ and $tloc(Xxy, W_2) = tloc(c^2, b^2 c^3) = 2 = tloc(Xx, W_2) - 1$.

Finally, as extensions of the tidset based on tail occurrences, we introduce an *occurrence set* $o_{\mathcal{D}}(X)$ and an *occurrence tidset* $o\tau_{\mathcal{D}}(X)$ of X in $\mathcal{D}$ as follows.

$$o_{\mathcal{D}}(X) = \{\langle t, tloc(X, W) \rangle \mid \langle t, W \rangle \in \mathcal{D}, X \sqsubseteq W, tloc(X, W) > 0\},$$
$$o\tau_{\mathcal{D}}(X) = \{t \in \mathcal{T} \mid \langle t, oc \rangle \in o_{\mathcal{D}}(X)\}.$$

We can always obtain $o\tau_{\mathcal{D}}(X)$ from just $o_{\mathcal{D}}(X)$. We will maintain the occurrence set $o_{\mathcal{D}}(X)$ of X to determine $X \sqsubseteq W$ for every $\langle t, W \rangle \in \mathcal{D}$.

4   Algorithm to Extract Mutually Dependent Multisets

The algorithms to extract all of the mutually dependent patterns [5,8] repeat to generate the candidates of mutually dependent patterns with growing the cardinality of patterns one by one and then add the mutually dependent patterns if they are mutually dependent until no mutually dependent pattern is extracted. In this paper, we extend their algorithms to those for mutually dependent multisets. Here, in the remainder of this paper, we omit the subscript $\mathcal{D}$ in the notations of $\tau_{\mathcal{D}}$, $freq_{\mathcal{D}}$, $supp_{\mathcal{D}}$, $P_{\mathcal{D}}$, $o_{\mathcal{D}}$ and $o\tau_{\mathcal{D}}$.

The difference between their algorithms and our algorithm is that the former is based on APRIORI [1] whereas the latter on APRIORITID [1] more efficient than APRIORI. Also the former maintains itemsets and their supports whereas the latter does multisets and their tail occurrences based on Theorems 4 and 5.

Our algorithm adopts a breadth-first generate-and-test strategy based on the cardinality of multisets same as APRIORITID [1]. Hence, throughout of our algorithm, we assume that every multiset X has its occurrence set $o(X)$ and our algorithm can always access $o(X)$ when X is generated. Then, after traversing a transaction database $\mathcal{D}$ once and storing the occurrence set for every item, we can compute the frequency $freq(X)$ of a multiset X by computing $|o\tau(X)|$ without traversing $\mathcal{D}$ again. Furthermore, our algorithm also checks whether or not X is a candidate of mutually dependent multisets by using $o(X)$.

The algorithm ISMDM in Algorithm 1 checks whether or not a multiset is mutually dependent. By Theorem 2, a multiset X is not mutually dependent iff there exists an item $x \in X$ such that $P(X|x) = freq(X)/freq(x) = |o\tau(X)|/|o\tau(x)| < \pi$ under the minimum dependency π.

---

procedure ISMDM(X, π)

    input   : A multiset X and the minimum dependency π.
    output: The truth value of *true* or *false*.

1   **foreach** $x \in X$ **do**
2   **if** $|o\tau(X)|/|o\tau(x)| < \pi$ **then**
3   **return** *false* **and halt**;

4   **return** *true*;

Algorithm 1. ISMDM

procedure $\text{OCCSET}(o(Xx), o(Xy))$

> **input** : Two occurrence sets $o(Xx)$ and $o(Xy)$.
> **output** : The occurrence set $o(Xxy)$ of Xxy.
> 1   **if** $x \prec y$ **then**
> 2 $o(Xxy) \leftarrow \{\langle t, oc \rangle \in o(Xy) \mid t \in o\tau(Xx) \cap o\tau(Xy)\};$
> 3   **else**
>   /* $x = y$ */
> 4 $o(Xxy) \leftarrow \{\langle t, oc - 1 \rangle \mid \langle t, oc \rangle \in o(Xx), oc > 1\};$
> 5   **return** $o(Xxy);$

Algorithm 2. OCCSET

The algorithm OCCSET in Algorithm 2 constructs the occurrence set $o(Xxy)$ of Xxy from two occurrence sets $o(Xx)$ and $o(Xy)$ based on Theorem 5.

The principle of the algorithm OCCSET is illustrated as follows.

1. When $x \prec y$, tids contained in $o(Xxy)$ are tids contained in both $o(Xx)$ and $o(Xy)$, that is, $o\tau(Xx) \cap o\tau(Xy)$. For every tid in $o\tau(Xx) \cap o\tau(Xy)$, the tail occurrence of Xxy is set to that of Xy by Theorem 5. In this case, $o(Xxy) = \emptyset$ if $o\tau(Xx) \cap o\tau(Xy) = \emptyset$. This procedure is realized as line 2 in the algorithm OCCSET.
2. When $x = y$, tids contained in $o(Xxy)$ constitute a subset of those in $o(Xx)$. For every tid t in $o(Xx)$, the tail occurrence of t in $o(Xxy)$ is that in $o(Xx)$ minus 1 by Theorem 5. In this case, $o(Xxy) = \emptyset$ if every element in $o(Xx)$ is of the form $\langle t, 1 \rangle$. This procedure is realized as line 4 in the algorithm OCCSET.

By using the algorithms ISMDM and OCCSET, we design the algorithm EXTMDM in Algorithm 3 to extract all of the mutually dependent multisets from a transaction database $\mathcal{D}$ under the minimum dependency π.

The algorithm EXTMDM first traverses a transaction database $\mathcal{D}$ just once and constructs the occurrence set $o(x)$ for $x \in C_1$ in lines 2 and 3. Then, it repeats the following procedures until $C_k = \emptyset$.

1. Check whether or not a multiset X is mutually dependent by using the algorithm ISMDM in Algorithm 1 for every $X \in C_k$ and add X to L_k if so in line 4.
2. For every pair of Xx and Xy in L_k such that $x \preceq y$, add Xxy to C_{k+1} if the occurrence set $o(Xxy)$ is not empty in lines from 5 to 8.

Theorems 3 and 4 guarantee that it is not necessary to construct multisets by adding an item y to the tail of Xx, once $tloc(Xx, W) = 0$ for every $\langle t, W \rangle \in \mathcal{D}$. Finally, when $C_k = \emptyset$, the algorithm EXTMDM returns all of the mutually dependent multisets as $\bigcup_{i=1}^{k-1} L_i$.

Example 4. Consider the transaction database $\mathcal{D}$ in Table 1 and set π to $2/3$.

The algorithm EXTMDM first traverses $\mathcal{D}$ once and construct the set C_1 of occurrence sets as Table 1. Here, for a multiset $X \in C_1$, the number in t_i is the

procedure EXTMDM($\mathcal{D}, \pi$)

  input : A transaction database $\mathcal{D}$ and the minimum dependency π.

  output : All of the mutually dependent multisets in $\mathcal{D}$.

1 $C_1 \leftarrow \{\{x\} \mid x \in \mathcal{X}\}; k \leftarrow 1;$

2 *Traverse* $\mathcal{D}$; /* Construct the occurrence sets for L_1 and C_2 */

3 **while** $C_k \neq \emptyset$ **do**

4 $\quad L_k \leftarrow \{X \in C_k \mid \text{ISMDM}(X, \pi) = true\}; C_{k+1} \leftarrow \emptyset;$

5 $\quad$ **foreach** $Xx \in L_k$ **do**

6 $\quad\quad$ **foreach** $Xy \in L_k$ s.t. $x \preceq y$ **do**

      /* Construct the occurrence set $o(Xxy)$ from already
       constructed occurrence sets $o(Xx)$ and $o(Xy)$ */

7 $\quad\quad\quad$ **if** OCCSET($o(Xx), o(Xy)) \neq \emptyset$ **then**

8 $\quad\quad\quad\quad$ $C_{k+1} \leftarrow C_{k+1} \cup \{Xxy\};$

9 $\quad k \leftarrow k + 1;$

10 **return** $\bigcup_{i=1}^{k-1} L_i;$

Algorithm 3. EXTMDM

Table 1. The transaction database $\mathcal{D}$, C_1, C_2 and C_3 in Example 4.

$\mathcal{D}$		C_1	t_1	t_2	t_3	C_2	t_1	t_2	t_3	C_2	t_1	t_2	t_3	C_3	t_1	t_2	t_3	C_3	t_1	t_2	t_3
t_1	a^2bc	a	2	−	2	aa	1	−	1	bb	−	1	1	aaa	−	−	−	bbb	−	−	−
t_2	b^2c^3	b	1	2	2	ab	1	−	2	bc	1	3	−	aab	1	−	2	bbc	−	3	−
t_3	a^2b^2	c	1	3	−	ac	1	−	−	cc	−	2	−	abb	−	−	1	bcc	−	2	−

value of $tloc(X, W)$ for $\langle t_i, W \rangle \in \mathcal{D}$ and "−" denotes no entry. Since all of the element in C_1 is mutually dependent, it extracts L_1 as C_1.

Consider every pair of Xx and Xy such that $x \preceq y$ in L_1. Then, there exists a case that $X = \emptyset$ and (x, y) is one of (a, a), (a, b), (a, c), (b, b), (b, c) and (c, c).

When $x \prec y$, the algorithm OCCSET works for the multiset ab that $o(ab)$ is set to $\langle t, oc \rangle \in o(b)$ such that $t \in o\tau(a) \cap o\tau(b)$ in line 2. Since $o\tau(a) \cap o\tau(b) = \{t_1, t_3\}$ and $o(b) = \{\langle t_1, 1 \rangle, \langle t_2, 2 \rangle, \langle t_3, 2 \rangle\}$, the algorithm OCCSET sets $o(ab)$ to $\{\langle t_1, 1 \rangle, \langle t_3, 2 \rangle\}$. Similarly, since $o\tau(a) \cap o\tau(c) = \{t_1\}$, $o\tau(b) \cap o\tau(c) = \{t_1, t_2\}$ and $o(c) = \{\langle t_1, 1 \rangle, \langle t_2, 3 \rangle\}$, the algorithm OCCSET sets $o(ac)$ and $o(bc)$ to $\{\langle t_1, 1 \rangle\}$ and $\{\langle t_1, 1 \rangle, \langle t_2, 3 \rangle\}$, respectively.

When $x = y$, the algorithm OCCSET works for the multiset aa that $o(aa)$ is set to $\langle t, oc - 1 \rangle$ such that $\langle t, oc \rangle \in o(a)$ and $oc > 1$ in line 4. Since $o(a) = \{\langle t_1, 2 \rangle, \langle t_3, 2 \rangle\}$, the algorithm OCCSET sets $o(aa)$ to $\{\langle t_1, 1 \rangle, \langle t_3, 1 \rangle\}$. Similarly, since $o(b) = \{\langle t_1, 1 \rangle, \langle t_2, 2 \rangle, \langle t_3, 2 \rangle\}$ and $o(c) = \{\langle t_1, 1 \rangle, \langle t_2, 3 \rangle\}$, the algorithm OCCSET sets $o(bb)$ and $o(cc)$ to $\{\langle t_2, 1 \rangle, \langle t_3, 1 \rangle\}$ and $\{\langle t_2, 2 \rangle\}$, respectively.

As a result, the algorithm EXTMDM constructs the set C_2 with occurrence sets as Table 1. Since $P(ac|c) = P(cc|c) = 1/2 < \pi$ in checking whether or not "ISMDM$(X, \pi) = true$," it extracts L_2 as $\{aa, ab, bb, bc\}$ in line 4.

Consider every pair of Xx and Xy such that $x \preceq y$ in L_2. Then, there exist two cases that (1) $X = a$ and (x, y) is one of (a, a), (a, b) and (b, b) and (2) $X = b$ and (x, y) is one of (b, b), (b, c) and (c, c).

When $x \prec y$, since $o\tau(aa) \cap o\tau(ab) = \{t_1, t_3\}$ and $o\tau(bb) \cap o\tau(bc) = \{t_2\}$, the algorithm OCCSET sets $o(aab)$ and $o(bbc)$ to $\{\langle t_1, 1\rangle, \langle t_3, 2\rangle\}(= o(ab))$ and $\{\langle t_2, 3\rangle\}$, respectively. When $x = y$, since $o(aa) = \{\langle t_1, 1\rangle, \langle t_3, 1\rangle\}$, $o(ab) = \{\langle t_1, 1\rangle, \langle t_3, 2\rangle\}$, $o(bb) = \{\langle t_2, 1\rangle, \langle t_3, 1\rangle\}$ and $o(cc) = \{\langle t_2, 2\rangle\}$, the algorithm OCCSET sets $o(aaa)$, $o(abb)$, $o(bbb)$ and $o(bcc)$ to $\emptyset$, $\{\langle t_3, 1\rangle\}$, $\emptyset$ and $\{\langle t_2, 1\rangle\}$, respectively.

As a result, the algorithm EXTMDM constructs the set C_3 with occurrence sets as Table 1. Since $P(abb|a) = P(bbc|c) = P(bcc|c) = 1/2 < \pi$ in checking whether or not "ISMDM$(X, \pi) = true$," it extracts L_3 as $\{aab\}$ in line 4.

Since the algorithm EXTMDM cannot construct C_4, it returns mutually dependent multisets $L_1 \cup L_2 \cup L_3 = \{a, b, c, aa, ab, bb, bc, aab\}$.

In order to extract all of the *frequent mutually dependent multisets*, by Theorem 1, it is sufficient to input the minimum support σ, store $|\mathcal{D}|$ when traversing $\mathcal{D}$ and replace the construction of L_k in line 4 in the algorithm EXTMDM with the following statement.

$$L_k \leftarrow \{X \in C_k \mid (\text{ISMDM}(X, \pi) = true) \wedge (|o\tau(X)|/|\mathcal{D}| \geq \sigma)\}.$$

Since just the algorithm ISMDM is essential to extract mutually dependent multisets in the algorithm EXTMDM, it is sufficient to replace the above statement with the following statement when we extract all of the *frequent multisets*.

$$L_k \leftarrow \{X \in C_k \mid |o\tau(X)|/|\mathcal{D}| \geq \sigma\}.$$

5 Experimental Results

In this section, we apply the algorithm EXTMDM to real data as antibiograms provided from Osaka Prefectural General Medical Center and artificial data obtained by repeating items from transaction data in FIMI repository [2].

In the following, we denote the support and the dependency by percentage (%). Also the computer environment is that OS is Ubuntu Linux (64bit), CPU is Xeon CPU E5-1650 v3 (3.50 GHz) and RAM is 1024 MB.

5.1 Antibiogram

In this section, we apply the algorithm EXTMDM to antibiograms provided from Osaka Prefectural General Medical Center in years from 1999 to 2012, consisting of a patient ID, a date, a sample, a detected bacterium and results of susceptibility test for 108 antibiotics, and so on, with total 295,031 records [6].

From this antibiogram, after setting detected bacteria as items, we convert two kinds of transaction data, *data on dates* and *data on patients* by changing tids. The tids in data on dates are dates, whereas the tids in data on patients are patients' IDs. We use just transactions with more than two items as data. Then, the number of transactions and the average number of items in transactions in data on dates are 3,338 and 34.1042, respectively, whereas those in data on patients are 15,596 and 6.7066, respectively.

Tables 2 and 3 illustrate, when varying the minimum dependency π from 10% to 100% under the minimum support 0.1%, the number n and the maximum cardinality c of extracted mutually dependent multisets and the computation time t (sec.) for the data on dates and the data on patients, respectively.

Tables 2 and 3 show that the computation time for the data on patients in Table 3 does not change well, which is larger than that for the data on dates in Table 2 when π is greater than or equal to 30% whereas smaller when π is less than or equal to 20%. The algorithm ExtMDM first constructs C_1, L_1 and C_2, whose size depends on the number of transactions of data, where the number of transactions in the data on dates is 3,338, whereas that in the data on patients is 15,596. Since the number of transactions of the data on patients is much larger than that on dates, we can conclude that the occurrence sets in the data on patients are larger than those on dates and the running time depends on the process of occurrence tidsets.

On the other hand, the maximum cardinality for the data on patients is not changed, whereas that on dates grows large when π is 10%, which is the reason that the computation time that $\pi = 10\%$ in Table 2 is very large.

Next, we analyze the extracted mutually dependent multisets from the viewpoint of *microbial substitution*, which is one of the causes of hospital acquired infection.

Table 4 illustrates the extracted mutually dependent multisets whose cardinality is 7 under the minimum support 0.1% and the minimum dependency 50%, from the data on dates. Here, the items of (S. aureus), (P. aeruginosa),

Table 2. The number and the maximum cardinality of extracted mutually dependent multisets and the computation time for the data on dates.

π	100	90	80	70	60	50	40	30	20	10
n	463	466	471	488	533	638	935	1,918	6,910	68,519
c	4	4	4	5	6	7	9	11	13	17
t	3.9166	3.9045	3.9253	3.9241	3.9719	4.1261	4.2289	4.8282	7.3796	71.0553

Table 3. The number and the maximum cardinality of extracted mutually dependent multisets and the computation time for the data on patients.

π	100	90	80	70	60	50	40	30	20	10
n	467	467	467	467	472	497	509	590	708	1,142
c	5	5	5	5	5	5	5	8	8	8
t	7.4423	7.4370	7.4451	7.4455	7.2321	7.2598	7.2434	7.1902	7.2680	7.3376

Table 4. The extracted mutually dependent multisets whose cardinality is 7 from the data on dates.

Mutually dependent multisets	Support	Dependency
(S. aureus)^6(P. aeruginosa)	56.68%	58.68%
(E. coli)(S. aureus)^5(P. aeruginosa)	54.76%	56.70%
(E. coli)(S. aureus)^6	52.67%	54.53%
(S. aureus)^5(P. aeruginosa)^2	51.56%	53.38%
(S. aureus)^7	51.47%	53.29%
(S. aureus)^5(P. aeruginosa)(E. faecalis)	50.36%	52.14%
(yeast)(S. aureus)^5(P. aeruginosa)	50.21%	51.99%

(E. coli) and (E. faecalis) denote Staphylococcus aureus, Pseudomonas aeruginosa, Escherichia coli and Enterococcus faecalis, respectively.

Table 4 shows that all of the extracted mutually dependent multisets whose cardinality is 7 contain the item (S. aureus) at least 5 times. Also they contain the item (P. aeruginosa) at most twice and just (E. coli)(S. aureus)^6 and (S. aureus)^7 do not contain (P. aeruginosa). Furthermore, both the support and the dependency of all of the multisets are more than 50%, which is high.

Furthermore, from the medical viewpoint, it is well-known that the occurrence of (P. aeruginosa) after the occurrence of (S. aureus) is regarded as *microbial substitution*, which Table 4 suggests.

Table 5 illustrates the extracted mutually dependent multisets whose cardinality is at least 3 under the minimum support 0.1% and the minimum dependency 20% from the data on patients. Here, the items of (S. maltophilia), (S. marcescens) and (E. aerogenes) denote Stenotrophomonas maltophilia, Serratia marcescens and Enterobacter aerogenes, respectively. Also "sp." denotes "species."

Table 5 shows that the extracted mutually dependent multisets whose cardinality is 4 are (S. aureus)^4 and (P. aeruginosa)^4. Also, in the extracted mutually dependent multisets, (S. aureus)^4 and (S. aureus)^3 have the support greater than 10%. Also (P. aeruginosa)^3, (P. aeruginosa)^4 and (yeast)(P. aeruginosa)^2 have the support greater than 4%. The others have the support smaller than 2%.

Whereas the supports of the extracted mutually dependent multisets in Table 5 are much smaller than those in Table 4, the extracted multisets in Table 4 contain several items not occurring in Table 5 and not concerned with the items (S. aureus) and (P. aeruginosa).

5.2 Artificial Data

In this section, we use artificial data obtained by repeating items from transaction data in FIMI repository [2]. For data D in FIMI repository, the data "$D \bmod k$" denote the data obtained by repeating every item i (as an integer) at $(i \bmod k) + 1$ times in D, where $k = 2, 5$ and 10. We use "accidents," "chess,"

Table 5. The extracted mutually dependent multisets whose cardinality is at least 3 from the data on patients.

Mutually dependent multisets	Support	Dependency
$(S.\ aureus)^4$	11.17%	24.14%
$(P.\ aeruginosa)^4$	4.13%	20.82%
$(S.\ aureus)^3$	15.88%	34.34%
$(P.\ aeruginosa)^3$	5.60%	30.10%
$(yeast)(P.\ aeruginosa)^2$	4.08%	20.42%
$(\alpha\text{-hemolysis sp.})(Neisseria\ sp.)(Haemophilus\ sp.)$	1.90%	28.03%
$(Enterobacter\ sp.)(\alpha\text{-hemolysis sp.})(Neisseria\ sp.)$	1.53%	22.63%
$(\alpha\text{-hemolysis sp.})(Neisseria\ sp.)(non\text{-hemolysis sp.})$	1.42%	20.93%
$(S.\ maltophilia)^3$	1.15%	21.77%
$(S.\ marcescens)^3$	0.97%	23.35%
$(E.\ aerogenes)^3$	0.56%	22.37%

"kosarak," "mushroom," "retail," "T10I4D100K" and "T40I10D100K" as data D. Then, we apply the algorithm EXTMDM to the data "D mod k," and evaluate how efficient it extracts mutually dependent multisets by comparing with mutually dependent patterns extracted by the algorithm EXTMDM from D.

Table 6 illustrates, for the artificial data "D mod k" whose number of transactions is "tran." and average number of items in transactions is "ave.", the running time t (sec.) and the number n and the maximum cardinality c of extracted mutually dependent multisets from "D mod k" under the minimum support σ and the minimum dependency π (75% or 50%).

For the data "accidents mod 10," our algorithm returns no mutually dependent multisets by memory overflow. The reason is that the information of storing candidate sets in the algorithm EXTMDM is too large to extract mutually dependent multisets, since both the number of transactions and the number of items in transactions are too large.

Table 6 shows that, when the running time is small like "chess" and "mushroom," the number of extracted mutually dependent multisets and the running time increase rapidly. On the other hand, when the running time is large like "T10I4D100K" and "T40I10D100K," those increase slowly.

Note that, by using the property that a singleton is always mutually dependent, we implement the algorithm EXTMDM to skip the construction of C_1 and to construct L_1 directly. This is the reason that the running time of the data D is sometimes larger than that of the data "D mod k."

In Sect. 5.1, the number of transactions and the average number of items in transactions are 3,338 and 34.1042 in data on dates and 15,596 and 6.7066 in data on patients. As a result, by comparing with those in Table 6, antibiograms have more appropriate sizes than the artificial data for our implementation.

Table 6. The running time t (sec.) and the number n and the maximum cardinality c of extracted mutually dependent multisets from the artificial data "D mod k" under the minimum support σ and the minimum dependency π.

data "D mod k"	tran.	ave.	σ	π	t	n	c
accidents	340,183	33.81	95%	75%	61.12	15	4
accidents mod 2		49.11			66.77	23	5
accidents mod 5		96.87			233.69	239	12
accidents mod 10		178.77			308.28	–	–
chess	3,196	37.00	95%	75%	0.41	77	5
chess mod 2		50.93			0.61	131	6
chess mod 5		105.77			2.63	1,719	15
chess mod 10		191.91			17.86	14,039	30
kosarak	990,002	8.10	1.5%	50%	4,011.16	38	2
kosarak mod 2		10.07			4,050.99	45	4
kosarak mod 5		18.79			7,820.18	86	6
kosarak mod 10		34.39			14,336.74	221	16
mushroom	8,124	23.00	20%	75%	7.90	75	5
mushroom mod 2		32.00			6.72	113	6
mushroom mod 5		66.81			7.42	403	11
mushroom mod 10		121.01			14.08	5,763	26
retail	88,162	10.31	0.1%	50%	854.69	959	2
retail mod 2		13.49			611.00	1,181	3
retail mod 5		27.18			811.99	2,342	9
retail mod 10		50.68			1,159.04	4,387	19
T10I4D100K	10,000	10.10	0.02%	75%	1,666.91	866	2
T10I4D100K mod 2		12.96			1,266.54	1,237	3
T10I4D100K mod 5		25.39			1,286.18	2,498	6
T10I4D100K mod 10		45.22			1,331.93	4,553	11
T40I10D100K	10,000	39.61	1%	75%	1,739.45	755	1
T40I10D100K mod 2		56.61			1,675.40	1,089	2
T40I10D100K mod 5		113.84			1,790.28	2,212	5
T40I10D100K mod 10		208.10			1,973.44	4,037	10

6 Conclusion and Future Works

In this paper, we have formulated mutually dependent multisets as extensions of mutually dependent patterns [5] or hyperclique patterns [8] as itemsets. Then, by improving the algorithms [5,8] based on APRIORI to the algorithm based on APRIORITID and introducing *tail occurrences*, we have designed a new algorithm to extract all of the mutually dependent multisets by traversing a transaction database just once. Finally, we have applied the algorithm to the real data as antibiograms and the artificial data. For antibiograms, we have succeeded in extracting the mutually dependent multisets suggesting *microbial substitution*.

Concerned with Sect. 5.1, it is a future work to analyze the extracted mutually dependent multisets from the medical viewpoint in more detail. In particular, it is a future work to incorporate the extraction of mutually dependent multisets from antibiograms with the methods of the event summarization [4] to extract time-related patterns.

Concerned with Sect. 5.2, the implementation of our algorithm is not fast enough to apply to large data, because the implementation of the algorithm EXTMDM in Algorithm 3 is not robust to store the information of candidate sets. Then, it is a future work to implement it robustly for a larger data. It is also a future work to investigate whether or not we can improve our algorithm based on the breadth-first search algorithms as APRIORI and APRIORITID [1, 9] to the depth-first search algorithm as FPGROWTH [3, 9].

In this paper, we have ignored the number of occurrences of multisets in transactions. We have formulated that a multiset is included in a transaction if it occurs at least once, which is the reason to determine such an inclusion by using just a tail occurrence. On the other hand, it is possible to be necessary to deal with the number of occurrences of multisets in transactions. Hence, it is a future work to extend multisets to deal with the number of occurrences.

Acknowledgment. The authors would like to thank anomymous refrees of DS2017 for valuable comments to revise the submitted version of this paper.

References

1. Agrawal, R., Srikant, R.: Fast algorithm for mining association rules. In: Proceedings of VLDB 1994, pp. 487–499 (1994)
2. Frequent itemset mining implementation repository (FIMI repository). http://www.fimi.ua.ac.be/data
3. Han, J., Pei, J., Yin, Y., Mao, R.: Mining frequent patterns withoout candidate generation: a frequent-pattern tree approach. Data Min. Knowl. Discov. **8**, 53–87 (2004)
4. Li, T. (ed.): Event Mining: Algorithms and Applications. CRC Press, Boca Raton (2015)
5. Ma, S., Hellerstein, J.L.: Mining mutually dependent patterns. In: Proceedings of ICDM 2001, pp. 409–416 (2001)
6. Nagayama, K., Hirata, K., Yokoyama, S., Matsuoka, K.: Extracting propagation patterns from bacterial culture data in medical facility. In: Otake, M., Kurahashi, S., Ota, Y., Satoh, H., Bekki, D. (eds.) New Frontiers in Artificial Intelligence. LNCS, vol. 10091, pp. 409–417. Springer, Cham (2017)
7. Omiecinski, E.: Alternative interest measures for mining associations. IEEE Trans. Knowl. Data Eng. **15**, 57–69 (2003)
8. Xiong, H., Tan, P.-N., Kumer, V.: Hyperclique pattern discovery. Data Min. Knowl. Discov. **13**, 219–242 (2006)
9. Zaki, H.J., Meira Jr., W.: Data Mining and Analysis. Cambridge University Press, New York (2014)

Bioinformatics

LOCANDA: Exploiting Causality in the Reconstruction of Gene Regulatory Networks

Gianvito Pio[(⊠)], Michelangelo Ceci, Francesca Prisciandaro,
and Donato Malerba

Department of Computer Science, University of Bari Aldo Moro,
Via Orabona, 4, 70125 Bari, Italy
{gianvito.pio,michelangelo.ceci,donato.malerba}@uniba.it,
f.prisciandaro1@studenti.uniba.it

Abstract. The reconstruction of gene regulatory networks via link prediction methods is receiving increasing attention due to the large availability of data, mainly produced by high throughput technologies. However, the reconstructed networks often suffer from a high amount of false
positive links, which are actually the result of indirect regulation activities. Such false links are mainly due to the presence of common cause and
common effect phenomena, which are typically present in gene regulatory
networks. Existing methods for the identification of a transitive reduction
of a network or for the removal of (possibly) redundant links suffer from
limitations about the structure of the network or the nature/length of
the indirect regulation, and often require additional pre-processing steps
to handle specific peculiarities of the networks at hand (e.g., cycles).

In this paper, we propose the method LOCANDA, which overcomes
these limitations and is able to identify and exploit indirect relationships
of arbitrary length to remove links considered as false positives. This is
performed by identifying indirect paths in the network and by comparing their reliability with that of direct links. Experiments performed on
networks of two organisms (*E. coli* and *S. cerevisiae*) show a higher accuracy in the reconstruction with respect to the considered competitors, as
well as a higher robustness to the presence of noise in the data.

Keywords: Causality · Bionformatics · Gene network reconstruction

1 Introduction

Recent studies in biology have been significantly supported by high throughput
technologies and by computational methods, which led to an improved understanding of the working mechanisms in several organisms. Such mechanisms can
be usually modeled through biological networks, which are able to easily describe
the considered biological entities as well as their relationships and interactions.
On the basis of the phenomenon under study, different types of biological networks can be considered. The most prominent example is that of networks modeling the control of transcription into messenger RNAs or proteins [2,13]. In these

© Springer International Publishing AG 2017
A. Yamamoto et al. (Eds.): DS 2017, LNAI 10558, pp. 283–297, 2017.
DOI: 10.1007/978-3-319-67786-6_20

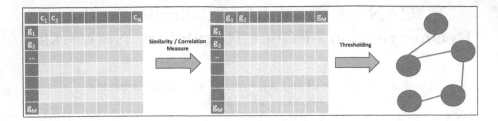

Fig. 1. Network reconstruction from expression data. On the left, a matrix of M genes, each associated to a vector containing the expression level measured under N different conditions. In the middle, the gene-gene matrix obtained by pair-wisely computing a similarity/correlation measure between the vectors. On the right, the reconstructed network obtained by imposing a threshold on the values of the gene-gene matrix.

networks, called Gene-Regulatory Networks (GRNs), nodes represent molecular entities, such as transcription factors, proteins and metabolites, whereas edges represent interactions, such as protein-protein and protein-DNA interactions.

The direct observation of the real structure of these interaction networks would require expensive in-lab experiments, usually performed through the so-called epistasis analysis. Although in the literature we can find some computational approaches which support such an analysis [17], gene expression data are much easier to obtain, therefore most of computational approaches proposed in the literature focused on predicting the existence of interactions from gene expression data, mainly on the basis of link prediction methods. These approaches analyze the expression level of the genes under different conditions (e.g., with a specific disease or after a treatment with a specific drug) or, alternatively, under a single condition in different time instants. The expression levels observed for each gene are represented as a feature vector and a gene-gene matrix is built by pair-wisely computing a similarity, correlation or information-theory-based measure between the vectors associated to genes [6]. Finally, the existence of links is inferred by imposing a threshold on the obtained score (see Fig. 1), where the direction is inferred only if the considered measure is asymmetric.

However, except for those based on clustering [15], these methods generally assume the independence among the interactions, i.e., they focus on each pair of genes separately, disregarding possible dependencies or indirect influences among them. This assumption leads to predict false positive interactions, which are usually due to causality phenomena: *(i)* common regulator genes (also referred to as *common cause* in the literature [9]) or *(ii)* commonly regulated genes (also referred to as *common effect* in the literature [9]). In the first case (see Fig. 2(a)), the feature vector associated to a gene C which exhibits a regulatory activity on two genes A and B will presumably be similar to the feature vectors associated to A and B. However, even if there is no interaction between the genes A and B, their feature vectors will appear similar, therefore a link between them could possibly be detected. Analogously, in the second case (see Fig. 2(b)), a gene C which is regulated by two genes A and B will presumably have a feature vector

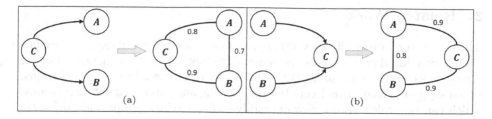

Fig. 2. Issues in the network reconstruction due to common cause (a) or common effect (b) phenomena. The direction of the interactions does not appear in the reconstructed networks if we consider the case of a symmetric similarity/correlation measure.

which is similar to the feature vectors associated to A and B. Therefore, even if A and B do not interact, their feature vectors will be similar and a link between them will possibly appear in the reconstructed network. Such issues are even more evident when data are affected by noise. Indeed, possible measurement errors can lead to a significant increment of false positives due to common cause and common effect phenomena, compromising the quality of the reconstruction.

The presence of these phenomena in the reconstruction of gene regulatory networks has been largely recognized in the literature, also considering possible hidden common causes and hidden common effects [10], and several approaches for post-processing the gene-gene matrix have been proposed. These methods, usually called *scoring schemes* [6], analyze large sets of genes simultaneously, in order to catch more global interaction activities and possibly reduce false positives due to the presence of common cause and common effect phenomena. One of the most popular scoring scheme is ARACNE [11], which evaluates all the possible connected gene triplets and removes the edge with the lowest score. ARACNE is limited to undirected networks and is not able to analyze more global indirect interactions (i.e., involving more than three genes). However, although the idea of removing the weaker edge is very simple, the intuition of considering the score as an indication of the reliability of the interaction is reasonable, and has been exploited by other works in the literature (e.g., [3]).

In this paper, inspired by the same idea, we introduce a new method, called LOCANDA, which is able to identify interaction chains of arbitrary length and is able to remove false positive interactions working on the identified chains. It is noteworthy that the approach we propose in this paper has its roots in methods for the analysis of graphs and, in particular, in works for the transitive reduction [1,7]. However, differently from existing methods, LOCANDA is able to handle weighted, directed and possibly cyclic networks without any pre-processing step.

In the Sect. 2, we briefly describe existing methods which exploit causality in the analysis or in the reconstruction of networks, giving emphasis to those tailored for the identification and removal of indirect interactions in (biological) networks. In Sect. 3, we describe our method LOCANDA, while in Sect. 4 we describe the experiments we performed and comment the obtained results. Finally, in Sect. 5, we draw some conclusions and outline possible future works.

2   Related Work

In the literature, we can find several approaches which catch and exploit causality phenomena for different goals. A general framework for the identification of causal links between variables [12] consists in *(i)* the analysis of correlations, which suggest possible (undirected) links, *(ii)* the analysis of partial correlations, which can be exploited to remove possibly indirect relationships and *(iii)* some assumptions on the structure of the network, such as acyclicity, which can suggest the possible direction of links. It is noteworthy that such assumptions can be easily violated in specific domains, such as biology, leading to an inaccurate reconstruction. An example of application of such a framework can be found in algorithms for learning the structure of Bayesian networks [9], which identify causalities between variables by analyzing the *d-separation* among them, which is based on the common cause and effect phenomena described in Sect. 1.

Other approaches exploit the concept of causality to identify a transitive reduction of a graph [1, 7]. These methods analyze a graph and produce a new graph containing a subset of links, which guarantees to convey the same information of the original graph. This means that, analogously to the method proposed in this paper, these approaches aim at removing edges that can be considered the result of an indirect relationship. Specifically, the method proposed in [1] finds a transitive reduction G' of the initial graph G, where G' has a directed path from vertex u to vertex v if and only if G has a directed path from vertex u to vertex v and there is no graph with such a property having fewer edges than G'. In other words, the obtained graph G' is the smallest graph (in terms of edges) such that given any pair of nodes $\langle u, v \rangle$, if v is (respectively, is not) reachable from u in the initial graph G, then v is (respectively, is not) reachable in the reduced graph G'. This means that the information conveyed by the graph, in this work, is associated to the reachability of nodes. Although based on the same principles of LOCANDA, this approach requires the identification of an equivalent acyclic graph before performing the analysis and is limited to unweighted graphs. Therefore, it can not exploit information about the reliability commonly associated to each edge in biological networks.

Analogously, in [7] the authors propose the identification of a Minimal Equivalent Graph (MEG), whose definition is the same as the transitive reduction proposed in [1]. The method consists of several steps, that are: *(i)* the identification of strongly connected components, *(ii)* the removal of cycles from each component, *(iii)* the identification of the minimal equivalent graph for each component and *(iv)* the reintroduction of the edges removed in the step *(i)*. Even if more sophisticated, this approach suffers from the same limitations described for [1].

Focusing on biological networks, in the literature, several approaches have been proposed to consider specific issues as well to exploit specific characteristics of such an application domain. In particular, it is possible to exploit the causality to infer the directionality of the interactions by exploiting time-series gene expression data [6]. In this case, the regulator gene (the cause), by definition, should act before the regulated gene (the effect). Therefore, a common strategy consists in computing the similarity between two genes u and v, by performing a progressive shifting forward in time of the time-series associated to the first gene u. If the similarity increases, then it is possible to conclude that u acts before v, therefore u regulates v. More sophisticated approaches exploit Granger causality [8] or hidden (i.e., unobserved, latent) common causes and common effects [10]. All these methods, however, are applicable only when analyzing time-series data, while they cannot be applied when each gene is associated with a vector representing its expression values in different (steady) conditions. In [11], the authors propose the (already mentioned) method ARACNE which exploits causality phenomena to identify and remove indirect relationships. Analogously to the approach presented in this paper, the method acts as a post-processing phase of the network reconstruction, aiming at removing interactions considered as the indirect effect of other interactions. This is performed by analyzing all the triplets of connected genes and by removing the weakest interaction, i.e. the edge with the lowest score. As already clarified, although based on the same principle of LOCANDA, this approach is limited to indirect interactions involving only three genes, thus it cannot identify interaction chains of arbitrary length.

In a more recent work [3], the authors propose a method for the identification of the transitive reduction of biological networks. This method is able to analyze both unweighted networks and possibly cyclic weighted networks. In this last case, however, following the approach adopted in [14], it requires to pre-process the network in order to make it acyclic. In detail, the method (i) identifies and shrinks the strongly connected components into single nodes, (ii) applies the reduction on the resulting acyclic graph, and (iii) re-expands the components. It is noteworthy that this procedure assumes that genes within each component are fully connected and do not perform any reduction within each component, since the results would strongly depend on the order of the analysis. Moreover, it assumes that the graph resulting from the step (i) is acyclic, i.e., there is no cycle among the components. However, the reduction phase is based on an idea which is similar to that adopted in LOCANDA, i.e. on the computation of an uncertainty score for paths connecting nodes, and on the removal of direct links having a higher uncertainty with respect to the identified indirect paths.

In summary, with respect to existing works in the literature, the method LOCANDA proposed in this paper identifies and removes links which are considered as the result of indirect regulation activities, exploiting common cause and common effect phenomena. LOCANDA has the following distinguishing characteristics: (i) unlike classical methods for the identification of a transitive reduction of networks [1,7], it is able to work on weighted networks, which is relevant when dealing with reconstructed biological networks where edges are associated to a score/reliability; (ii) unlike [11], it is able to work on directed networks, which (if available) becomes important to correctly consider causality phenomena; (iii) similar to [3] and unlike [11], it is able to catch indirect relationships of arbitrary length by comparing the reliability of direct links to that of identified indirect relationships; (iv) contrary to [1,3,7] it is able to directly work on cyclic networks, without any pre-processing steps and by guaranteeing the same result independently on the order of analysis.

3   The Method LOCANDA

In this section, we describe the method LOCANDA for the identification and removal of false positive links in a reconstructed gene network. The method is based on the concepts of common cause and common effect already introduced in Sect. 1. We remind that LOCANDA is not limited to the simple cases depicted in Fig. 2, but is able to detect and exploit indirect relationships of arbitrary length. In the following, before describing our approach, we introduce some useful notions and formally define the task we solve. Let:

- V be the set of genes, i.e., nodes in the reconstructed network.
- $E \subseteq (V \times V \times \mathbb{R})$ be the set of interactions in the reconstructed network, i.e., weighted edges in the form $\langle source\_node, destination\_node, edge\_weight \rangle$.
- P be a generic path between two nodes v_1 (*source node*) and v_k (*destination node*) in the network, defined as a sequence of nodes $[v_1, v_2, \ldots, v_k]$, such that $\forall_{i=1,2,\ldots,k-1}, \exists w_i : \langle v_i, v_{i+1}, w_i \rangle \in E$.
- $f(P)$ be a function that measures the reliability of the path P according to the edges involved in its sequence of nodes.

A path P between u and v is considered more reliable than the edge $\langle u, v, w \rangle$ if $f(P) > w$. According to such an assumption, the task we solve consists in the identification of a reduced set of edges $\widetilde{E} \subseteq E$, satisfying the following properties:

- the *reachability* of nodes is preserved. Formally, given two nodes u, v, there exists at least a path P connecting them through the edges in $\widetilde{E}$ if and only if there exists at least a path P connecting them through the edges in E.
- an edge $\langle u, v, w \rangle$ is removed, i.e., it does not belong to the reduced set $\widetilde{E}$, if there exists a path P from u to v which is more reliable than $\langle u, v, w \rangle$.

Note that, contrary to [1,7], we do not require the minimality of the number of edges in the reduced network, since we are not interested in *pure* transitive reduction, but in removing possible false positive edges identified during the reconstruction of the network. Indeed, in the case of reconstructed gene networks, the fact that the information conveyed by a link can be represented by a sequence of nodes (a path) is not a sufficient condition to consider the link as a false positive due to the presence of common cause or common effect phenomena. For this reason, we remove a link only if its reliability appears lower than the reliability of the identified path, measured by $f(\cdot)$. In this work, we take into account different possible measures to estimate the reliability of the path. In particular, being $w(v_i, v_j)$ the weight associated to the edge between v_i and v_j, we consider the following measures:

- *Minimum (Min)*, which corresponds to the lowest edge weight in the path, following the principle of the "weakest link in the chain".
  Formally, $f([v_1, v_2, \ldots, v_k]) = \min\limits_{i=1,2,\ldots,k-1} w(v_i, v_{i+1})$.
- *Product (Prod)*, i.e., the product of the edge weights involved in the path. This approach is motivated by the common strategy adopted for the combination of probabilities of (naively independent) events.
  Formally, $f([v_1, v_2, \ldots, v_k]) = \prod_{i=1}^{k-1} w(v_i, v_{i+1})$.

- *Average (Avg)*, i.e., the average of the edge weights involved in the path. Formally, $f([v_1, v_2, \ldots, v_k]) = \frac{1}{k} \cdot \sum_{i=1}^{k-1} w(v_i, v_{i+1})$.
- *Weighted Average (WAvg)*, i.e., the average of the edge weights involved in the path, linearly weighted on the basis of their closeness to the source node. This approach can be motivated by the assumption that the influence of the source node on the other nodes in the path fades linearly on the basis of their distance. Formally, $f([v_1, v_2, \ldots, v_k]) = \frac{1}{\sum_{i=1}^{k-1} \frac{1}{i}} \cdot \sum_{i=1}^{k-1} \left[\frac{1}{i} \cdot w(v_i, v_{i+1}) \right]$.

The pseudo-code of the algorithm LOCANDA is reported in Algorithm 1. We also report a running example in Fig. 3. Before describing LOCANDA, we remind that the method is able to analyze both undirected and directed networks, weighted according to a score representing the reliability about the existence of the interaction (computed by any method for network reconstruction). Here we assume to work with a weighted directed network (the most general case), since an unweighted network can be always mapped into a directed network by introducing an edge for each direction, with the same reliability score. The first step of LOCANDA consists in the removal of self-edges (line 2), since some methods for network reconstruction identify them erroneously. Although self-regulation activities are possible in biology, in reconstructed networks such links are due to errors in the computation of similarity/correlation measures on the vector associated to a single gene. In our example, the self-edge on the node E (Fig. 3(b)) is removed, leading to the network in Fig. 3(c). Then the algorithm analyzes each node (that we call *source node*) aiming at identifying all the reachable nodes and a path to reach them. Note that the visit of the network is performed according to a depth-first and best-first strategy, based on the reliability of the edges. The algorithm works in a greedy fashion, since an exhaustive exploration of all the possible paths would lead to an exponential time complexity. When there are several edges to follow, LOCANDA considers the path that locally (i.e., by observing only the neighborhood) appears the most reliable.

LOCANDA exploits three data structures: the set of visited nodes (*visited*), the current sequence of nodes (*path*) and a *stack*, according to which nodes are explored. Moreover, it exploits a structure (*RT*) similar to the routing table used by routing algorithms, which keeps information about the nodes reachable from the source node. In particular, for each reachable node (*destination*), it stores:

- the **next-hop**, i.e., the node adjacent to the source node that we need to follow to reach it, according to the current path.
- the **path score** associated to the current path, on which is based the choice of the optimal path to keep. LOCANDA will prefer a new path with respect to a previously identified path if this value is higher.
- the **path weight**, which represents the reliability associated to the current path according to $f(\cdot)$, that will be exploited to remove links.

Note that we prefer to consider two different criteria for the choice of the optimal path to consider (path score) and for the estimation of the reliability of the path (path weight), since they could not be generally based on the same assumptions. In particular, the path score will correspond to the sum of edges in

Algorithm 1. Pseudo-code of the method LOCANDA.

Data:
- V: the set of genes (nodes in the network)
- $E \in (V \times V \times \mathbb{R})$: the set of interactions (edges in the network), represented as $\langle source\_node, destination\_node, edge\_weight \rangle$
- $f(\cdot)$: the measure for the reliability of a path

Result:
- $\widetilde{E}$: the updated (reduced) set of interactions

```
1  begin
2  │   Ẽ ← E \ E.getSelfEdges();
3  │   foreach src ∈ V do
       │       /* Structures initialization. Records in the routing table RT are in the form
       │          ⟨dest_node, next_hop, path_score, path_weight⟩. Operations on RT are based on
       │          dest_node. Updates are considered as a new record if it does not exist.    */
4      │       visited ← {src}; path ← [src]; path_score ← 0; stack ← [ ]; RT ← [ ];
       │       /* Initialize the routing table for adjacents of src                          */
5      │       foreach ⟨src, adj, w⟩ ∈ Ẽ in ascending order w.r.t. w do
6      │       │   RT.update(adj, adj, w, f([adj]));
7      │       └   stack.push(adj);

8      │       while stack is not empty do
9      │       │   current_node ← stack.pop();
10     │       │   visited ← visited ∪ {current_node};
11     │       │   edge_weight ← Ẽ.getEdgeWeight(path.getLast(), current_node);
12     │       │   old_path_score ← RT.getPathScore(current_node);
13     │       │   new_path_score ← path_score + edge_weight;
       │       │   /* Update the RT if the route does not exist or if the new path has a
       │       │      higher score than the previous path                                    */
14     │       │   if old_path_score = null or old_path_score < new_path_score then
15     │       │   │   next_hop ← path.getFirst();
16     │       │   └   RT.update(current_node, next_hop, new_path_score, f(path));
       │       │   /* Push non-visited adjacent nodes of the current node into the stack,
       │       │      ordered by weight                                                      */
17     │       │   foreach ⟨current_node, adj, w⟩ ∈ Ẽ in ascending order w.r.t. w do
18     │       │   │   if adj ∉ visited then
19     │       │   └   └   stack.push(adj);
       │       │   /* Update the current path                                                */
20     │       │   if some nodes were added to stack then
21     │       │   │   path.add(current_node);
22     │       │   │   path_score ← new_path_score;
23     │       │   else if stack is not empty then
24     │       │   │   next ← stack.top();
25     │       │   │   while ⟨path.getLast(), next⟩ ∉ Ẽ do
26     │       │   │   │   last ← path.getLast();
27     │       │   │   │   path.removeLast();
28     │       │   └   └   path_score ← path_score − Ẽ.getEdgeWeight(path.getLast(), last);

       │       │   /* Remove a direct link if it is not used to reach other nodes and its less
       │       │      reliable than the indirect link (path)                                 */
29     │       all_next_hops ← RT.getAllNextHops();
30     │       foreach ⟨src, adj, w⟩ ∈ Ẽ do
31     │       │   if adj ∉ all_next_hops and w < RT.getPathWeight(adj) then
32     │       └   └   Ẽ ← Ẽ \ {⟨src, adj, w⟩};

33 │   return Ẽ;
```

the path, since, combined with the adopted strategy for the choice of the edge to follow (i.e., the highest), leads to the identification of long and reliable paths. On the contrary, the estimation of the path weight will be based on several different measures, that we will describe later.

The analysis of a source node is performed as follows. First, the data structures are initialized (line 4), by considering the source node as already expanded and by adding it to the current path. Second, we analyze all its adjacent nodes, i.e., we push them into the stack, ordered in ascending order according to the edge weight, and initialize the routing table by setting themselves as their next-hop (lines 5–7). Then, the main part of the algorithm (lines 8–28) iterates until the stack still has some nodes to analyze. In particular, LOCANDA pops a node (*current_node*) from the stack (see Fig. 3(d)), marks it as visited (lines 9–10), and computes the score associated with the current path to reach the *current_node* from the source (lines 11–13). If the current path is the first identified path to reach *current_node* or it has a higher score with respect to the previous path in the routing table, LOCANDA updates the routing table (lines 14–16).

Then the algorithm expands the current node, by pushing its adjacent nodes into the stack in ascending order with respect to the edge weight, if not already visited (lines 17–19). If at least a node was pushed (see Figs. 3(e),(f),(g)), the current path is updated to follow *current_node* (lines 20–22), otherwise (see Fig. 3(h)) the algorithm steps back, until it can find an existing edge between the last node in the path and the next node in the stack (lines 23–28). In both cases, the path and its score are updated incrementally (lines 22 and 26–28).

When there is no more nodes in the stack, LOCANDA removes all the direct links such that the properties described before are satisfied. In particular, it removes a link between the source node u and an its adjacent v if v is never used as next-hop to reach other nodes and if the path identified to reach v from u appears more reliable then the direct link (lines 29–32). The algorithm then proceeds with the next source node. It is noteworthy that the removed links will never be considered again from the algorithm. This can be done without any risk to lose relevant paths, since those edges would never be considered in any case, even analyzing the nodes of the networks in a different order. As an example, the removed edge between A and B in Fig. 3(i) would not be followed in any case during the analysis of the node G as source node. Therefore, the order of analysis of source nodes does not affect the resulting reduced network.

The immediate removal of such links also improves the algorithm time complexity. Indeed, although in the pessimistic case LOCANDA has a time complexity of $O(|V| \cdot |E|)^1$, this choice decreases the number of edges at each iteration.

4   Experiments

We performed our experiments on the datasets considered in [4]. These datasets consist of steady-state expression data (10 conditions), generated by the tool

---

[1] For space constraint, we do not prove formally the time complexity of the algorithm.

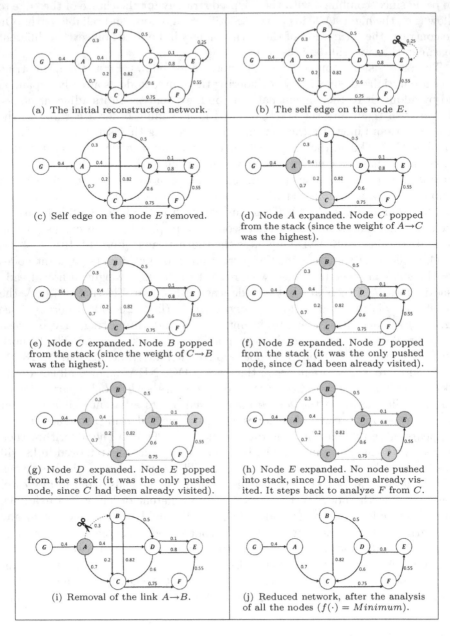

(a) The initial reconstructed network.

(b) The self edge on the node E.

(c) Self edge on the node E removed.

(d) Node A expanded. Node C popped from the stack (since the weight of $A{\rightarrow}C$ was the highest).

(e) Node C expanded. Node B popped from the stack (since the weight of $C{\rightarrow}B$ was the highest).

(f) Node B expanded. Node D popped from the stack (it was the only pushed node, since C had been already visited).

(g) Node D expanded. Node E popped from the stack (it was the only pushed node, since C had been already visited).

(h) Node E expanded. No node pushed into stack, since D had been already visited. It steps back to analyze F from C.

(i) Removal of the link $A{\rightarrow}B$.

(j) Reduced network, after the analysis of all the nodes ($f(\cdot) = Minimum$).

Fig. 3. An example of execution of LOCANDA and the analysis of the *source node A*. Grey nodes: already expanded; blue node: the current node to analyze, extracted from the stack; black edges: not seen yet; grey edges: already seen, but still not followed; blue edges: belonging to the current path; red edges: will not be followed, since would bring to already expanded nodes; black-dashed edges: to be removed. (Color figure online)

SynTReN [16] on the basis of the well-defined regulatory networks of the organisms *E. coli* and *S. cerevisiae* (henceforth Yeast) [6]. SynTReN selects connected sub-networks of the input networks and generates gene expression data which best describe the network structure. We consider sub-networks of 100 and 200 genes, characterized by 121 and 303 links, with an average node degree of 2.42 and 3.03, respectively. In order to evaluate the robustness to noise, coherently to [4], we consider three versions of each dataset, with different levels of (additive, lognormally-distributed) noise, i.e., 0.0 (without noise), 0.1 and 0.5, introduced by SynTReN. Gene regulatory networks were reconstructed by adopting the system GENERE [4], which, according to the experiments, obtains state-of-the-art results in terms of Area Under the ROC Curve (AUC). In particular, we selected the parameter configuration of GENERE which led to the best results.

We considered as a competitor the system ARACNE [11], that we already described in Sect. 2. Moreover, we considered, as a baseline, the original network reconstructed by GENERE. For all the systems, we performed the experiments by imposing a lower threshold on the weight of the edges in $\{0.0, 0.1, \ldots 1.0\}$. For LOCANDA, we performed the experiments with all the measures for the estimation of the reliability of the path proposed in Sect. 3, that are: minimum (Min), product (Prod), average (Avg) and weighted average (WAvg).

The evaluation measure that we consider is based on the Area Under the ROC Curve. It is noteworthy that the classical AUC evaluation focuses on known examples in the gold standard, disregarding all the predicted links for which the existence is unknown. This means that the obtained AUC value can be significantly distorted, since focused only on the small subset of known links in the reconstructed network. Since, in real scenarios, the biologists have to analyze the whole set of predicted links, possibly ranked in descending order with respect to their score, we define the weighted AUC as follows:

$$WAUC(V, \widetilde{E}) = \left(1 - \frac{sumOfWeights(\widetilde{E})}{|V| \cdot (|V| - 1)}\right) \cdot AUC(\widetilde{E}) \qquad (1)$$

where $sumOfWeights(\widetilde{E}) = \sum_{\langle u,v,w \rangle \in \widetilde{E}}(w)$ is the sum of edge weights in the reduced reconstructed network, $AUC(\widetilde{E})$ is the classical Area Under the ROC Curve and $|V| \cdot (|V| - 1)$ is the number of possible links in the network. It is noteworthy that this measure penalizes the original AUC score proportionally to the number (and the weight) of links in the reduced network. This is motivated by the fact that a large set of predicted links, all with a high score (i.e., without a clear indication about their rank) would require an extensive manual analysis performed by biologists. On the other hand, reconstructed networks with many links will not be penalized significantly if a large set of links has a very low score, since they would be probably disregarded by biologists during their analysis. Note that, due to the weighting defined in Eq. 1, WAUC values near to 0.5 do not correspond to a random prediction as in the standard AUC evaluation.

The obtained results are plotted in the box plots depicted in Fig. 4. Box plots are drawn by considering the different values for the input threshold on

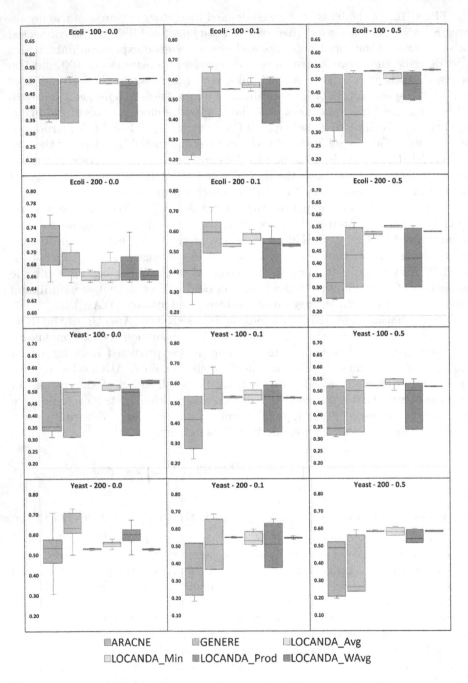

Fig. 4. Box plots depicting the results. On the X-axis there are the different methods; on the Y-axis there is the WAUC obtained by varying the threshold on the edge weight.

the edge weight. This allows us to evaluate the stability of the results with respect to such a parameter. First, we can observe that ARACNE, GENERE and LOCANDA_Prod obtain unstable results with respect to the input threshold, whereas the other variants of LOCANDA obtain very stable results. Moreover, we can observe that the networks reconstructed by GENERE appear, in general, accurate and often lead to the highest WAUC value (see the datasets Ecoli-100-0.1, Ecoli-200-0.1, Ecoli-200-0.5, Yeast-100-0.1, Yeast-100-0.5, Yeast-200-0.0 and Yeast-200-0.1). However, such a result can be obtained with a specific value of the input threshold and a wrong decision can lead to very poor results. On the contrary, a non-optimal choice of the value for the input threshold does not affect significantly the results obtained by LOCANDA_Min, LOCANDA_Avg and LOCANDA_WAvg, that lead to stable and high WAUC values in almost all the cases. ARACNE generally obtains lower WAUC values, which also appear highly dependent on the value of the input threshold. An exception can be observed in the dataset Ecoli-200-0.0 in which ARACNE obtains the best result.

Analyzing the influence on the results caused by the presence of noise in the data, we can observe that, without noise or with a low amount of noise, GENERE and ARACNE obtain acceptable (although unstable) results. However, when the amount of noise increases, their average WAUC values decrease significantly. On the contrary, LOCANDA, especially with the variants based on Min, Avg and WAvg, generally shows good and stable results, even in the case of the datasets with the highest noise. This proves that the proposed method is actually very robust to the possible presence of noise in the data.

Finally, we performed the Friedman test with the Nemenyi post-hoc test, with $\alpha = 0.05$, in order to evaluate whether the obtained results appear significant from a statistical viewpoint. Following [5], we plot a graph which summarizes the results in Fig. 5. Observing the graph, we can conclude that, although LOCANDA_Min generally leads to the best results, the difference among the three variants based on Min, Avg and WAvg is not statistically significant. However, the difference between the results obtained by these three variants and by the other approaches, including ARACNE and GENERE, is statistically significant.

The non-optimal results obtained by the variant based on product can be motivated by the fact that it is based on assumptions that are often violated in biological networks (i.e., the independence between the events). On the contrary, the very good results obtained by the variants Min and WAvg can be motivated by the fact that their assumptions correctly reflect the real interactions among genes. At this respect, we can conclude that: (i) the variants based on Min and WAvg are the most appropriate for the reconstruction of gene networks, and (ii) LOCANDA can be easily adapted to analyze networks representing data about other application domains, by identifying a proper function $f(\cdot)$ able to catch specific assumptions of the domain at hand.

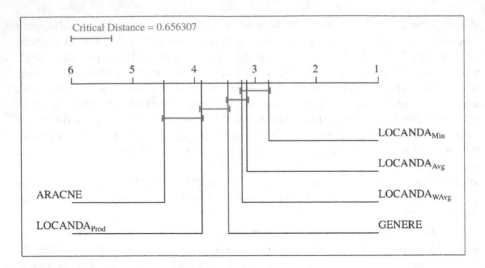

Fig. 5. Results of the Friedman test and Nemenyi post-hoc test with $\alpha = 0.05$.

5 Conclusions and Future Work

In this work, we proposed the method LOCANDA for the analysis of reconstructed biological networks, which identifies and exploits causality phenomena to remove links which can be considered the result of indirect regulation activities. Contrary to existing methods for the identification of a transitive reduction of a network or for the identification of redundancies in reconstructed biological networks, LOCANDA simultaneously offers all the following characteristics: (i) it is able to analyze directed weighted networks, fully exploiting the weights on the edges which represent their reliability; (ii) it does not require any preprocessing step on the network in order to handle the possible presence of cycles; (iii) it is able to identify indirect relationships of arbitrary length and to exploit them to remove direct links considered as false positives. The estimation of the reliability of a path is guided by a function, which can be tuned according to specific underlying phenomena and assumptions with respect to the application domain at hand. Focusing on biological networks, the obtained results show that LOCANDA, especially in its variant based on minimum, is able to obtain better and more stable results with respect to the considered competitors, even with highly noisy data. Moreover, according to the Friedman test and Nemenyi post-hoc test, such difference appears statistically significant.

As future works, we plan to compare LOCANDA with additional competitor systems, also in the analysis of a larger network about the Homo Sapiens. We will also perform a qualitative analysis of the results, guided by experts in biology. Moreover, we will evaluate the effectiveness of LOCANDA in the analysis of networks representing data about other domains, focusing on the influence of the function $f(\cdot)$ when different assumptions on the network are verified.

Acknowledgements. We would like to acknowledge the support of the European Commission through the projects MAESTRA - Learning from Massive, Incompletely annotated, and Structured Data (Grant Number ICT-2013-612944) and TOREADOR - Trustworthy Model-aware Analytics Data Platform (Grant Number H2020-688797).

References

1. Aho, A.V., Garey, M.R., Ullman, J.D.: The transitive reduction of a directed graph. SIAM J. Comput. **1**(2), 131–137 (1972)
2. Atias, N., Sharan, R.: Comparative analysis of protein networks: hard problems, practical solutions. Commun. ACM **55**(5), 88–97 (2012)
3. Bošnački, D., Odenbrett, M.R., Wijs, A., Ligtenberg, W., Hilbers, P.: Efficient reconstruction of biological networks via transitive reduction on general purpose graphics processors. BMC Bioinform. **13**(1), 281 (2012)
4. Ceci, M., Pio, G., Kuzmanovski, V., Džeroski, S.: Semi-supervised multi-view learning for gene network reconstruction. PLOS ONE **10**(12), 1–27 (2015)
5. Demšar, J.: Statistical comparisons of classifiers over multiple data sets. J. Mach. Learn. Res. **7**, 1–30 (2006)
6. Hempel, S., Koseska, A., Nikoloski, Z., Kurths, J.: Unraveling gene regulatory networks from time-resolved gene expression data - a measures comparison study. BMC Bioinform. **12**(1), 292 (2011)
7. Hsu, H.T.: An algorithm for finding a minimal equivalent graph of a digraph. J. ACM **22**(1), 11–16 (1975)
8. Itani, S., Ohannessian, M., Sachs, K., Nolan, G.P., Dahleh, M.A.: Structure learning in causal cyclic networks. In: Proceedings of the International Conference on Causality: Objectives and Assessment, COA 2008, vol. 6, pp. 165–176. JMLR.org (2008)
9. Korb, K.B., Nicholson, A.E.: Bayesian Artificial Intelligence, 2nd edn. CRC Press Inc., Boca Raton (2010)
10. Lo, L., Wong, M., Lee, K., Leung, K.: Time delayed causal gene regulatory network inference with hidden common causes. PLOS ONE **10**(9), 1–47 (2015)
11. Margolin, A., Nemenman, I., Basso, K., Wiggins, C., Stolovitzky, G., Favera, R., Califano, A.: ARACNE: an algorithm for the reconstruction of gene regulatory networks in a mammalian cellular context. BMC Bioinform. **7**(Suppl 1), S7 (2006)
12. Pearl, J.: Causality: Models, Reasoning, and Inference. Cambridge University Press, New York (2000)
13. Penfold, C.A., Wild, D.L.: How to infer gene networks from expression profiles, revisited. Interface Focus **1**(6), 857–870 (2011)
14. Pinna, A., Soranzo, N., de la Fuente, A.: From knockouts to networks: establishing direct cause-effect relationships through graph analysis. PLOS ONE **10**(5), e12912 (2010)
15. Pio, G., Ceci, M., Malerba, D., D'Elia, D.: ComiRNet: a web-based system for the analysis of miRNA-gene regulatory networks. BMC Bioinform. **16**(9), S7 (2015)
16. Van den Bulcke, T., Van Leemput, K., Naudts, B., van Remortel, P., Ma, H., Verschoren, A., De Moor, B., Marchal, K.: SynTReN: a generator of synthetic gene expression data for design and analysis of structure learning algorithms. BMC Bioinform. **7**, 43 (2006)
17. Zitnik, M., Zupan, B.: Data imputation in epistatic MAPs by network-guided matrix completion. J. Computat. Biol. **22**(6), 595–608 (2015)

Discovery of Salivary Gland Tumors' Biomarkers via Co-Regularized Sparse-Group Lasso

Sultan Imangaliyev[1,6](✉), Johannes H. Matse[2,3], Jan G.M. Bolscher[3],
Ruud H. Brakenhoff[1], David T.W. Wong[4], Elisabeth Bloemena[2],
Enno C.I. Veerman[3], and Evgeni Levin[5,6]

[1] Department of Otolaryngology/Head and Neck Surgery,
VU University Medical Center Amsterdam/Cancer Center Amsterdam,
Amsterdam, The Netherlands
s.imangaliyev@vumc.nl
[2] Department of Pathology, VU University Medical Center Amsterdam,
Amsterdam, The Netherlands
[3] Department of Oral Biochemistry, Academic Centre for Dentistry Amsterdam,
Amsterdam, The Netherlands
[4] Center for Oral/Head and Neck Oncology Research,
University of California Los Angeles, Los Angeles, USA
[5] Department of Experimental Vascular Medicine, Academic Medical Center,
Amsterdam, The Netherlands
[6] Horaizon BV, Rotterdam, The Netherlands

Abstract. In this study, we discovered a panel of discriminative microR-NAs in salivary gland tumors by application of statistical machine learning methods. We modelled multi-component interactions of salivary microRNAs to detect group-based associations among the features, enabling the distinction of malignant from benign tumors with a high predictive performance utilizing only seven microRNAs. Several of the identified microRNAs are separately known to be involved in cell cycle regulation. Integrated biological interpretation of identified microRNAs can provide potential new insights into the biology of salivary gland tumors and supports the development of non-invasive diagnostic tests to discriminate salivary gland tumor subtypes.

Keywords: Cancer biomarkers · Salivary gland tumors · Feature selection · Stability selection · Co-regularization · microRNA

1 Background and Motivation

Statistical machine learning techniques have recently received significant attention due to their outstanding performance in various data analysis tasks. One of the most common tasks in machine learning is the feature selection, frequently referred to as biomarker selection when applied to biomedical datasets [13]. In this study, we used a recently constructed database [10] to analyze the microRNA (miRNA) expression profiles of head and neck cancer samples via a statistical

© Springer International Publishing AG 2017
A. Yamamoto et al. (Eds.): DS 2017, LNAI 10558, pp. 298–305, 2017.
DOI: 10.1007/978-3-319-67786-6_21

machine learning algorithm which can detect group-based associations. We modelled multi-component interactions of salivary miRNAs enabling the distinction of malignant from benign salivary gland tumors with a high predictive performance. Moreover, stability selection [11] and randomization test [4] procedures were conducted on the collected data to select the most robust miRNA biomarkers.

Head and neck cancer is the sixth most common type of cancer, and patients' overall five-year survival rate is only around 50% [7]. As one of the subtypes of head and neck cancer, salivary gland tumors are a rare but very heterogeneous set of tumors comprised of 13 benign and 24 malignant subtypes. Early diagnosis is a key factor in contributing to a positive outcome of salivary gland tumor treatment. Clinical examination, with or without fine needle aspiration cytology, preoperative CT-scan, and MRI are the most commonly used methods for diagnosing these rare tumor types [15]. More recently, a non-invasive and inexpensive salivary diagnostic tools are being developed to assist in the diagnosis of these and other types of oral tumors. In particular, salivary miRNAs have attracted much attention for non-invasive diagnostic applications since altered levels of salivary miRNA expression have been implicated in the etiology of cancer [10,23].

2   Previous Work

In our recent study [10], we profiled miRNA expression in saliva from oral cancer patients and produced a database that can be used for studying differences in whole saliva from patients with a malignant or a benign parotid gland tumor. One of the main limitations of the previous study, however, was the use of a univariate statistical approach to analyze the generated high-dimensional data. The tacit assumption of such an approach is that the underlying biological process is dominated by a few miRNAs acting as isolated entities. While this approach is attractive due to its relative simplicity, biological processes such as tumor development are in reality much more complex, and are driven by multi-component interactions among miRNA, mRNA, DNA, peptides, signaling molecules, and drug activities. Furthermore, it is known that some biological processes are controlled by activating a hierarchical cascade of interdependent miRNA regulators [2,17]. Multi-component group-based interactions, such as hierarchal miRNA systems, are difficult to detect using standard statistical tests.

To exploit existing hierarchical relationship among features, one should use a model which incorporates this domain-specific knowledge explicitly in its objective function. Many existing hierarchical feature selection models [9,12] are not directly applicable to genomics data, because they exploit ontology-based semantic relationships and do not use sparsity-inducing norms. To a lesser extent, a similar problem also applies to sparsity inducing methods, such as group lasso [22] and sparse-group lasso [16], because they implement group information but ignore its hierarchical properties. We suggest using a co-regularized sparse-group lasso algorithm, which does not have above mentioned limitations and it works under the assumption that features are divided into hierarchically related groups. Such a feature 'grouping effect' is useful when interrogating the miRNA dataset in order to detect group-based miRNA interactions. Co-regularized sparse-group lasso was evaluated in our recent study [14] on a synthetic dataset, and its performance was examined under various settings. The results of the study supported conclusion that its application is beneficial in comparison to the standard lasso [19], elastic net [24], group lasso [22], and sparse-group lasso [16] techniques.

3 Methodology

3.1 Co-Regularized Sparse-Group Lasso

Let $S = (\mathbf{X}, \mathbf{y})$ denote a dataset where $\mathbf{y}$ is the $n \times 1$ label vector and $\mathbf{X}$ is the $n \times p$ input dataset matrix, where n is the number of examples and p the number of features. If the features are divided into M groups, we use the following notations: $\mathbf{X} = (\mathbf{X}^1|...|\mathbf{X}^M)$ is an input dataset matrix where each $\mathbf{X}^{(v)}$ is a $n \times p^{(v)}$ sub-matrix, where $p^{(v)}$ is the number of features in group v. $\boldsymbol{\beta} = (\beta_1, ..., \beta_p) = (\boldsymbol{\beta}^{(1)}|...|\boldsymbol{\beta}^{(M)})$ is a vector of weights, where $\boldsymbol{\beta}^{(v)}$ $(p^{(v)} \times 1)$ is the vector of weights of the corresponding features of group v. Using these notations, co-regularized sparse-group lasso model [14] computes weights that minimize the following objective function:

$$L_{crSGL}(\alpha, \lambda, \boldsymbol{\beta}) = \frac{1}{2n} \left\| \mathbf{y} - \sum_{v=1}^{M} \mathbf{X}^{(v)} \boldsymbol{\beta}^{(v)} \right\|_2^2 \tag{1}$$

$$+ \alpha\lambda \|\boldsymbol{\beta}\|_1 + (1 - \alpha)\lambda \sum_{v=1}^{M} \sqrt{p^{(v)}} \left\| \boldsymbol{\beta}^{(v)} \right\|_2$$

$$+ \frac{1}{2n} \sum_{l,v=1}^{M} \gamma_{lv} (\mathbf{X}^{(l)} \boldsymbol{\beta}^{(l)} - \mathbf{X}^{(v)} \boldsymbol{\beta}^{(v)})^2,$$

where γ_{lv} is the co-regularization coefficient between groups l and v. Further details of the co-regularized sparse-group lasso, such as weight update procedure, soft-thresholding operator and a step-wise sketch of the algorithm are provided in our previous work [14].

3.2 Stability Selection and Randomization Test

The regularized feature selection model is parameterized by the hyper-parameter $\alpha \in [0, 1]$, which modulates the emphasis of either the L_1- or L_2-norm regularizations. This, and two other hyper-parameters λ, γ, influence the number of features selected [14,16]. To minimize this influence on the model performance, a stability selection [11] procedure was introduced. In this study, the stability selection procedure was performed by running the feature selection model multiple times on randomly subsampled data, and choosing only those features that are selected most frequently across all sample partitions. Furthermore, while solving a penalized classification problem, we assume that due to biological reasons there is an association between the miRNA expression value matrix and the classification labels vector. This assumption can be tested by conducting modeling on the dataset with reshuffled class labels, while keeping the corresponding miRNA expression value indices fixed. Such a procedure is often used to test the statistical significance of results, and is known as a randomization test [4].

4 Experiments

The miRNA dataset obtained from the discovery phase [10] was used as an input for the machine learning algorithm. Specifically, the miRNA dataset contains data from ten patients diagnosed with malignant parotid gland tumors and ten patients diagnosed with benign parotid gland tumors. miRNA expression profiles were determined by real-time PCR encompassing 750 miRNAs. To compensate for variance in feature amplitudes, all input features were zero-mean unit-variance scaled.

Under a standard miRNA nomenclature system [5], general pattern of a miRNA name is a 'three letter species abbreviation–miR–number–qualifier'. Using miRNA names, the dataset can be split to a group of 17 miRNAs which originate from mice species, and the next group consists of 733 miRNAs which originate from human species. Within those human-originated miRNAs group, 15 belong to let-7 family discovered originally from *C. elegans*. Within the rest of 718 human non-let-7 miRNAs, further group division is also possible depending on the miRNA qualifier (*e.g.* 3p or 5p arm of a hairpin loop).

The features were selected based on their stability after 50 runs of feature selection model on random data shuffles using the stability selection procedure. Model hyper-parameters were estimated using an exhaustive grid search within a 10-fold stratified shuffled cross-validation procedure [6] on 80% of the training dataset, and the model's generalization error was assessed on the remaining 20% test dataset. The performance measure used for a binary classification task is a Receiver Operating Characteristics Area Under Curve (ROC AUC).

5 Results and Discussion

5.1 Model Performance Evaluation

Stability selection procedure conducted on the miRNA expression profiles allowed the identification of seven of the most stable discriminative miRNA biomarkers. These biomarkers' associated weights as well as their stability coefficients are presented in the right panel of Fig. 1. Out of the 750 analyzed miR-NAs, *hsa-miR-449b* had the highest stability coefficient (0.88). Other miRNAs with relatively high stability coefficients were *hsa-miR-374*, *hsa-miR-411*, *hsa-miR-599*, *hsa-miR-1285*, *hsa-miR-324-5p*, and *hsa-miR-449a*. ROC curves were constructed to determine the diagnostic/predictive values of the combined seven selected miRNAs, resulting in a ROC AUC score of 0.92 (Fig. 1, Left). We also illustrate the differences in expression values of the top seven biomarkers for the malignant and benign tumor samples in the left panel of Fig. 2. The p-value of the randomization test for 10,000 shuffles is 0.0097 < 0.01, thus assuming 1% significance level, we can conclude that the model prediction performance is statistically significant.

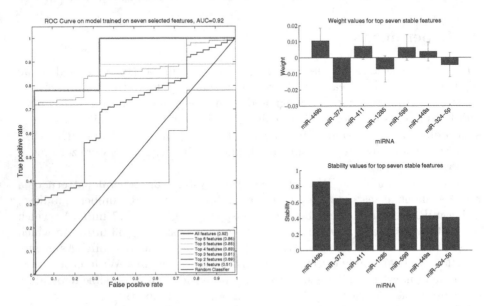

Fig. 1. Overview of predictive power of the selected seven features. *Left:* ROC curves for which corresponding ROC AUC scores were computed for models with sequentially added features. The best prediction model selected seven combined miRNAs with a ROC AUC value of 0.92. *Right-top:* Weights of the selected top seven biomarkers. *Right-bottom:* Stability coefficients of the selected top seven biomarkers.

5.2   Biological Interpretation

In this study, we have discovered a panel of seven miRNA biomarkers which enabled the differentiation between benign and malignant parotid salivary gland tumors with a high degree of accuracy. Four of the presently identified miRNAs (*hsa-miR-374*, *hsa-miR-1285*, *hsa-miR-449a* and *hsa-miR-449b*) have been implicated in the control of the cell cycle either as 'gas pedals' or as 'brake pedals', and their actions have already been validated *in vitro* in earlier studies [1,3,8,18,20,21]. These four validated 'gas pedal' and 'break pedal' miR-NAs coordinately accelerate the cell division cycle as presented in Fig. 2.

According to previous studies, *hsa-miR-374* and *hsa-miR-1285* promote cell cycle progression; *hsa-miR-374* suppresses the expression of *GADD45A*, an inducer of cell cycle arrest [1,20], while both *hsa-miR-374* and *hsa-miR-1285*

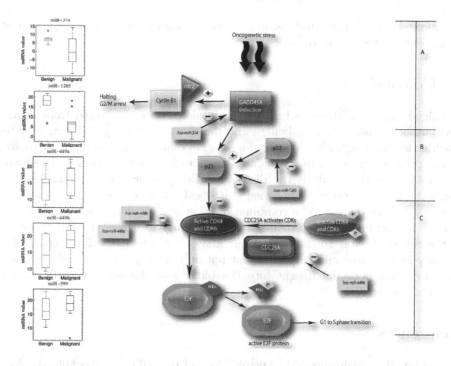

Fig. 2. A higher expression of *hsa-miR-374* and *hsa-miR-1285*, and a lower expression of *hsa-miR-449a* and *hsa-miR-449b* in saliva from patients with a malignant tumor, may result in an uncontrolled acceleration of the cell cycle in malignant tissues. *Section A: hsa-miR-374* inhibits the translation of *GADD45A*, encoding a DNA-damage-inducible protein involved in growth arrest [1, 20]. *Section B: hsa-miR-1285* inhibits the expression of *p21* directly by binding to its mRNA, but also indirectly by inhibiting the expression of *p21* activator *p53* [3, 18]. *Section 1: hsa-miR-449a* and *hsa-miR-449b* both can directly inhibit the expression of *CDK4* and *CDK6* [8, 20, 21]. '−' sign represents inhibition; '+' sign represents activation. *Left:* The zero-mean unit-variance scaled expression values of several stable biomarker miRNAs in saliva from patients with malignant or benign salivary gland tumors.

suppress the expression of *CDKN1A*, which encodes cell cycle regulator *p21* [3, 18]. This results in the suppression of the *p21*-mediated inhibition of *CDK6* which, in conjunction with *CDK4*, acts as a switch to direct the cell towards S phase. It can be envisaged that increased levels of these miRNAs, *e.g.* in malignant tumors, will enhance the rate of cell division.

On the other hand, *hsa-miR-449a* and *hsa-miR-449b* act as 'brake-pedals' for cell-division by directly and indirectly suppressing *CDK6* and *CDK4* [8, 20, 21]. We hypothesize that the increased expression levels of *hsa-miR-374* and *hsa-miR-1285* (the 'gas-pedals') in combination with the decreased expression levels of *hsa-miR-449a* and *hsa-miR-449b* (the 'brake-pedals') will contribute to an uncontrolled acceleration of the cell cycle in malignant tissues. To sum up, the set of statistical machine learning algorithms selected a group of miRNAs that coordinately target the cell division cycle and possibly result in development of malignant salivary gland tumors.

6   Conclusions

In this study, we analyzed a miRNA expression dataset collected from a limited number of patients with benign and malignant salivary gland tumors. The machine learning algorithm identified a panel of biomarkers of malignant tumors whose targets were separately validated *in vitro* in previous studies. The main contribution of this study is the discovery of diagnostic cancer biomarkers by application of our recently published statistical machine learning method to a cancer genomics dataset. The combination of sparse regularization and the stability selection procedures provided an opportunity to utilize the full potential of the collected data. To the best of our knowledge, this is the first report of a salivary gland tumor miRNA panel that not only distinguishes tumor subtypes, but also gives potential insight into the biology of the salivary gland tumor development process.

References

1. Bueno, M.J., Malumbres, M.: MicroRNAs and the cell cycle. Biochim. Biophys. Acta (BBA)-Mol. Basis Dis. **1812**(5), 592–601 (2011)
2. Dueck, A., Eichner, A., Sixt, M., Meister, G.: A miR-155-dependent microRNA hierarchy in dendritic cell maturation and macrophage activation. FEBS Lett. **588**(4), 632–640 (2014)
3. Fan, W., Richter, G., Cereseto, A., Beadling, C., Smith, K.A.: Cytokine response gene 6 induces p. 21 and regulates both cell growth and arrest. Oncogene **18**(47), 6573–6582 (1999)
4. Fisher, R.A.: The Design of Experiments. Oliver and Boyd, Edinburgh (1935)
5. Griffiths-Jones, S., Grocock, R.J., Van Dongen, S., Bateman, A., Enright, A.J.: miRBase: microRNA sequences, targets and gene nomenclature. Nucleic Acids Res. **34**(Suppl. 1), D140–D144 (2006)
6. Hastie, T., Tibshirani, R., Friedman, J.: The Elements of Statistical Learning. Springer, New York (2009)
7. Leemans, C.R., Braakhuis, B.J., Brakenhoff, R.H.: The molecular biology of head and neck cancer. Nat. Rev. Cancer **11**(1), 9–22 (2011)
8. Lize, M., Pilarski, S., Dobbelstein, M.: E2F1-inducible microrna 449a/b suppresses cell proliferation and promotes apoptosis. Cell Death Differ. **17**(3), 452–458 (2010)
9. Lu, S., Ye, Y., Tsui, R., Su, H., Rexit, R., Wesaratchakit, S., Liu, X., Hwa, R.: Domain ontology-based feature reduction for high dimensional drug data and its application to 30-day heart failure readmission prediction. In: 2013 9th International Conference on Collaborative Computing: Networking, Applications and Worksharing (Collaboratecom), pp. 478–484. IEEE (2013)
10. Matse, J.H., Yoshizawa, J., Wang, X., Elashoff, D., Bolscher, J.G., Veerman, E.C., Bloemena, E., Wong, D.T.: Discovery and prevalidation of salivary extracellular microRNA biomarkers panel for the noninvasive detection of benign and malignant parotid gland tumors. Clin. Cancer Res. **19**(11), 3032–3038 (2013)

11. Meinshausen, N., Bühlmann, P.: Stability selection. J. Roy. Stat. Soc. Ser. B (Stat. Methodol.) **72**(4), 417–473 (2010)

12. Ristoski, P., Paulheim, H.: Feature selection in hierarchical feature spaces. In: Džeroski, S., Panov, P., Kocev, D., Todorovski, L. (eds.) DS 2014. LNCS, vol. 8777, pp. 288–300. Springer, Cham (2014). doi:10.1007/978-3-319-11812-3_25

13. Saeys, Y., Inza, I., Larrañaga, P.: A review of feature selection techniques in bioinformatics. Bioinformatics **23**(19), 2507–2517 (2007)

14. Santos, P.L.A., Imangaliyev, S., Schutte, K., Levin, E.: Feature selection via co-regularized sparse-group lasso. In: Pardalos, P.M., Conca, P., Giuffrida, G., Nicosia, G. (eds.) MOD 2016. LNCS, vol. 10122, pp. 118–131. Springer, Cham (2016). doi:10.1007/978-3-319-51469-7_10

15. Seifert, G., Brocheriou, C., Cardesa, A., Eveson, J.: WHO international histological classification of tumours tentative histological classification of salivary gland tumours. Pathol.-Res. Pract. **186**(5), 555–581 (1990)

16. Simon, N., Friedman, J., Hastie, T., Tibshirani, R.: A sparse-group lasso. J. Comput. Graph. Stat. **22**(2), 231–245 (2013)

17. Tang, R., Li, L., Zhu, D., Hou, D., Cao, T., Gu, H., Zhang, J., Chen, J., Zhang, C.Y., Zen, K.: Mouse miRNA-709 directly regulates miRNA-15a/16-1 biogenesis at the posttranscriptional level in the nucleus: evidence for a microRNA hierarchy system. Cell Res. **22**(3), 504–515 (2012)

18. Tian, S., Huang, S., Wu, S., Guo, W., Li, J., He, X.: MicroRNA-1285 inhibits the expression of p53 by directly targeting its 3 untranslated region. Biochem. Biophys. Res. Commun. **396**(2), 435–439 (2010)

19. Tibshirani, R.: Regression shrinkage and selection via the lasso. J. Roy. Stat. Soc. Ser. B (Methodol.) **58**(1), 267–288 (1996)

20. Wang, X.W., Zhan, Q., Coursen, J.D., Khan, M.A., Kontny, H.U., Yu, L., Hollander, M.C., O'Connor, P.M., Fornace, A.J., Harris, C.C.: GADD45 induction of a G2/M cell cycle checkpoint. Proc. Natl. Acad. Sci. **96**(7), 3706–3711 (1999)

21. Yang, X., Feng, M., Jiang, X., Wu, Z., Li, Z., Aau, M., Yu, Q.: miR-449a and miR-449b are direct transcriptional targets of E2F1 and negatively regulate pRb-E2F1 activity through a feedback loop by targeting CDK6 and CDC25A. Genes Dev. **23**(20), 2388–2393 (2009)

22. Yuan, M., Lin, Y.: Model selection and estimation in regression with grouped variables. J. Roy. Stat. Soc. Ser. B (Stat. Methodol.) **68**(1), 49–67 (2006)

23. Zhang, X., Cairns, M., Rose, B., O'Brien, C., Shannon, K., Clark, J., Gamble, J., Tran, N.: Alterations in miRNA processing and expression in pleomorphic adenomas of the salivary gland. Int. J. Cancer **124**(12), 2855–2863 (2009)

24. Zou, H., Hastie, T.: Regularization and variable selection via the elastic net. J. Roy. Stat. Soc. Ser. B (Stat. Methodol.) **67**(2), 301–320 (2005)

Knowledge Discovery

Measuring the Inspiration Rate of Topics in Bibliographic Networks

Livio Bioglio, Valentina Rho, and Ruggero G. Pensa[✉] [iD]

Department of Computer Science, University of Turin, Turin, Italy
{livio.bioglio,valentina.rho,ruggero.pensa}@unito.it

Abstract. Information diffusion is a widely-studied topic thanks to its applications to social media/network analysis, viral marketing campaigns, influence maximization and prediction. In bibliographic networks, for instance, an information diffusion process takes place when some authors, that publish papers in a given topic, influence some of their neighbors (coauthors, citing authors, collaborators) to publish papers in the same topic, and the latter influence their neighbors in their turn. This well-accepted definition, however, does not consider that influence in bibliographic networks is a complex phenomenon involving several scientific and cultural aspects. In fact, in scientific citation networks, influential topics are usually considered those ones that spread most rapidly in the network. Although this is generally a fact, this semantics does not consider that topics in bibliographic networks evolve continuously. In fact, knowledge, information and ideas are dynamic entities that acquire different meanings when passing from one person to another. Thus, in this paper, we propose a new definition of influence that captures the diffusion of inspiration within the network. We propose a measure of the inspiration rate called inspiration rank. Finally, we show the effectiveness of our measure in detecting the most inspiring topics in a citation network built upon a large bibliographic dataset.

Keywords: Information diffusion · Topic modeling · Citation networks

1 Introduction

Information diffusion is a fundamental and widely-studied topic in many research fields, including computational social science, machine learning and network analytics, thanks to its applications to social media/network analysis [1], viral marketing campaigns [17], influence maximization [4] and prediction [6]. An information diffusion process takes place when some active nodes (e.g., customers, social profiles, scientific authors) influence some of their inactive neighbors in the network and turn them into active nodes with a certain probability, and the newly activated nodes, in their turn, can progressively trigger some of their neighbors into becoming active [12]. Information diffusion is similar to the spread of diseases in epidemiology and it has also been modeled as such [7] by considering influence as a contagion process. However the correct definition of "influence"

© Springer International Publishing AG 2017
A. Yamamoto et al. (Eds.): DS 2017, LNAI 10558, pp. 309–323, 2017.
DOI: 10.1007/978-3-319-67786-6_22

strongly depends on the application. In mouth-to-mouth viral campaign, a user who buys a product at time t influences their neighbors if they buy the same product at time $t + \delta$. In social media, influence is the process that enables the diffusion of memes, (fake) news, viral posts across the network through different social actions such as likes, shares or retweets. In bibliographic networks, author a influences author b when a and b are connected by some relationship (e.g., collaboration, co-authorship, citation) and either b cites one of the papers published by author a, or author b publishes in the same topic as author a [12]. The latter definition, however, does not consider that influence in bibliographic networks is a complex phenomenon involving several scientific and cultural aspects. For instance, in scientific citation networks, the most cited papers are often seminal papers that introduce some topics (or some new aspects of a topic) for the first time. They are often cited "by default" and thus they spread in the network for very long periods. Moreover, in most existing works, influential topics are simply those ones that spread most rapidly in the network. Although this is generally a fact, this semantics does not consider that topics in bibliographic networks evolve continuously. In fact, knowledge, information and ideas are dynamic entities that acquire different meanings when passing from one person to another. For instance, "deep learning", a term invented in early 2000s, has known a rapid development and evolution that has influenced many research fields including semiconductor technology and circuits [3,5,20].

In this paper we address the problem of information diffusion in a bibliographic network by using the notion of *inspiring topics*. According to our definition, the most inspiring topics are those that evolve rapidly in the network by triggering fast citation rates. As an example, consider an author a_0 that publish a paper p_0 covering a given topic X at initial time interval t_0 of width δ. In the following time interval t_1, the activated authors are those that publish a paper p_1 citing paper p_0. In the following time interval t_2, the authors that publish a paper p_2 citing paper p_1 are activated. In general, we only consider citations from papers published at time interval t_i to papers published at the previous time interval t_{i-1}. Moreover, differently from other state-of-the-art methods, we consider topics assigned to papers by an adaptive Latent Dirichlet Annotation (LDA) technique [15]. According to this method, a paper p is said to cover a topic X if the LDA model states that p is generated by X with a probability greater than a threshold. Therefore, for a given time interval width δ, our topic diffusion model enables the ranking of topics according to their inspiration rate: topics that rank high for small values of δ are the most inspiring ones.

The salient contributions of this paper can be summarized as follows: (1) we define *inspiration* as an alternative to influence in information diffusion; (2) we introduce the definition of *inspiration rank* as a measure of the topic inspiration rate: topics that trigger fast citation rates have a high inspiration rank; (3) we use an adaptive LDA technique for assigning topics to each paper; (4) we propose a topic analysis model enabling the ranking of topics according to their inspiration rate. By comparing our model to a standard diffusion model, we show the effectiveness of our framework on a large corpus consisting of about 155,000 scientific papers and 225,000 authors.

The remainder of the paper is organized as follows: related works are analyzed in Sect. 2; the topic diffusion model is presented in Sect. 3; Sect. 4 provides the report of our experiments; finally, we draw some conclusions in Sect. 5.

2 Related Works

Information diffusion has been first regarded as a derivation of the process of disease propagation in contact networks [14], a well-studied problem in epidemiology. An obvious application stands in the domain of marketing, where diffusion models are used to understand the process of information spread among potential customers with the goal of improving viral marketing campaigns [9]. In [17], the authors mathematically characterize the propagation of products recommendation in the network of individuals.

Besides viral marketing studies, the success of Web 2.0 and online social networks has also boosted researches on topic diffusion. In [10,11] the authors leverage the theory of infectious diseases to capture the structure of topics and analyze their diffusion in the blogsphere. In [25], Yang and Counts analyze Twitter by constructing a model that captures the speed, scale, and range of information diffusion. In [24], the same authors compare the diffusion patterns within Twitter and a weblog network, finding that Twitter's network is more decentralized and connected locally. In [2], a novel and more accurate information propagation model is defined: the authors propose a topic-aware extensions of the well-known Independent Cascade and Linear Threshold models [16] by taking into account authoritativeness, influence and relevance.

Digital libraries and bibliographic networks have also taken advantage of information diffusion studies. Thanks to the availability of data sets of unprecedented size many studies have analyzed citation, co-authorship or co-participation networks to identify patterns of diffusion and influence, and to rank authors. In [18], Radicchi et al. define an author ranking method based on a diffusion algorithm that mimics the spreading of scientific credits on the network. Shi et al., instead, study the structural features of the information paths in the citation networks of publications in computer science [21]. Among their findings, they discover that citing more recent papers corresponds to receiving more citations in turn. In [12], the authors propose to model information diffusion in multi-relational bibliographic networks, by distinguishing different types of relationships. In addition, they propose a method to learn the parameters of their model leveraging real publication logs.

Differently from all these works, we focus on topic diffusion and evolution by leveraging explicit citations in bibliographic networks. Topic evolution has already been regarded as extensions of the Latent Dirichlet Allocation (LDA) or the Probabilistic Latent Semantic Analysis algorithms [8]. In [13] the authors leverage citations to address the problem of topic evolution analysis on scientific literature. When detecting topics in a collection of new papers at a given time instant, they also consider citations to previously published papers and propose a novel LDA-based topic modeling technique named Inheritance Topic Model. In our work, we adopt a similar solution, but we look at topic evolution from the information diffusion perspective, by computing a ranking of most inspiring topics, defined as those topics for which we observe a rapid evolution and inspiration rate in the network.

3   Inspiration Propagation

In this section we introduce the mathematical background and the theoretical framework of our ranking method.

We consider a set of n documents $D = d_1, \ldots, d_n$ and a set of K topics $Z = z_1, \ldots, z_K$. Each document $d_i \in D$ is characterized by a distribution of topics $\Theta_i = <\theta_{i1}, \ldots, \theta_{iK}>$, where $\forall i, k, 0 \leq \theta_{ik} \leq 1$ and $\sum_{k=1}^{K} \theta_{ik} = 1$. Each document is authored by one or more authors belonging to the set $A = \{a_1, \ldots, a_N\}$ of all possible N authors. Moreover, each document d_i has a timestamp ts_i corresponding to the publication date.

Authors and papers are part of a *heterogenous information network*, i.e., a directed graph $\mathcal{G}(\mathcal{V}, \mathcal{E})$, where $\mathcal{V} = V^d \cup V^a$ and $\mathcal{E} = E^{ad} \cup E^{dd}$. Each $v_i^d \in V^d$ and $v_l^a \in V^a$ are, respectively, a vertex representing the i-th document $d_i \in D$ and a vertex representing the l-th author $a_l \in A$. Moreover, each $(v_l^a, v_i^d) \in E^{ad}$ is a directed edge meaning that author a_l has coauthored document d_i and each $(v_i^d, v_j^d) \in E^{dd}$ is a directed edge coding the fact that document d_i cites document d_j. Furthermore, E^{ad} is such that if $(v_l^a, v_i^d) \in E^{ad}$, then $(v_i^d, v_j^a) \in E^{ad}$ (i.e., each connection between documents and authors is reciprocal).

Within the heterogenous information network $\mathcal{G}(\mathcal{V}, \mathcal{E})$, we identify the *citation network* $G(V, E)$, where $V = V^a$ is the set of author vertices and $E = \{(v_h, v_l)\}$ is the set of directed citation edges. In particular, $(v_h, v_l) \in E$ iff there exists a path $path(v_h^a, v_l^a) = v_h^a \xrightarrow{ad} v_i^d \xrightarrow{dd} v_j^d \xrightarrow{ad} v_l^a$ within the information network $\mathcal{G}(\mathcal{V}, \mathcal{E})$. Roughly speaking, an edge (v_h, v_l) can be found in the citation network $G(V, E)$ iff author v_h has cited some (at least one) paper coauthored by

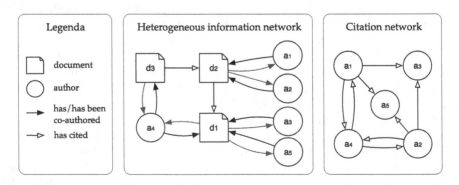

Fig. 1. A heterogenous information network and the corresponding citation network.

v_l in one of the papers she coauthored. An example of heterogenous information network and its corresponding citation network is given in Fig. 1.

In the following sections we describe the topic diffusion model adopted in our framework, as well as the topic modeling method used to associate topics with documents.

3.1   Topic Diffusion Model

Differently from most topic diffusion models that consider both co-authorship and citation links, our approach only considers explicit citations. In most existing approaches (such as the one presented in [12]), the influence process takes place when an author publishes some paper on a given topic at time t and some of her neighbors publish any paper on the same topic at time $t + \delta$. Usually explicit citations are simply ignored, but they are crucial to understand the evolution and transformation of a topic across the network during a time period. Moreover, when explicit citations are ignored and heterogeneous links between authors are considered, the true semantics of propagation is less clear: influence may occur because of some external factors, e.g., the topic is popular at publication time, the authors are part of the same consortium within a project, or they publish in the same topic just by chance. Instead, in our work, we propose to measure "inspiration" as an alternative to classic influence processes. Conversely speaking, inspiration takes place when an author cites another author explicitly in one of her papers, regardless of its topic. The general definition of inspiration is then as follows.

Definition 1 (inspiration). *Let $\mathcal{G}(\mathcal{V}, \mathcal{E})$ be a heterogenous information network. Author $a_h \in A$ **is inspired by** author $a_l \in A$ $(a_l \neq a_h)$ iff there is a path $v_h^a \xrightarrow{ad} v_i^d \xrightarrow{dd} v_j^d \xrightarrow{ad} v_l^a$ in $\mathcal{G}$ s.t. $ts_i \geq ts_j$.*

In the following we provide the theoretical details of our topic diffusion model. Let $\mathcal{T} = [T_0, T_n]$ be a time interval. We define a set $\Delta\mathcal{T} = \{\Delta T_0, \ldots, \Delta T_N\}$ of possibly overlapping time intervals over $\mathcal{T}$ s.t. $\forall t = 1 \ldots N$ $\Delta T_{t-1} \prec \Delta T_t$. We introduce the definitions of *initial topic-based inspiration* and *subsequent topic-based inspiration* for a given topic z_k.

Definition 2 (initial topic-based inspiration). *Let $\mathcal{G}(\mathcal{V}, \mathcal{E})$ be a heterogenous information network, $\Delta\mathcal{T} = \{\Delta T_0, \ldots, \Delta T_N\}$ a set of time intervals and Θ a topic distribution. For a given topic z_k and a given threshold $\tau \in [0, 1]$, author $a_h \in A$ **is initially inspired by** author $a_l \in A$ $(a_l \neq a_h)$ iff there is a path $v_h^a \xrightarrow{ad} v_i^d \xrightarrow{dd} v_j^d \xrightarrow{ad} v_l^a$ in $\mathcal{G}$ s.t. $ts_j \in \Delta T_0$, $ts_i \in \Delta T_1$ and $\theta_{jk} \geq \tau$.*

According to this definition, the initial inspiration takes place when an author a_l publishes a document d_j during ΔT_0 $(ts_j \in \Delta T_0)$ covering topic z_k $(\theta_{jk} \geq \tau)$, and another author a_h publishes a document d_i, during the following time interval ΔT_1 $(ts_i \in \Delta T_1)$. Notice that we do not impose any constraints on the topic covered by document d_i. Let us now introduce the definition of *subsequent topic-based inspiration*.

Definition 3 (subsequent topic-based inspiration). *Let $\mathcal{G}(\mathcal{V}, \mathcal{E})$ be a heterogenous information network and $\Delta T = \{\Delta T_0, \ldots, \Delta T_N\}$ a set of time intervals. For a given topic z_k, author $a_h \in A$ is **subsequently inspired** by author $a_l \in A$ ($a_l \neq a_h$) at time ΔT_t iff there is a path $v_h^a \xrightarrow{ad} v_i^d \xrightarrow{dd} v_j^d \xrightarrow{ad} v_l^a$ in $\mathcal{G}$ s.t. $ts_j \in \Delta T_{t-1}$, $ts_i \in \Delta T_t$ and a_l has been initially/subsequently inspired by another author $a_m \in A$ ($a_m \neq a_h$ and $a_m \neq a_l$) during ΔT_{t-1} for topic z_k.*

It can be noticed that this definition is recursive, meaning that the subsequent inspiration occurs when an author a_h has cited an author a_l that has been either subsequently inspired or initially inspired by a third author a_m in the previous time interval. Moreover, according to our diffusion model, inspiration takes place when a citation occurs between two consecutive time intervals. Even though this may appear a strong constraint, we recall that the definition of the set ΔT of time interval is very general. In particular, we introduce two parameters $\delta > 0$ and $\gamma \geq 0$ ($\gamma < \delta$), representing respectively the size of a sliding time window and the overlap between two consecutive time windows. Given these two parameters and a time interval $T = [T_0, T_n]$, we define $\Delta T = \{\Delta T_0, \ldots, \Delta T_N\}$ in such a way that $\Delta T_t = [T_0 + t(\delta - \gamma), T_0 + t(\delta - \gamma) + \delta)$, for $t = 0, \ldots, N$ with $N = \lceil \frac{T_n - (T_0 + \delta - 1)}{\delta - \gamma} \rceil$.

3.2   Computation of the Inspiration Rank

We now describe how to assign a rank value to each topic depending on its inspiration speed. To this purpose, for a given topic z_k and a given set of time intervals $\Delta T = \{\Delta T_0, \ldots, \Delta T_N\}$ we measure the cumulative number of new authors inspired at each time interval, according to the definitions of inspiration given in Sect. 3.1. In particular, given the heterogenous information network $\mathcal{G}(\mathcal{V}, \mathcal{E})$ and a threshold τ, we call $A_0 = \{a_h | \exists (v_h^a, v_i^d) \in E^{ad} \wedge ts_i \in \Delta T_0 \wedge \theta_{ik} > \tau\}$ the set of authors that publish a paper on topic z_k during ΔT_0. Then, we define $A_1 = \{a_h \mid \exists a_l \in A_0 \ s.t. \ a_h \text{ is initially inspired by } a_l\}$ and, $\forall t = 2, \ldots, N$, $A_t = \{a_h \mid \exists a_l \in A_{t-1} \ s.t. \ a_h \text{ is subsequently inspired by } a_l \text{ during } \Delta T_t\}$. In a nutshell, A_1 is the set of initially inspired authors, $A_2, \ldots, A_N$ are the sets of subsequently inspired authors.

Finally, we construct a set of two-dimensional points $\{(t, y_t)\}$, $t = 1, \ldots, N$ where $y_t = |A_t|$ for $t = 1$ and $y_t = |A_{t-1} \cup A_t|$ for $t = 2, \ldots, N$. We use this set to compute a linear function $y = \hat{\sigma} t + \hat{c}$ by solving the following simple linear regression problem

$$(\hat{\sigma}, \hat{c}) = \arg \min_{\sigma, c} \sum_{t=1}^{N} (y_t - c - \sigma t)^2 \tag{1}$$

using the least squares method.

The *inspiration rank* value is then defined as the slope $\hat{\sigma}$ of the linear function $y = \hat{\sigma} x + \hat{c}$ obtained by solving Eq. 1. More formally:

Definition 4 (inspiration rank). *Given a heterogenous information network* $\mathcal{G}(\mathcal{V}, \mathcal{E})$, *a topic* z_k *and a set of time intervals* $\Delta\mathcal{T} = \{\Delta T_0, \ldots, \Delta T_N\}$, *the inspiration rank of* z_k, *called* $IR(\mathcal{G}, \Delta\mathcal{T}, z_k)$ *is given by*

$$IR(\mathcal{G}, \Delta\mathcal{T}, z_k) = \hat{\sigma} \tag{2}$$

where $\hat{\sigma}$ *is the solution of the linear regression problem given in Eq. 1.*

Notice that, by varying parameters δ and γ, which define the width and overlap of time intervals in $\Delta\mathcal{T}$, different values of information rank can be obtained.

In order to compare our ranking method to the usual idea of topic diffusion, for each topic we also compute a *diffusion rank* value as follows. For each time interval $\Delta T_t \in \Delta\mathcal{T}$ we set $A_t' = \{a_h \mid \exists (v_h^a, v_i^d) \in E^{ad} \wedge ts_i \in \Delta T_t \wedge \theta_{ik} > \tau\}$, i.e., A_t' is the set of authors that have published a paper on topic z_k during time interval Δ_t. Then, we construct a set of two-dimensional points $\{(t, y_t')\}$, $t = 1, \ldots, N$ where $y_t' = |A_t'|$ for $t = 1$ and $y_t' = |A_{t-1}' \cup A_t'|$ for $t = 2, \ldots, N$. Again, we fit these values to a linear function $y' = \hat{\sigma}t + \hat{c}$ and set the *diffusion rank* $DR(\mathcal{G}, \Delta\mathcal{T}, z_k)$ equal to the slope $\hat{\sigma}$.

3.3   Topic Extraction

In this section, we introduce the topic modeling technique that we adopt to determine the distribution of topics for each document $d_i \in D$. Topic extraction is performed using Latent Dirichlet Allocation (LDA), a generative probabilistic model of a corpus, that aims at describing a set of observations, e.g. textual documents, using a set of unobserved latent elements, e.g. topics. LDA considers each document as a distribution over latent topics and each topic as a distribution over terms. Given α as prior knowledge about topics distribution, LDA assumes the following generative process for each document d of a corpus: (1) draw a distribution over topics $\theta_d \sim \text{Dirichlet}(\alpha)$, (2) for each word i in d draw a topic z_{di} from θ_d and draw the word w_{di} from z_{di}.

For our purposes we use a slightly modified version of LDA, named *Online LDA* [15]. In fact, traditional LDA implementations are based on either variational inference or collapsed Gibbs sampling; both methods require to process the entire corpus in order to compute the topic model, and it is not possible to query the model with previously unseen documents. In contrast, Online LDA replaces the previously used inference methods with the stochastic variational inference technique that allows *online* training, update of an existing model with new documents and query for unseen documents. Algorithm 1 shows the procedure to infer topics assignment on a new document. The document is represented by a vector of terms occurrences n of length N, K is the number of topics in the LDA model, α is the Dirichlet prior, β is the topic-term distribution matrix. The algorithm iteratively refines the variational parameters ϕ (line 4) that represents the word probability in each topic and ψ (line 5), which encodes the topics proportion within the document. When the procedure converges [15] ψ is returned as the topics assignment for the document represented by n (line 7).

Algorithm 1. Topic inference on unseen documents in Online LDA.

1 Initialize $\psi_k = 1$, $\forall k = 1, \ldots, K$.
2 **repeat**
3   **for** $k = 1, \ldots, K$ **do**
4   Set $\phi_{wk} \propto \exp\{\mathbb{E}_q[\log \theta_k] + \mathbb{E}_q[\log \beta_w]\}$ $\forall w = 1, \ldots, N$
5   Set $\psi_k = \alpha + \sum_{w=1}^{N} \phi_{wk} n_w$
6 **until** $\frac{1}{K} \sum_k |$change in $\psi_k| < \epsilon$;
7 **return** ψ

---

4   Experiments

In this section, we present the results of our experiments conducted on a large corpus of scientific documents. In particular, we analyze the outcomes of our measure of inspiration in terms of effects on topic ranking. We compare our results with the standard diffusion approach, broadly adopted in most research works dealing with topic diffusion. In the following, we first describe the dataset used in our experiments and how we construct it; then, we provide some insights on the topic extraction and labeling tasks; finally, we give the details of the experimental protocol and report the results returned by our ranking-by-inspiration method in comparison with the standard ranking-by-diffusion approach.

4.1   Dataset

The dataset used in our experiments is a subset of the Computer Science paper citation network. This dataset is created by automatically merging two datasets originally extracted through ArnetMiner [23]: the *DBLP* and *ACM* citation networks[1]. The merge procedure is necessary because both datasets lack some information: the *ACM* dataset contains many abstracts and citations between documents, but venues do not follow any naming convention and authors are ambiguous; In *DBLP*, venues and authors are clearly identified, but abstracts

Table 1. Datasets statistics.

	ACM-v8	DBLP-v8	Merged	Selected
No. of papers	2,381,674	3,272,990	1,373,202	154,947
No. of complete papers	1,668,246	3,241,890	1,143,443	154,947
No. of venues names	265,149	11,553	6,959	153
No. of authors	1,508,051	1,752,440	903,771	225,559
No. of out-citations	8,650,089	8,466,858	6,513,765	1,321,905
No. of in-citations	-	-	5,365,753	1,000,657

---

[1] https://aminer.org/citation.

are missing and citations contain repetitions. Some statistics on the datasets are shown in Table 1. Papers are considered *complete* if all basic information are present, i.e. title, abstract (ACM only), year, venue and at least one outgoing or incoming citation. The *merged* dataset has been obtained by matching ACM and DBLP entries as follows: two papers match if both title and list of authors are the same. Then, abstracts and citations are extracted from ACM data; authors, title and venue are extracted from DBLP data. Finally, the *selected* dataset considers only papers published in the context of a set of manually preselected venues in the period from 2000 to 2014, covering the following research area: *artificial intelligence, machine learning, pattern recognition, data mining, information retrieval, database* and *information management*. The *selected* dataset is available online[2].

4.2   Text Processing and Topic Extraction

The input data given to the topic extraction algorithm is obtained as the result of a cleaning and vectorization process performed on the concatenation of paper title and abstract. In particular, the cleaning module ignores terms that appears only once in the dataset and in more than 80% of the documents. A domain dependent stop words list is also excluded from topic computation. First, documents are pre-processed with NLP techniques that perform tokenization, lemmatization, stop words removal and term frequency computation in order to prepare the corpus for the topic modeling algorithm. For performing this task, we adopt a scalable and robust topic modeling library [19] that enables the extraction of an adaptive set of topics using an online learning version of Latent Dirichlet Allocation [15].

Topic modeling is performed on all papers published between 2000 and 2004 that appear within the *selected* dataset using Latent Dirichlet Allocation, searching for $K = 50$ topics. The extracted topic model is then used to assign a weighted list of topics to all papers published between 2005 and 2014. We perform LDA on a time interval preceding the one used for analysis, instead of the whole corpus, because in this way we focus on well-established topics rather than on emerging ones. However this choice does not limit our findings: in fact, many research topics investigated during the last ten years (including, e.g., *deep learning*) have been faced for the first time in the first half decade of the 21st century.

Topics Labeling. For improving the readability of our model, we introduce a simple topics labeling step that associates, to each topic z_k represented by a weighted list of words, up to three labels. The labels are computed as the first three results obtained by querying Wikipedia with the set of most representative words for z_k. We identify as *most representative* the 6 words having a weight greater than 0.01 or, if the first set is empty, the top 3 words. An example of labels extracted with this method is shown in Table 2.

---

[2] Dataset encoded in ArnetMiner V8 format, https://github.com/rupensa/tranet.

Table 2. Example of extracted topic description and associated labels.

Topic description	Labels
0.091*network + 0.058*neural + 0.025*input + 0.021*learning + 0.021*adaptive + 0.020*neuron + 0.017*dynamic + 0.014*function + ...	Artificial Neural Network, Artificial Neuron, -

4.3   Results

In our experiments, we calculate the ranking of topics according to their *inspiration rank* and diffusion rank in the time interval from 2005 to 2014 for $1 \leq \delta \leq 6$ and $0 \leq \gamma \leq \delta$. In all our experiments $\tau = 0.2$. Algorithms and scripts are implemented in Python, and data are stored in a MongoDB[3] database server. The source code and the dataset are available online[4]: the whole analysis process can be driven within an interactive Jupyter notebook[5]. The experiments are performed on a server with two 3.30 GHz Intel Xeon E5-2643 CPUs, 128 GB RAM, running Linux.

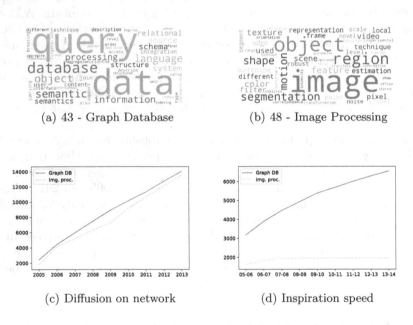

(a) 43 - Graph Database   (b) 48 - Image Processing

(c) Diffusion on network   (d) Inspiration speed

Fig. 2. Diffusion and word clouds of the selected topics.

---

[3]   https://www.mongodb.com/.

[4]   https://github.com/rupensa/tranet.

[5]   https://jupyter.org/.

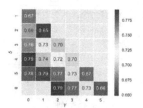 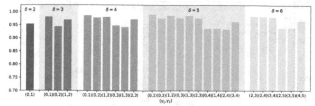

(a) Spearman correlation values between inspiration and diffusion rank for several values of δ and γ.

(b) Spearman correlation values of inspiration ranks computed with several γ values and same δ.

Fig. 3. Correlation computed between inspiration and diffusion ranks. (Color figure online)

Examples of Inspiration Trend. As a first illustrative result, we show the inspiration trend of two topics, and compare it to their diffusion trend. The topics selected for this analysis are *Graph DB* (topic-43) and *Image Processing* (topic-48). They are intrinsically described in word clouds shown in Fig. 2 by means of their more representative terms. These two topics have been selected due to their similarity in terms of number of assigned papers (8, 969 for topic 43 and 8, 646 for topic 48), authors (24, 143 for topic 43 and 23, 056 for topic 48) and distribution of papers in the considered time frame. According to Fig. 2c, which shows the diffusion trend as computed by the method in [12], these topics have very close diffusion trends in the bibliographic network. However, there is a strong difference in the inspiration trend, as shown in Fig. 2d: in fact, topic 43 (graph databases) evolves more rapidly than topic 48 (image processing). This behavior can be explained by the increasing and fast research results obtained by the database community, also boosted by the research on semantic queries and triplestores. Image processing, in contrast, appears as an evergreen albeit not particularly evolving research field in the time frame considered here. In this experiment, we employ $K = 50$, $\delta = 2$ and $\gamma = 1$.

Inspiration Rank vs. Diffusion Rank. In order to study the difference between the proposed ranking and the usual one, we measure the Spearman's rank correlation coefficient [22] between the the *inspiration rank* and *diffusion rank*. The Spearman's rank coefficient assesses monotonic relationships between two series of values. It basically captures the correlation between the two rankings and ranges between -1 (for inversely correlated sets of values) and $+1$ (for the maximum positive correlation).

Figure 3a shows the Spearman's rank correlation coefficient between inspiration rank IR and diffusion rank DR for several values of δ and γ. The empty tiles on the bottom left are due to lack of data: since our dataset covers only 10 years, when $\delta \gg \gamma$ there is only one time interval valuable for calculating the rank, that is not sufficient for fitting a linear function.

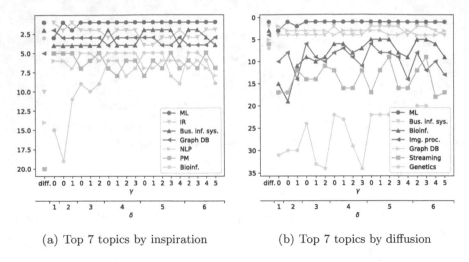

(a) Top 7 topics by inspiration   (b) Top 7 topics by diffusion

Fig. 4. Ranking by inspiration and diffusion for several values of δ and γ. Diffusion rank are not affected by parameter variation.

In general, it can be noticed that the two ranks are always positively correlated. However, for lower values of δ (i.e., for small time windows), the correlation is sensibly slighter (only 0.67 for $\delta = 1$ and 0.65 for $\delta = 2$, with $\gamma = 1$). This can be explained by the fact that topics that diffuse faster are not necessarily the most inspiring ones, according to our definition. When the inspiration rank is high for small time windows, it means that citations occur very fast. The fact that the two rank values get more similar when δ increases, also confirms our intuition. In fact, it is more likely that papers are cited after four or five years, rather than the year following its publication. When this occurs, it means that this topic is evolving very fast, inspiring plenty of new research works. Another noticeable result is that correlation decreases when γ increases. This is due to the fact that larger overlap values allow to capture more citations to papers published in the previous interval. However, its effect is weaker than the one of parameter δ, as shown in Fig. 3b, where the correlation between any pair of γ values for the same value of δ are illustrated. This particular results also shows that our method is rather stable toward variations of parameter γ.

Ranking Comparison. Here, we analyze the ranking of the top 7 topics based on the average inspiration rank, compared to the ranking of the same topics based on diffusion rank. The results are depicted in Fig. 4a (notice that diffusion ranks are not affected by parameter variation). We notice that the best 4 topics are almost the same ones for all values of δ and γ, then the ranking becomes more chaotic. More interestingly, topic IR (Information retrieval) is always ranked in the top 3 positions for inspiration, while it is ranked 10*th* according to diffusion. Our measure capture a real trend in Computer Science: the increasing research efforts in information retrieval have been driven by search engine and social media applications, as well as by Semantic Web technologies. Topic Graph DB (graph databases) is also ranked higher by our technique. Research on this topic has been boosted by semantic database achievements in the last 15 years. Notice that our techniques also ranks NLP (natural language processing) and PM (pattern mining) among the top 7 topics, coherently with the actual efforts in these domains pushed by the advances in sentiment analysis and other Semantic Web applications as well as in frequent itemset and sequence mining in the considered period. These topics are only ranked 13th and 20th according to standard diffusion metrics.

It is worth noting that, by analyzing the ranking of the top 7 topics based on the average diffusion rank, and their respective ranking based on inspiration (Fig. 4b), we observe that some of the topics that have a relatively lower rank in the ranking-by-inspiration approach can be considered as application of Computer Science techniques. For instance, it is a fact that Bioinformatics (ranked third) has spread rapidly in the last 10 years. However, in our approach this topics gets a lower rank: this can be explained by the fact that, in the research areas under investigation, covering data mining and machine learning, papers in this multidisciplinary field are more likely to be inspired by (rather than to inspire) other research topics (such as, clustering, machine learning or pattern mining). The same observation applies to genetics and image processing.

Finally, we explore the ranking provided by the two methods for a set of topics of interest, namely: Deep learning (topic-3), Clustering (topic-22), Information retrieval (topic-26), Neural networks (topic-33) and Pattern mining (topic-41). The ranking positions are shown for several values of δ and γ in Fig. 5. The same conclusions drawn in the previous experiments hold here, in particular for pattern mining (PM). Interestingly, deep learning is ranked low, despite its objective success in the last five years. This may be explained by the fact that the topic model has been trained on papers published from 2000 to 2004 when deep

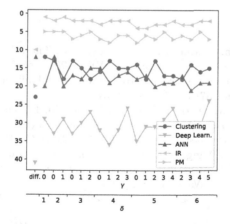

Fig. 5. Rankings on inspiration and diffusion speeds for several values of δ and γ on a set of selected topics.

learning was beginning to be recognized as a research field itself. Furthermore, since we only consider Computer Science venues, the broad influence of deep learning on other research areas can not be captured. Notice that, however, it is always ranked far higher by our method. This is another result indicating that inspiration capture a more realistic influence semantics than simple topic diffusion.

5   Conclusions

We have proposed a new definition of influence that takes into account the inspiration of a given topic within a citation network. We have defined a new influence measure, called *inspiration rank*, that captures the inspiration rate of topics extracted by an adaptive LDA technique, within a given time interval. The *inspiration rank* allows the discovery of the most inspiring topics according to different levels of speed. We have shown experimentally the effectiveness of our measure in detecting the most inspiring topics in a citation network built upon a large bibliographic dataset. Although the core application is the analysis of topic diffusion in citation networks, our methods can be also applied on other information networks, including patent and news, provided that a link between two documents can be inferred directly or indirectly.

As future work, we will define new author and paper ranking methods based on our inspiration measure. Furthermore, we will investigate new algorithms to learn the topic diffusion parameters under different diffusion models adopting our definition of inspiration.

Acknowledgments. This work is partially funded by project MIMOSA (MultIModal Ontology-driven query system for the heterogeneous data of a SmArtcity, "Progetto di Ateneo Torino_call2014_L2_157", 2015–17).

References

1. Bakshy, E., Rosenn, I., Marlow, C., Adamic, L.A.: The role of social networks in information diffusion. In: Proceedings of WWW 2012, pp. 519–528. ACM (2012)
2. Barbieri, N., Bonchi, F., Manco, G.: Topic-aware social influence propagation models. Knowl. Inf. Syst. **37**(3), 555–584 (2013)
3. Boguslawski, B., Sarhan, H., Heitzmann, F., Seguin, F., Thuries, S., Billoint, O., Clermidy, F.: Compact interconnect approach for networks of neural cliques using 3D technology. In: Proceedings of IFIP/IEEE VLSI-SoC 2015, pp. 116–121 (2015)
4. Chen, W., Wang, Y., Yang, S.: Efficient influence maximization in social networks. In: Proceedings of ACM SIGKDD 2009, pp. 199–208. ACM (2009)
5. Coates, A., Huval, B., Wang, T., Wu, D.J., Catanzaro, B., Ng, A.Y.: Deep learning with COTS HPC systems. In: Proceedings of ICML 2013, pp. 1337–1345. JMLR.org (2013)
6. Cui, P., Wang, F., Liu, S., Ou, M., Yang, S., Sun, L.: Who should share what?: item-level social influence prediction for users and posts ranking. In: Proceeding of ACM SIGIR 2011, pp. 185–194. ACM (2011)

7. Daley, D.J., Kendall, D.G.: Epidemics and rumours. Nature **208**, 1118 (1964)
8. Gohr, A., Hinneburg, A., Schult, R., Spiliopoulou, M.: Topic evolution in a stream of documents. In: Proceedings of SIAM SDM 2009, pp. 859–870. SIAM (2009)
9. Goldenberg, J., Libai, B., Muller, E.: Talk of the network: a complex systems look at the underlying process of word-of-mouth. Market. Lett. **12**(3), 211–223 (2001)
10. Gruhl, D., Guha, R.V., Liben-Nowell, D., Tomkins, A.: Information diffusion through blogspace. In: Proceedings of WWW 2004, pp. 491–501. ACM (2004)
11. Gruhl, D., Liben-Nowell, D., Guha, R.V., Tomkins, A.: Information diffusion through blogspace. SIGKDD Explor. **6**(2), 43–52 (2004)
12. Gui, H., Sun, Y., Han, J., Brova, G.: Modeling topic diffusion in multi-relational bibliographic information networks. In: Proceedings of CIKM 2014, pp. 649–658. ACM (2014)
13. He, Q., Chen, B., Pei, J., Qiu, B., Mitra, P., Giles, C.L.: Detecting topic evolution in scientific literature: how can citations help? In: Proceedings of ACM CIKM 2009, pp. 957–966. ACM (2009)
14. Hethcote, H.W.: The mathematics of infectious diseases. SIAM Rev. **42**(4), 599–653 (2000)
15. Hoffman, M.D., Blei, D.M., Bach, F.R.: Online learning for latent dirichlet allocation. In: Proceedings of NIPS 2010, pp. 856–864 (2010)
16. Kempe, D., Kleinberg, J.M., Tardos, É.: Maximizing the spread of influence through a social network. In: Proceedings of ACM SIGKDD 2003, pp. 137–146. ACM (2003)
17. Leskovec, J., Adamic, L.A., Huberman, B.A.: The dynamics of viral marketing. TWEB **1**(1), 5 (2007)
18. Radicchi, F., Fortunato, S., Markines, B., Vespignani, A.: Diffusion of scientific credits and the ranking of scientists. Phys. Rev. E **80**, 056103 (2009)
19. Řehůřek, R., Sojka, P.: Software framework for topic modelling with large corpora. In: Proceedings of LREC 2010 Workshop on New Challenges for NLP Frameworks, pp. 45–50 (2010)
20. Seo, J., Seok, M.: Digital CMOS neuromorphic processor design featuring unsupervised online learning. In: Proceedings of IFIP/IEEE VLSI-SoC 2015, pp. 49–51. IEEE (2015)
21. Shi, X., Tseng, B.L., Adamic, L.A.: Information diffusion in computer science citation networks. In: Proceedings of ICWSM 2009. The AAAI Press (2009)
22. Spearman, C.: The proof and measurement of association between two things. Am. J. Psychol. **15**(1), 72–101 (1904)
23. Tang, J., Zhang, J., Yao, L., Li, J., Zhang, L., Su, Z.: Arnetminer: extraction and mining of academic social networks. In: Proceedings of KDD 2008, pp. 990–998 (2008)
24. Yang, J., Counts, S.: Comparing information diffusion structure in weblogs and microblogs. In: Proceedings of ICWSM 2010. The AAAI Press (2010)
25. Yang, J., Counts, S.: Predicting the speed, scale, and range of information diffusion in twitter. In: Proceedings of ICWSM 2010. The AAAI Press (2010)

Discovering Minority Sub-clusters and Local Difficulty Factors from Imbalanced Data

Mateusz Lango, Dariusz Brzezinski$^{(\boxtimes)}$, Sebastian Firlik, and Jerzy Stefanowski

Institute of Computing Science, Poznan University of Technology,
ul. Piotrowo 2, 60–965 Poznan, Poland
{mlango,dbrzezinski,jstefanowski}@cs.put.poznan.pl

Abstract. Learning classifiers from imbalanced data is particularly challenging when class imbalance is accompanied by local data difficulty factors, such as outliers, rare cases, class overlapping, or minority class decomposition. Although these issues have been highlighted in previous research, there have been no proposals of algorithms that simultaneously detect all the aforementioned difficulties in a dataset. In this paper, we put forward two extensions to popular clustering algorithms, ImKmeans and ImScan, and one novel algorithm, ImGrid, that attempt to detect minority sub-clusters, outliers, rare cases, and class overlapping. Experiments with artificial datasets show that ImGrid, which uses a Bayesian test to join similar neighboring regions, is able to re-discover simulated clusters and types of minority examples on par with competing methods, while being the least sensitive to parameter tuning.

Keywords: Class imbalance · Minority class categorization · Data difficulty factors · Class overlapping · Minority sub-clusters

1 Introduction

Improving classifiers learned from class-imbalanced data has been a topic of growing research in recent decades and several specialized algorithms have been introduced [2,6]. However, less effort has been put into studying the *characteristics* of imbalanced data, which make learning from imbalanced data so difficult.

It has been shown that neither the global imbalance ratio between the minority class and majority class nor the cardinality of the minority class are the main sources of difficulty. Other *data difficulty factors*, referring to internal characteristics of class distributions, are usually more influential. Several studies have demonstrated the high impact of the following factors: decomposition of the minority class into many sub-concepts [7,9], overlapping between the classes [5,16], and presence of many minority class examples inside the majority class region [13]. When these data difficulty factors occur *together* with class imbalance, they may seriously deteriorate the recognition of the minority class [12,13].

Identification of data difficulty factors may help in distinguishing different categories of imbalanced datasets (easier or more difficult to learn from).

© Springer International Publishing AG 2017
A. Yamamoto et al. (Eds.): DS 2017, LNAI 10558, pp. 324–339, 2017.
DOI: 10.1007/978-3-319-67786-6_23

Consequently, specialized classifiers and preprocessing methods are more sensitive to certain data categories [11,12]. Therefore, such an analysis of data characteristics and data difficulty factors may be important, on the one hand, to better understand the nature of class imbalanced data and, on the other, to aid the development of new classification methods.

Nevertheless, automatic discovery of the aforementioned factors in real world datasets is not an easy task and may not give unique results. Most known studies on difficulty factors have been carried out on synthetic data with ground truth knowledge. Discovering *sub-concepts* of the minority class is usually done with clustering algorithms such as k-means [9,14]. However, tuning parameters of clustering, the number of expected sub-concepts, dealing with complex, non-spherical shapes and outliers is problematic in real imbalanced datasets. Therefore, discovering minority sub-concepts still constitutes a research challenge.

Some other difficulty factors may be linked to *different types of examples* forming the minority class distribution with respect to their relative position. This view has led Napierala and Stefanowski to differentiate between safe and unsafe examples for recognizing minority instances [11]. The unsafe examples are further categorized into borderline, rare cases, and outliers. These authors have also introduced an approach to identify these types of examples by analyzing class label distributions in the neighborhood of minority examples [11,12]. The results of these works have been useful for constructing new preprocessing methods and specialized classifier ensembles for imbalanced data [19]. However, this approach is unable to detect sub-concepts inside the minority class.

Therefore, an open research question is: whether it is possible to construct a clustering approach that simultaneously discovers sub-concepts in complex imbalanced data and categorizes types of examples inside discovered clusters?

The main aim of this paper is to solve this research problem by introducing new specialized clustering algorithms. For this purpose, we put forward two extensions of popular clustering algorithms, ImKmeans and ImScan, as well as propose a novel approach, called ImGrid, dedicated to discovering minority sub-concepts and categorizing examples simultaneously. ImGrid uses spatial, density, and statistical characteristics of the attribute space, to detect and analyze minority class regions. The algorithms are experimentally evaluated on a comprehensive set of synthetic datasets with hidden sub-concept structures and various proportions of data difficulty factors.

The paper is organized as follows: related literature is discussed in Sect. 2; the proposed ImKmeans, ImScan, and ImGrid algorithms are described in Sect. 3; experimental results are discussed in Sect. 4; and finally conclusions and lines of future research are drawn in Sect. 5.

2   Related Work

2.1   Characteristics of Imbalanced Data

A dataset is considered to be imbalanced when the cardinality of the minority class is much smaller than the majority class (which is expressed by the global imbalanced ratio between these two classes). In this paper we consider a standard formulation of the binary class or binarized multi-class imbalance problem [6].

It is worth noting that the global imbalance between classes may not pose difficulty for learning accurate classifiers by itself. Some, even highly, imbalanced data can be accurately learned by standard algorithms if the classes are well separated. When the rarity of the minority class is combined with other data difficulty factors concerning instance distributions in the attribute space [19], then it has a stronger negative impact on the recognition of the minority class.

Although many of the considered data factors are also known to affect learning in balanced domains, when they occur *together* with class imbalance the deterioration of classification performance is amplified and affects mostly (or sometimes only) the minority class. In this study, we focus on the following data difficulty factors: decomposing the minority class into sub-concepts, overlapping between classes, presence of outliers, and rare instances.

The influence of *class decomposition* into smaller sub-parts has been noticed by Japkowicz et al. [9]. Their experimental studies have demonstrated that the degradation of classification performance has resulted from the fragmentation of the minority class, rather than from changing the global imbalance ratio. Such sub-clusters of minority examples, surrounded by majority examples correspond to, so called, *small disjuncts*, which are harder to learn and cause more classification errors than larger sub-concepts [16]. Other experiments [13] showed that classification performance drops when decision boundaries around sub-clusters are non-linear and overlap with majority class examples. Finally, a visual analysis of projections of popular imbalanced UCI data [12] confirmed that the minority class often does not form a compact homogeneous distribution, but is scattered into many smaller sub-clusters surrounded by majority examples.

High *overlapping* between regions of different classes in the attribute space has already been recognized as particularly influential for standard, balanced, classification problems. However, its impact is even stronger when recognizing minority examples, see e.g. experimental studies [5]. The authors of these studies have also shown that the local imbalance ratio inside the overlapping region is more influential than the global ratio.

Other researchers characterize difficulty factors by considering mutual positions of a minority example with respect to other examples. One of the first studies in this direction [10] distinguished between *safe* and *unsafe* examples. More precisely, examples located in homogenous sub-regions populated by examples from the same class were called safe, whereas all other examples were denoted as unsafe. Napierala and Stefanowski proposed to further categorize unsafe examples into *borderline*, *rare cases*, and *outliers* [11]. Borderline examples can either be located in the overlapping between classes or positioned very close to complex non-linear decision boundaries. The two other types of examples occur deeper inside the safe region of the majority class. Outliers are isolated minority class singletons, whereas rare examples correspond to very small groups (pairs or triples) of examples. Comprehensive experiments [11,12] have shown that most benchmark imbalanced datasets contain mainly unsafe minority examples and the categorization of unsafe data correlates with the performance of classifiers.

2.2 Identification of Sub-concepts and Types of Minority Examples

Nearly all approaches to identify sub-concepts apply clustering algorithms that are run on examples of a single class, without analyzing their relation to remaining classes. Japkowicz et al. [7,14] proposed a k-means based oversampling method, where random oversampling is applied to majority and minority class clusters until the global class distribution becomes balanced. Other researchers discover within-class sub-concepts while constructing a classifier, by exploiting classifier predictions to tune the number of clusters k [18]. Nevertheless, the use of clustering algorithms for real-world datasets is still a non-trivial task. In case of k-means, the main difficulty is to tune the number of clusters k. It is also not obvious which optimization criteria should be considered as most clustering evaluation metrics were proposed for purely unsupervised frameworks. Moreover, existing works focus on the minority class without taking into account its local relationship with majority examples and challenges, such as class overlapping, rare cases, and outliers.

In an attempt to address these issues, as one of the contributions of this paper, we verify the utility of density-based clustering algorithms for the task of detecting sub-concepts. One of the analyzed algorithms is DBSCAN [4], which is capable of finding clusters of any shape and does not require the specification of the number of clusters. Nevertheless, DBSCAN requires specification of the following parameters: the minimal number of data points min_points and the maximal distance among those points ϵ in order to begin the formation of a new cluster. Just as finding a suitable k in k-means, the tuning of min_points and $epsilon$ is not trivial.

Additionally, the proposed ImGrid algorithm is inspired by grid-based clusterers [3]. Algorithms from this group divide the attribute space into a set of cells, which are later joined in order to form clusters. To the best of our knowledge, grid-based clusterers have not been applied to imbalanced data analysis.

Concerning the identification of types of minority class examples, Napierala and Stefanowski [11] proposed to identify four types of examples (safe, borderline, rare, outlier) by analyzing class label distributions inside the neighborhood of each minority class example. The authors considered two ways of modeling the neighborhood, either with k-nearest neighbors or kernels. Depending on the number of examples from the majority class inside the neighborhood, it is estimated how safe or unsafe a minority example is. If all, or nearly all, its neighbors belong to the minority class, this example is treated as a safe one, otherwise it is categorized as one of three unsafe types: borderline, rare, outlier. The decision which of the four types should be assigned to a given example can either depend on the parameter k or thresholds on the within-region class probabilities.[1]

---

[1] Details on tuning the size of the neighborhood and a comparison between the k-NN and kernel-based approach can be found in [12].

3   Clustering and Categorizing Minority Examples

Following the critical discussion on limitations of existing approaches in the previous section, the main goal of our work is to create a clustering algorithm that is capable of not only discovering minority class sub-concepts, but also revealing their underlying example types. For this purpose, we put forward a novel learning approach, called *ImGrid*, and devise modifications of k-means and DBSCAN, called *ImKmeans* and *ImScan*.

ImGrid (Imbalanced Grid) is inspired by grid clustering algorithms [3]. The main steps of the algorithm involve: (1) dividing the attribute space into grid cells, (2) joining similar adjacent cells taking into account their minority class distributions, (3) labeling examples according to difficulty factors, (4) forming minority sub-clusters.

Since cells are joined based on example distributions, each cell should contain enough examples to make the estimation of the example density feasible. Hence, the presented algorithm divides each dimension of the attribute space into $\lceil \sqrt[m]{|D|/10} \rceil$ equally wide intervals, where $|D|$ is the size of the dataset and m is the number of dimensions of the attribute space. This formula, inspired by histogram bin count heuristics, ensures that, on average, we have 10 data points in each cell. The value 10 was chosen to make the cell as small as possible, while retaining a reasonable amount of data for statistical comparisons of cell distributions.

The second step of ImGrid requires a method for joining adjacent cells. The joining mechanism takes into account the distribution of minority and majority class examples in grid cells and combines them only if the distribution of the classes is similar. In particular, the algorithm aims at connecting cells that contain examples of similar difficulty, and one way of achieving this goal is to use the statistical hypothesis testing framework. The most popular tests for the comparison of discrete distributions are Pearson's chi-squared test and its exact alternatives, such as Fisher's exact test or Barnard's test [1]. However, since those tests cannot directly state that the distributions are identical (they can only fail to reject the null hypothesis), we decided to use a Bayesian test [8], which allows to calculate the hypotheses' probability. Using this test, ImGrid joins adjacent cells when the data statistically shows that the distributions are similar. The level of required confidence in order to merge cells is a parameter of the algorithm α; note that in this case α should be always greater than 0.5 (probability that the distributions are similar should be higher than they are not). Since in binary classification the comparison of the class distribution reduces to the analysis of two proportions, we have chosen a test constructed on the Bayes factor for the beta-binomial model. As prior distribution of the classes we use a non-informative Jeffreys prior [8].

In the third step of ImGrid, one of four difficulty labels (safe, borderline, rare, or outlier) is assigned to each cluster based on the ratio of minority and majority class examples. We refer to previous studies on modeling data difficulties [12] and use the following thresholds to assign the labels: if the proportion of minority examples p is greater than 0.7, the safe label is assigned; if $0.7 \geq p > 0.3$ then borderline label is attached; the rare or outlier label is assigned if $0.3 \geq p > 0.1$ or $0.1 \geq p > 0$, respectively [12].

Finally, having a clustering that divides the data into regions of different difficulties, adjacent cells containing minority examples are joined. By joining minority examples regardless of their difficulty labels, the algorithm forms minority sub-clusters. The pseudocode of ImGrid is presented in Algorithm 1.

Algorithm 1. ImGrid

Input: D: m-dimensional dataset, α: threshold for statistical test
Output: *types_grid*: grid with detected types, *clustering_grid*: grid with clustering

1: $grid \leftarrow$ split dimensions into $\left\lceil \sqrt[m]{|D|/10} \right\rceil$ equi-width intervals   ▷ 1) Create grid
2: to each *cell* $\in grid$ assign corresponding data points
3: **while** $[cell_1, cell_2] \leftarrow$ FIND_CELLS_TO_JOIN($grid$) **do**   ▷ 2) Join similar cells
4:     $grid.join(cell_1, cell_2)$
5: **for** *cell* $\in grid$ **do**   ▷ 3) Assign type to examples in cells
6:     $p \leftarrow cell.minority\_num/cell.example\_num$
7:     $cell.assign\_label(p)$
8: *types_grid* $\leftarrow grid.copy()$
9: **for** *cell* $\in grid$ **do**   ▷ 4) Form minority clusters
10:     **for** *neighbor* $\in cell.get\_neighbors()$ **do**
11:         **if** $cell.has\_minority()$ **and** $neighbor.has\_minority()$ **then**
12:             $grid.join(cell, neighbor)$
13: *clustering_grid* $\leftarrow grid$
14: **return** $[types\_grid, clustering\_grid]$

1: **function** FIND_CELLS_TO_JOIN($grid$)
2:     sort cells in $grid$ by the prevalence of minority class in descending order
3:     **for** *cell* $\in grid$ **do**
4:         **for** *neighbor* $\in cell.get\_neighbors()$ **do**
5:             $neighbor.p \leftarrow$ probability that the distribution of examples in *neighbor* and *cell* is the same
6:         $[p, best\_neighbor] \leftarrow neighbor$ with the highest $neighbor.p$
7:         **if** $p > \alpha$ **then return** $[cell, best\_neighbor]$
8:     **return false**

We also put forward extensions of the k-means and DBSCAN clustering algorithms. The proposed approaches consist of: (1) dividing the whole dataset into two datasets with examples of one class only, (2) performing standard clustering on the dataset with minority examples, (3) incorporating majority examples into the clustering result, and finally, (4) assigning difficulty labels to each cluster.

In the third step of the extensions we attempted to imitate the philosophy of the original clustering algorithms. ImKmeans assigns each majority example to the minority cluster with the nearest centroid. This naive approach relies solely on the global clustering information available in k-means, thus, local difficulty type categorization produced ImKmeans may be very imprecise. Conversely, ImScan attaches majority examples to the cluster of its nearest minority example, but only if the distance to the nearest minority example does not exceed ϵ. Both ImScan and ImKmeans assign difficulty labels using the same rules as ImGrid. Algorithms 2 and 3 present the pseudocodes of the proposed extensions.

Algorithm 2. ImKmeans

Input: D: dataset, k: number of clusters
Output: $clustering$: a set of clusters with detected types

1: $[D_+, D_-] \leftarrow$ split D based on class labels   ▷ 1) Divide dataset
2: $clustering \leftarrow$ K-MEANS(D_+, k) ▷ 2) Cluster minority examples
3: **for** $maj\_example \in D_-$ **do**   ▷ 3) Add majority examples to clusters
4:      $c \leftarrow$ cluster with the nearest centroid to $maj\_example$
5:      $c.add\_example(maj\_example)$
6: **for** $c \in clustering$ **do**   ▷ 4) Assign type to examples in clusters
7:      $p \leftarrow c.minority\_num/c.example\_num$
8:      $c.assign\_label(p)$
9: **return** $clustering$

Algorithm 3. ImScan

Input: D: dataset, ϵ: radius of considered neighborhood, min_points: minimum number of points required to form a dense region
Output: $clustering$: a set of cluster with detected types

1: $[D_+, D_-] \leftarrow$ split D based on class labels   ▷ 1) Divide dataset
2: $clustering \leftarrow$ DBSCAN$(D_+, \epsilon, min\_points)$ ▷ 2) Cluster minority examples
3: **for** $maj\_example \in D_-$ **do**   ▷ 3) Add majority examples to clusters
4:      $nearest \leftarrow$ minority example closest to $maj\_example$
5:      **if** $distance(nearest, maj\_example) < \epsilon$ **then**
6:          $c \leftarrow$ cluster of $nearest$
7:          $c.add\_example(maj\_example)$
8: **for** $c \in clustering$ **do**   ▷ 4) Assign type to examples in clusters
9:      $p \leftarrow c.minority\_num/c.example\_num$
10:      $c.assign\_label(p)$
11: **return** $clustering$

To the best of our knowledge, there have been no previous proposals of algorithms that simultaneously detect minority sub-concepts and identify local data difficulty factors. In the following section, we examine the utility of the proposed algorithms in terms sub-concept discovery and minority example categorization.

4   Experimental Study

In this section, we experimentally evaluate ImGrid, ImScan, and ImKmeans, on 78 synthetic datasets with controlled proportions and placement of data difficulty factors. The proposed algorithms are analyzed in terms of their ability to discover class sub-concepts and detect different difficulty labels. Hence, the clustering performance and the accuracy of difficulty type categorization is measured. Finally, we asses how well the algorithms balance clustering and categorization tasks, and compare their processing time.

4.1 Experiment Setup

In our experiments, we compare the proposed three algorithms (ImGrid, ImScan, ImKmeans) and the algorithm for the identification of example difficulty type by Napierala and Stefanowski [11] (Napierala) with the following parameters:

- ImGrid: $\alpha \in \{0.75, 0.80, 0.85, 0.90, 0.95\}$;
- ImScan: $\epsilon \in \{10, 30, 50, 70, 90\}$, $min\_points \in \{2\}$;
- ImKmeans: $k \in [1, 9]$;
- Napierala: number of neighbors $k \in \{5, 7, 9, 11\}$.

The algorithm of Napierala and Stefanowski was chosen as a baseline for categorizing minority class examples, however, this algorithm is not applicable to clustering tasks and will not be compared with the remaining approaches in terms of detecting sub-concepts. On the other hand, when analyzing clustering performance, ImKmeans and Imscan default to standard definitions of k-means and DBSCAN, and, therefore, can be considered as baseline approaches in terms of detecting minority sub-concepts. We note that the min_points parameter in ImScan was set to 2 to ensure that rare cases can be identified as separate clusters and distinguished from outliers.

Clustering was evaluated using Adjusted Mutual Information [17] (AMI) which takes into account not only the total number of clusters but also the correctness of example assignation to sub-concepts. Categorization was treated as a classification task and assessed using G-mean [10] over four difficulty types (safe, borderline, rare, outlier). These measures were selected, as they are deemed suitable for imbalanced data [6,17]. We note that traditionally if at least one class is unrecognized by the classifier, G-mean resolves to zero. To alleviate this property, we changed the recall of unrecognized classes from zero to 0.001. To differentiate from traditional G-mean, we denote the used measure as G-mean$^{0.001}$.

All the algorithms and evaluation methods were implemented in Python using the *scikit-learn* library [15].[2] Experiments were conducted on a machine equipped with an Intel i7-5500U 2.4 Ghz processor and 8 GB of RAM.

4.2 Datasets

In our experiments, we used 78 synthetic binary classification datasets with six basic shapes of minority sub-clusters and varying proportions of data difficulty

---

[2] Source code, datasets, and reproducible test scripts available at: https://github.com/langus0/imgrid.

factors. The datasets were created with the imbalanced dataset generator of Wilk et al. [20], which provides the ground truth for sub-concept and difficulty type detection. As this study focuses on local data difficulty factors, all the datasets have a constant 1:5 global class imbalance ratio. Table 1 presents the main characteristics of each dataset.

Table 1. Dataset characteristics; superscripts denote versions of datasets with different proportions of minority example types: ^{u}-unsafe, ^{b}-borderline, ^{r}-rare; subscript $_s$ denotes "sparse" versions of datasets, with much less examples

Dataset	Inst.	Attr.	Clust.	Safe	Border	Rare	Outlier
clover	1500	2/3	1	100 %	0 %	0 %	0 %
dis	1500	2/3/5	3	100 %	0 %	0 %	0 %
hyp	1500	2/3/5	1	100 %	0 %	0 %	0 %
joined	1500	2	4	100 %	0 %	0 %	0 %
normal	1500	2/3/5	1	100 %	0 %	0 %	0 %
rothyp	1500	2	1	100 %	0 %	0 %	0 %
<dataset>^{u}	...	...	+13	80 %	12 %	6 %	2 %
<dataset>^{b}	...	...	...	40 %	60 %	0 %	0 %
<dataset>^{r}	...	...	+50	30 %	40 %	20 %	10 %
<dataset>$_s$	250	...	...	100 %	0 %	0 %	0 %
<dataset>$_s^u$	250	...	+2	80 %	12 %	6 %	2 %

The clover dataset resembles a clover (Fig. 2a) with five prolonged leaves in 2d- (clover) and 3d-attribute space (clover₃). dis constitutes an example of spherical minority class sub-clusters in 2, 3, or 5 dimensions. Datasets hyp and rothyp exemplify a simple and rotated hyperplane decision boundary between two classes. The joined dataset allows to test the algorithms on overlapping sub-clusters, whereas normal is a uniformly distributed sphere in 2, 3 or 5 dimensions. Each of the six basic ("safe") datasets (clover, dis, hyp, joined, normal, rothyp) has five additional versions:

– ^{u}-**unsafe**: where the minority class also contains borderline instances, rare cases, and outliers;
– ^{b}-**borderline**: where the minority class is surrounded by a thick border of examples overlapping with the majority class;
– ^{r}-**rare**: where the number of safe examples is smaller than the number of unsafe examples;
– $_s$-**sparse**: where the dataset has much less examples, which introduces sparsity to the attribute space;
– $_s^u$-**sparse**: where the dataset is sparse and contains unsafe examples.

It is important to note that rare cases and outliers introduce additional minority sub-clusters to the dataset.

4.3   Minority Class Sub-clusters

Due to the large number of datasets and space limitations, detailed tabular results for each algorithm can be found in online supplementary materials.[3] In this and the following section, we summarize the results by means of selected plots, tabular summaries, and statistical hypothesis tests.

To compare the minority clustering performance of the proposed algorithms, we calculated Adjusted Mutual Information (AMI) for each clustering. However, since clustering is an unsupervised learning task and results can strongly depend on algorithm parameters (e.g. k in k-means), we compare the algorithms on two levels. The first level involves comparing *best models*, i.e., we choose the best parametrization of a given algorithm for each dataset separately, and report this "best" value for each dataset. This level corresponds to assessing algorithms, as if we explicitly knew how to tune them (which is usually not true). The second level involves comparing *mean models*, i.e., reporting the algorithms performance averaged over all parameterizations. This scenario corresponds to comparing algorithms as if they were parametrized by chance.

Table 2 shows detailed results for best models in terms of AMI on three selected datasets: $clover^u$, dis^b, and $rothyp^r$. Additionally, upper panels of Fig. 2 show sub-concept clusterings for the selected datasets. The results confirm commonly known, complementary, characteristics of k-means and density-based

Table 2. Clustering and categorization results for three selected datasets: $clover^u$, dis^b, and $rothyp^r$. Napierala added for comparison on categorization.

	G-mean$^{0.001}$	AMI	Time	Clusters	Safe	Border	Rare	Outlier
$clover^u$								
ImKmeans	0.006	0.038	0.640	2	0	0	250	0
ImScan	0.150	0.613	0.483	11	0	235	8	7
ImGrid	0.530	0.486	0.142	9	140	58	44	8
Napierala	0.758	-	0.032	—	105	117	23	5
dis^b								
ImKmeans	0.002	0.980	0.699	4	0	0	500	0
ImScan	0.032	0.577	0.438	2	0	250	0	0
ImGrid	0.526	0.577	0.156	2	66	121	57	6
Napierala	0.526	-	0.031	—	60	123	66	1
$rothyp^r$								
ImKmeans	0.117	0.167	2.183	9	111	0	82	57
ImScan	0.106	0.183	0.428	43	82	10	123	35
ImGrid	0.134	0.000	0.165	1	134	8	59	49
Napierala	0.171	-	0.024	—	127	6	61	56

[3] http://www.cs.put.poznan.pl/dbrzezinski/software/MinorityAnalysis.html.

clustering algorithms. On the dis^b dataset, ImKmeans is capable of finding the perfect clustering, as this dataset has only homogenous sub-concepts. On $clover^u$ and $rothyp^r$, however, ImKmeans has trouble with noisy examples. Conversely, ImScan and ImGrid perform quite well on noisy datasets, but fail to detect the right number of clusters, when the sub-concepts overlap.

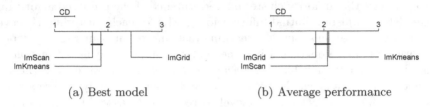

(a) Best model   (b) Average performance

Fig. 1. Performance ranking using AMI. Algorithms that are not significantly different according to the Nemenyi test (at $\alpha = 0.05$) are connected.

Figure 1 graphically presents the results of the Friedman test with Nemenyi post-hoc analysis for both levels of comparison. The null-hypothesis that best models for each algorithm perform similarly was rejected with $p < 0.001$. Ideally tuned versions of ImKmeans and ImScan are significantly better in terms of AMI than ImGrid. However, the null-hypothesis of the Friedman test for comparing mean models cannot be rejected ($p = 0.245$). Moreover, ImGrid obtains the highest mean rank in this comparison. This shows how crucial parameter tuning is to the performance of k-means and DBSCAN.

4.4   Minority Example Categorization

We also compared the ability of the algorithms to detect difficulty types of minority examples. Lower panels of Fig. 2 show example categorization corresponding to clusterings from the upper panels.

One can notice that ImScan and ImGrid obtain quite accurate difficulty type predictions. It is worth noting, however, that on the presented plots ImGrid produces more accurate predictions. This is due to the fact that the plots present best models in terms of the clustering performance (AMI). Contrary to ImGrid, ImScan's results are not robust to the change of parameters, hence its best parametrization for clustering did not usually correspond with the best model for categorization. It is also worth noting that ImGrid achieves G-mean$^{0.001}$ values fairly close to those obtained by Napierala, which is an algorithm designed strictly for detecting example difficulty types and has no clustering capabilities.

As it was done for clustering performance, we also performed the Friedman test to assess the significance of differences in performance for best and mean models. However, in this comparison we additionally analyze the performance of the algorithm of Napierala and Stefanowski. As Fig. 3 shows, in terms of G-mean$^{0.001}$, Napierala is the best categorization algorithm, but it is not significantly better than ImGrid when looking at mean models. Moreover, once again it can be noticed that ImScan is highly sensitive to parameter tuning.

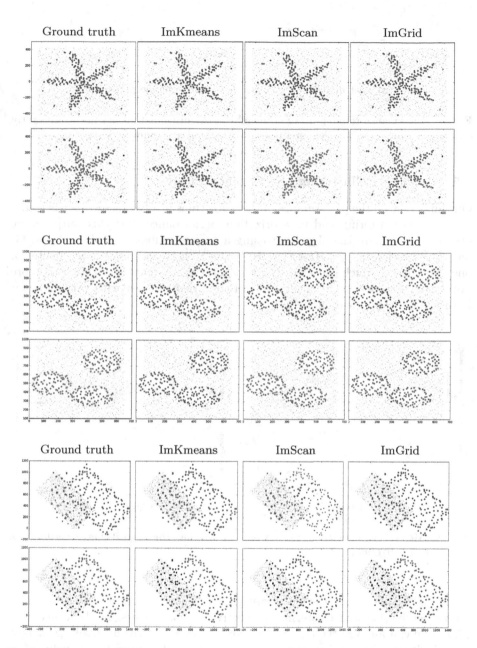

Fig. 2. Comparison of clustering results (upper panel of each pair) and minority type identification (lower panels) on the `clover`[u], `dis`[b], and `rothyp`[r] datasets. Clusters in upper panels are differentiated using shapes and colors. Types of minority class examples in lower panels are color-coded as follows: safe - green, borderline - orange, rare - red, outlier - black. Figure should be read in color. (Color figure online)

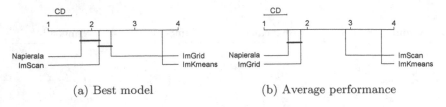

(a) Best model   (b) Average performance

Fig. 3. Performance ranking using G-mean$^{0.001}$. Algorithms that are not significantly different according to the Nemenyi test (at $\alpha = 0.05$) are connected.

4.5   Balancing Clustering and Categorization

Our final view on the performance of the algorithms involved assessing the trade-off between clustering and categorization performance. For this purpose, we decided to evaluate the algorithm using a linear combination of AMI and G-mean$^{0.001}$, as follows: $\beta AMI + (1 - \beta) Gmean^{0.001}$. By varying the parameter β, one can control which aspect of the task, clustering or categorization, is more important. Figure 4 shows mean ranks of the Friedman test with varying β, for both the best and mean models.

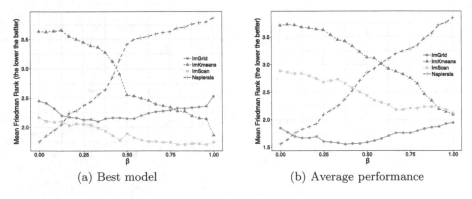

(a) Best model   (b) Average performance

Fig. 4. Mean Friedman test ranks when evaluating the algorithms according to: $\beta AMI + (1 - \beta) Gmean^{0.001}$

For $\beta = 0$ and $\beta = 1$, the mean ranks in Fig. 4 correspond to results presented for categorization and clustering, respectively. Nonetheless, in the range $\beta \in (0, 1)$ one can analyze the trade-off offered by each algorithm. Looking at best models, it can be seen that ImScan can be successfully tuned to any value of β, with ImGrid being usually second. However, when comparing mean models, it can be noticed that on average ImGrid produces better results, suggesting that is much less prone to parameter tuning. This is an important finding, as the goal of this study was to simultaneously detect minority sub-clusters and data difficulty factors, without prior knowledge about how to cluster examples.

Furthermore, we measured the running time of each algorithm and performed a Friedman test (Fig. 5). One can notice that Napierala is the fastest approach, however, it is an algorithm for the recognition of difficulty types only. Among algorithms which provide information about both minority sub-concepts and examples difficulty, ImGrid is significantly the fastest method. In terms of concrete values, ImGrid is on average almost two times faster than ImScan, its best competitor.

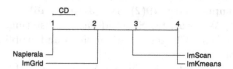

Fig. 5. Running time ranking. Algorithms that are not significantly different according to the Nemenyi test (at $\alpha = 0.05$) are connected.

5   Conclusions and Future Research

The main aim of this study was to find novel ways of discovering local data difficulty factors from imbalanced data. Up till now, efforts in this field have concentrated, *separately*, on detecting sub-concepts of the minority class and detecting local relationships between minority and majority examples. We argue that existing approaches to identifying minority sub-concepts are impractical as they heavily rely on non-trivial parameter tuning and are sensitive to outliers and other difficulty factors.

In this paper, we put forward ImGrid, an algorithm that *simultaneously* detects minority sub-clusters, outliers, rare cases, and class overlapping in imbalanced data. Additionally, we proposed two extensions to popular clustering algorithms, ImKmeans and ImScan, that incorporate knowledge about relationships between minority and majority examples. A comprehensive series of experiments characterized the strengths and weaknesses of each algorithm, showing that, depending on parameter tuning, each of the proposed algorithms is capable of successfully detecting sub-concept or characterizing difficulty types. However, the results highlighted ImGrid as a fast and easily parametrized trade-off between minority class clustering and example categorization.

Due to its small dependency on parameter tuning, ImGrid could be used to analyze real world datasets. Nevertheless, as future work the topic of defining its grid space for real data may be revisited, as more flexible approaches to dividing the attribute space can still be proposed. Moreover, the combination of clustering and example categorization gives the user two layers of information about an imbalanced dataset. These layers could be combined in a data difficulty metric, which would inform the user about the main difficulties in the dataset and suggest possible actions. Finally, it would be very interesting to use the gathered information to improve specialized algorithms for imbalanced data.

Acknowledgments. The authors' research was partly funded by the Polish National Science Center under Grant No. DEC-2013/11/B/ST6/00963. D. Brzezinski acknowledges the support Institute of Computing Science Statutory Funds.

References

1. Barnard, G.: A new test for 2×2 tables. Nature **156**, 177 (1945)
2. Branco, P., Torgo, L., Ribeiro, R.P.: A survey of predictive modeling on imbalanced domains. ACM Comput. Surv. **49**(2), 31:1–31:50 (2016)
3. Cheng, W., Wang, W., Batista, S.: Grid-based clustering. In: Aggarwal, C.C., Reddy, C.K. (eds.) Data Clustering: Algorithms and Applications, pp. 127–148. CRC Press, London (2013)
4. Ester, M., Kriegel, H., Sander, J., Xu, X.: A density-based algorithm for discovering clusters in large spatial databases with noise. In: Simoudis, E., Han, J., Fayyad, U.M. (eds.) Proceedings of the 2nd International Conference on Knowledge Discovery and Data Mining, pp. 226–231 (1996)
5. García, V., Sánchez, J., Mollineda, R.: An empirical study of the behavior of classifiers on imbalanced and overlapped data sets. In: Rueda, L., Mery, D., Kittler, J. (eds.) CIARP 2007. LNCS, vol. 4756, pp. 397–406. Springer, Heidelberg (2007). doi:10.1007/978-3-540-76725-1_42
6. He, H., Ma, Y. (eds.): Imbalanced Learning: Foundations, Algorithms, and Applications. Wiley-IEEE Press, Hoboken (2013)
7. Japkowicz, N., Stephen, S.: The class imbalance problem: a systematic study. Intell. Data Anal. **6**(5), 429–449 (2002)
8. Jeffreys, H.: Some tests of significance, treated by the theory of probability. Proc. Camb. Philos. Soc. **31**, 203–222 (1935)
9. Jo, T., Japkowicz, N.: Class imbalances versus small disjuncts. SIGKDD Explor. **6**(1), 40–49 (2004)
10. Kubat, M., Matwin, S.: Addressing the curse of imbalanced training sets: one-sided selection. In: Proceedings of the International Conference on Machine Learning, pp. 179–186 (1997)
11. Napierala, K., Stefanowski, J.: Identification of different types of minority class examples in imbalanced data. In: Corchado, E., Snášel, V., Abraham, A., Woźniak, M., Graña, M., Cho, S.-B. (eds.) HAIS 2012. LNCS, vol. 7209, pp. 139–150. Springer, Heidelberg (2012). doi:10.1007/978-3-642-28931-6_14
12. Napierala, K., Stefanowski, J.: Types of minority class examples and their influence on learning classifiers from imbalanced data. J. Intell. Inf. Syst. **46**(3), 563–597 (2016)
13. Napierała, K., Stefanowski, J., Wilk, S.: Learning from imbalanced data in presence of noisy and borderline examples. In: Szczuka, M., Kryszkiewicz, M., Ramanna, S., Jensen, R., Hu, Q. (eds.) RSCTC 2010. LNCS, vol. 6086, pp. 158–167. Springer, Heidelberg (2010). doi:10.1007/978-3-642-13529-3_18
14. Nickerson, A., Japkowicz, N., Milios, E.E.: Using unsupervised learning to guide resampling in imbalanced data sets. In: Proceedings of the 8th International Conference on Artificial Intelligence and Statistics, pp. 261–265. Society for Artificial Intelligence and Statistics (2001)
15. Pedregosa, F., et al.: Scikit-learn: machine learning in Python. J. Mach. Learn. Res. **12**, 2825–2830 (2011)

16. Prati, R.C., Batista, G.E.A.P.A., Monard, M.C.: Class imbalances *versus* class overlapping: an analysis of a learning system behavior. In: Monroy, R., Arroyo-Figueroa, G., Sucar, L.E., Sossa, H. (eds.) MICAI 2004. LNCS, vol. 2972, pp. 312–321. Springer, Heidelberg (2004). doi:10.1007/978-3-540-24694-7_32
17. Romano, S., Vinh, N.X., Bailey, J., Verspoor, K.: Adjusting for chance clustering comparison measures. J. Mach. Learn. Res. **17**(134), 1–32 (2016)
18. Sobhani, P., Viktor, H., Matwin, S.: Learning from imbalanced data using ensemble methods and cluster-based undersampling. In: Appice, A., Ceci, M., Loglisci, C., Manco, G., Masciari, E., Ras, Z.W. (eds.) NFMCP 2014. LNCS, vol. 8983, pp. 69–83. Springer, Cham (2015). doi:10.1007/978-3-319-17876-9_5
19. Stefanowski, J.: Dealing with data difficulty factors while learning from imbalanced data. In: Matwin, S., Mielniczuk, J. (eds.) Challenges in Computational Statistics and Data Mining. SCI, vol. 605, pp. 333–363. Springer, Cham (2016). doi:10.1007/978-3-319-18781-5_17
20. Wojciechowski, S., Wilk, S.: Difficulty factors and preprocessing in imbalanced data sets: an experimental study on artificial data. Found. Comput. Decis. Sci. **42**(2), 149–176 (2017)

Fusion Techniques for Named Entity Recognition and Word Sense Induction and Disambiguation

Edmundo-Pavel Soriano-Morales[✉], Julien Ah-Pine, and Sabine Loudcher

Université de Lyon, Lyon 2, ERIC EA 3083, Bron, France
{edmundo.soriano-morales,julien.ah-pine,sabine.loudcher}@univ-lyon2.fr

Abstract. In this paper we explore the use of well-known multimodal fusion techniques to solve two prominent Natural Language Processing tasks. Specifically, we focus on solving Named Entity Recognition and Word Sense Induction and Disambiguation by applying feature-combination methods that have already shown their efficiency in the multimedia analysis domain. We present a series of experiments employing fusion techniques in order to combine textual linguistic features. Our intuition is that by combining different types of features we may find semantic relatedness among words at different levels and thus, the combination (and recombination) of these levels may yield gains in terms of metrics' performance. To our knowledge, employing these techniques has not been studied for the tasks we address in this paper. We test the proposed fusion techniques on three datasets for named entity recognition and one for word sense disambiguation and induction. Our results show that the combination of textual features indeed improves the performance compared to single feature representation and the trivial feature concatenation.

1 Introduction

Named Entity Recognition (NER) and Word Sense Induction and Disambiguation (WSI/WSD) requires textual features to represent the similarities between words in order to discern between different words' meanings. NER goal is to automatically discover, within a text, mentions that belong to a well-defined semantic category. The classic task of NER involves detecting entities of type Location, Organization, Person and Miscellaneous. The task is of great importance for more complex NLP systems, e.g., relation extraction, opinion mining. Common solutions to NER consist on one of the following: via matching patterns created manually or extracted semi-automatically; or by training a supervised machine learning algorithm with large quantities of annotated text. The latter being the currently more popular solution to this task.

Word Sense Induction and Disambiguation involves two closely related tasks[1]. WSI aims to automatically discover the set of possible senses for a target word given a text corpus containing several occurrences of said target word.

---

[1] Even though these tasks are related, they are independent from one another. Still, in this paper we consider them to be a single one.

© Springer International Publishing AG 2017
A. Yamamoto et al. (Eds.): DS 2017, LNAI 10558, pp. 340–355, 2017.
DOI: 10.1007/978-3-319-67786-6_24

Meanwhile, WSD takes a set of possible senses and determines the most appropriate sense for each instance of the target word according to the instance's context. WSI is usually approached as an unsupervised learning task, i.e., a cluster method is applied to the words occurring in the instances of a target word. The groups found are interpreted as the senses of the target word. The WSD task is usually solved with knowledge-based approaches, based on WordNet; or more recently with supervised models which require annotated data.

As stated before, both tasks rely on features extracted from text. Usually, these representations are obtained from the surrounding context of the words in the input corpus. Mainly two types of representations are used. According to their nature we call these features lexical and syntactical. The first type requires no extra information than that contained already in the analyzed text itself. It consists merely on the tokens surrounding a word, i.e., those tokens that come before and after within a fixed window. The second type, syntactical features, is similar to the lexical representation in that we also consider as features the tokens that appear next to the corpus' words. Nonetheless, it requires a deeper degree of language understanding. In particular, these features are based on part of speech tags, phrase constituents information, and syntactical functionality between words, portrayed by syntactical dependencies. Likewise, specific features, particular to a task are also employed. These features later on become standard features in the literature.

Most of the approaches in the literature dealing with these tasks use each of these features independently or stacked together, i.e., different feature columns in an input representation space matrix. In the latter case, features are usually combined without regards to their nature.

The main intuition of the present work is that word similarities may be found at different levels according to the type of features employed. In order to exploit these similarities, we look into multimedia fusion methods. In order to better perform an analysis task, these techniques combine multimodal representations, their corresponding similarities, or the decisions coming from models fitted with these features. In this paper, we try to mutually complement independent representations by utilizing said fusion techniques to combine (or fuse) features in the hope of improving the performance of the tasks at hand, specially compared to the use of features independently.

Fusion techniques have previously shown their efficiency, mainly on text and image related tasks, where there is a need to model the relation between images and text extracts. Here, in order to apply multimedia fusion techniques, we consider textual features as different modalities, i.e., instead of having textual and image features we have lexical and syntactical features. The main contribution of this work is to assess the effectiveness of simple yet untested techniques to combine classical and easy to obtain textual features. As a second contribution, we propose a series of feature combination and recombination to attain better results. We test our intuitions on both NER and WSI/WSD tasks and over four different corpora: CoNLL-2003 [17], WikiNER and Wikigold [4] for NER; Semeval-2007 [1] for WSI/WSD.

The rest of the paper is organized as follows: in Sect. 2, we go into further details about fusion techniques. We introduce the fusion operators that we use in our experiments in Sect. 3. Then, in Sect. 4 we show the effectiveness of the presented methods by testing them on NER and WSI/WSD and their respective datasets. Finally, in Sect. 5 we present our conclusions and future directions to explore.

2   Background and Related Work

In this section, we describe the fusion techniques we use in our methodology as well as relevant use-cases where they have been employed.

2.1   Multimodal Fusion Techniques

Multimodal fusion is a set of popular techniques used in multimedia analysis tasks. These methods integrate multiple media features, the affinities among these attributes or the decisions obtained from systems trained with said features, to obtain rich insights about the data being used and thus to solve a given analysis task [2,3]. We note that these techniques come at the price of augmenting the training time of a system by increasing both the dimension space and/or the density of a given feature matrix.

In the multimodal fusion literature we can discern two main common types of techniques: early fusion and late fusion.

Early Fusion. This technique is the most widely used fusion method. The principle is simple: we take both modal features and concatenate them into a single representation matrix. More formally, we consider two matrices that represent different modality features each over the same set of individuals. To perform early fusion we concatenate them column-wise, such that we form a new matrix having the same number of lines but increasing the number of columns to the sum of the number of columns of both matrices. The matrices may also be weighted as to control the influence of each modality.

The main advantage of early fusion is that a single unique model is fitted while leveraging the correlations among the concatenated features. The method is also easy to integrate into an analysis system. The main drawback is that we increase the representation space and may make it harder to fit models over it.

Late Fusion. In contrast to early fusion, in late fusion the combination of multimodal features are generally performed at the decision level, i.e., using the output of independent models trained each with an unique set of features [5]. In this setting, decisions produced by each model are combined into a single final result set. The methods used to combine preliminary decisions usually involve one of two types: rule-based (where modalities are combined according to domain-specific knowledge) or linear fusion (e.g., weighting and then adding or multiplying both matrices together). This type of fusion is very close to the so-called ensemble methods in the machine learning literature. Late fusion combines both modalities in the same semantic space. In that sense, we may also combine modalities via an affinity representation instead of final decision sets. In other words, we can combine two modality matrices by means of their respective similarities. A final representation is then usually obtained by adding the weighted similarity matrices.

The advantages of late fusion include the combination of features at the same level of representation (either the fusion of decisions or similarity matrices). Also, given that independent models are trained separately, we can chose which algorithm is more adequate for each type of features.

Cross-Media Similarity Fusion. A third type of fusion technique, cross-media similarity fusion (or simply cross fusion), introduced in [2,5], is defined and employed to propagate a single similarity matrix into a second similarity matrix. In their paper, the authors propagated information from textual media towards visual media. In our case, we transfer information among textual features. For example, to perform a cross fusion between lexical and syntactical features, we perform the following steps:

1. Compute the corresponding similarity matrices for each type of feature.
2. Select only the k-nearest neighbor for each word within the lexical similarity matrix. These neighbors are to be used as lexical representatives to enrich the syntactical similarities.
3. Linearly combine both similarity matrices (lexical k-nearest lexical neighbors with the syntactical features) via a matrix product.

Cross fusion aims to bridge the semantic gap between two modalities by using the most similar neighbors as proxies to transfer valuable information from one modality onto another one. Usually, the result of a cross fusion is combined with the previous techniques, early and late fusion. In this paper we perform experiment in that sense.

Hybrid Fusion. We may leverage the advantages of the previous two types of fusion techniques by combining them once more in a hybrid setting. As described in [3,18], the main idea is to simultaneously combine features at the feature level, i.e., early fusion, and at the same semantic space or decision level. Nonetheless, they define a specific type of hybrid fusion. In this paper, we adopt a looser definition of hybrid fusion. That is, we perform hybrid fusion by leveraging the combination of the fusion strategies described before.

We consider the first three types of fusion techniques (early fusion, late fusion and cross fusion) as the building blocks to the experiments we conduct. While we work with a single modality, i.e., textual data, we consider the different kinds of features extracted from it as distinct modalities. Our intuition being that the semantic similarities among words in these different spaces can be combined in order to exploit the latent complementarity between the lexical and syntactical representations. The fusion should therefore improve the performance of the NLP tasks at hand, NER and WSI/WSD.

Our first goal is to assess the effectiveness of the classic fusion methods and then, as a second goal, to propose new combinations that yield better outcomes in terms of performance than the simpler approaches. The new combinations are found empirically. Nonetheless, as we will show, their effectiveness replicates across different datasets and NLP tasks.

2.2 NER and WSI/WSD

To the best of our knowledge, there is no work that addresses both NER and WSI/WSD explicitly while using fusion techniques from the multimedia analysis domain. Still, we base our experiments on those carried on in [6,8,10] using well-known supervised (structured perceptron) and unsupervised (spectral clustering) learning algorithms. A thorough review on NER and WSI/WSD can be found in [13,14], respectively.

3 Methodology

In the present section we address the core of the work performed in this paper. We formally describe the fusion techniques we employ in the next section. Also, we delineate the procedure followed in our experiments.

The experiments we carry on consist in generating fusion matrices that will serve as input to a learning algorithm in order to solve NER and WSI/WSD. These input feature matrices are based upon lexical, syntactical, or other types of representation. The procedure can be seen in Fig. 1.

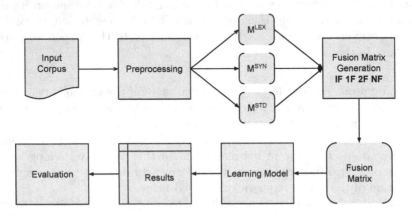

Fig. 1. Steps followed on our experiments. First the corpus is preprocessed, then features are extracted from the text. A fusion matrix is generated, which in turn is used as input to a learning algorithm. Finally, the system yields its results and to be analyzed.

3.1   Fusion Strategies

We begin by presenting a formal definition of the fusion techniques employed and described in the previous sections. We define (weighted) early fusion, late fusion and cross fusion as follows:

Early Fusion

$$E(A, B) = \mathbf{hstack}(A, B) \tag{1}$$

Weighted Early Fusion

$$wE_\alpha(A, B) = \mathbf{hstack}(\alpha \cdot A, (1 - \alpha) \cdot B) \tag{2}$$

Late Fusion

$$L_\beta(A, B) = \beta \cdot A + (1 - \beta) \cdot B \tag{3}$$

Cross fusion

$$X_\gamma(A, B) = \mathbf{K}(A, \gamma) \times B \tag{4}$$

Parameters A and B are arbitrary input matrices. They may initially represent, for example, the lexical (M^{LEX}) or syntactical based (M^{SYN}) features matrix, or their corresponding similarity matrices, S^{LEX} and S^{SYN}, respectively. In a broader sense, matrices A and B may represent any pair of valid[2] fusion matrices.

In early fusion, $E(A, B)$, the matrices A and B are combined together via a function called **hstack** which concatenates, column-wise, both matrices A and B. Weighted early fusion represents the same operation as before with an extra parameter: α, which controls the relative importance of each matrix. In the following, we refer to both operations as early fusion. When α is determined, we refer to weighted early fusion.

Regarding late fusion $L_\beta(A, B)$, the β parameter determines again the importance of the matrix A, and consequently also the relevance of matrix B.

In cross fusion $X_\gamma(A, B)$, the $\mathbf{K}(\cdot)$ function keeps the top-γ closest words (columns) to each word (lines) while the rest of the values are set to zero.

Using the previously defined operators, we distinguish four levels of experiments:

1. **Single Features:** in this phase we consider the modalities independently as input to the learning methods. For instance, we may train a model for NER using only the lexical features matrix M^{LEX}.
2. **First Degree Fusion:** we consider the three elementary fusion techniques by themselves (early fusion, late fusion, cross fusion) without any recombination. These experiments, as well as those from the previous level, serve as the baselines we set to surpass in order to show the efficacy of the rest of the fusion approaches. As an example, we may obtain a representation matrix by performing an early fusion between the lexical matrix and the syntactical features matrix: $E(M^{LEX}, M^{SYN})$. In this level we distinguish

---

[2] Valid in terms of having compatible shapes while computing a matrix sum or multiplication.

two types of cross fusion: Cross Early Fusion (XEF) and Cross Late Fusion (XLF). The first one combines a similarity matrix with a feature matrix: $X(S^{LEX}, M^{SYN})$. The second one joins a similarity matrix with a similarity matrix: $X(S^{SYN}, S^{LEX})$.

3. **Second Degree Fusion:** we recombine the outputs of the previous two levels with the elementary techniques. This procedure then yields a recombination of "second-degree" among fusion methods. We introduce the four types of second degree fusions in the following list. Each one is illustrated with an example:

   (a) Cross Late Early Fusion (XLEF): $X(X(S^{STD}, S^{SYN}), M^{STD})$
   (b) Cross Early Early Fusion (XEEF: $X(S^{STD}, X(S^{STD}, S^{SYN}))$
   (c) Early Cross Early Fusion (EXEF): $E(M^{STD}, X(S^{LEX}, M^{STD}))$
   (d) Late Cross Early Fusion (LXEF): $L(M^{STD}, X(S^{STD}, M^{STD}))$

4. **N-Degree Fusion:** in this last level we follow a similar approach to the previous level by combining the output of the second-degree fusion level multiple times (more than two times) with other second-degree fusion outputs. Again, in this level we test the following two fusion operations:

   (a) Early Late Cross Early Fusion (ELXEF): $E(M^{STD}, L(M^{STD}, X(S^{STD}, M^{STD})))$
   (b) Early ELXEF (EELXEF): $E(M^{LEX}, E(E(M^{STD}, L(M^{STD}, X(S^{STD}, M^{STD}))), L(M^{LEX}, X(S^{SYN}, M^{LEX}))))$

3.2 Feature Matrices

In the previous subsection we presented the fusion operators used in our experiments. Below we detail the three types of features used to describe the words of each of the tested corpus.

Lexical Matrix (LEX). For each token in the corpus, we use a lexical window of two words to the left and two words to the right, plus the token itself. Specifically, for a target word w, its lexical context is $(w_{-2}, w_{-2}, w, w_{+1}, w_{+2})$. This type of context features is typical for most systems studying the surroundings of a word, i.e., using a distributional approach [11].

Syntactical Matrix (SYN). Based on the syntactic features used in [11,15], we derive contexts based on the syntactic relations a word participates in, as well as including the part of speech (PoS) of the arguments of these relations. Formally, for a word w with modifiers $m_1, \ldots, m_k$ and their corresponding PoS tags $p_1^m, \ldots, p_k^m$; a head h and its corresponding PoS tag p_h, we consider the context features $(m_1, p_{m_1}, lbl_1), \ldots, (m_k, p_{m_k}, lbl_k), (h, p_h, lbl\_inv_h)$. In this case, lbl and lbl_{inv} indicate the label of the dependency relation and its inverse, correspondingly. Using syntactic dependencies as features should yield more specific similarities, closer to synonymy, instead of the broader topical similarity found through lexical contexts.

NER Standard Features Matrix (STD). The features used for NER are based on those used in [4,8]. The feature set consists of: the word itself, whether the word begins with capital letter, prefix and suffix up to three characters (within a window of two words to the left and two words to the right), and the PoS tag of the current word. These features are considered to be standard in the literature. We note that the matrix generated with these features is exclusively used in the experiments regarding NER.

3.3   Learning Methods

We use supervised and unsupervised learning methods for NER and WSI/WSD respectively. On the one hand, for NER, as supervised algorithm, we use an averaged structured perceptron [6,8] to determine the tags of the named entities. We considered Logistic Regression and linear SVM. We chose the perceptron because of its performance and its lower training time.

On the other hand, for WSD/WSI, specifically for the induction part, we applied spectral clustering, as in [10], on the input matrices in order to automatically discover senses (a cluster is considered a sense). Regarding disambiguation, we trivially assign senses to the target word instances according to the number of common words in each cluster and the context words of the target word. In other words, for each test instance of a target word, we select the cluster (sense) with the maximum number of shared words with the current instance context.

4   Experiments and Evaluation

We experiment with four levels of fusion: Single Features (SF), First-degree Fusion (1F), Second-degree Fusion (2F) and N-degree Fusion (NF). The representation matrices for NER come from lexical context features M^{LEX}, syntactical context features M^{SYN} or standard features M^{STD}. On the other hand, experiments on WSI/WSD exclusively employ matrices M^{LEX} and M^{SYN}.

Our first goal is to compare the efficiency of the basic multimedia fusion techniques applied to single-modality multi-feature NLP tasks, namely NER and WSI/WSD. A second goal is to empirically determine a fusion combination setting able to leverage the complementarity of our features.

To this end, we evaluate the aforementioned 4 fusion levels. We note that the fusion combinations in the third and fourth level (2F and NF) are proposed based on the results obtained in the previous levels. In other words, in order to reduce the number of experiments, we restrict our tests to the best performing configurations. This is due to the large number of possible combinations (an argument to a fusion operation may be any valid output of a second fusion operation).

4.1   Named Entity Recognition

Pre-processing. As is usual when preprocessing text before performing named entity recognition, [16], we normalize tokens that include numbers. This allows a degree of abstraction to tokens that contain years, phone numbers, etc.

Features. The linguistic information we use are extracted with the Stanford's CoreNLP parser [12]. Again, the features used for these experiments on NER are those described before: lexical, syntactic and standard features, i.e., M^{LEX}, M^{SYN}, and M^{STD}, respectively.

Test Datasets. We work with three corpora coming from two different domains:

(1) CoNLL-2003 (CONLL): This dataset was used in the language-independent named entity recognition CoNLL-2003 shared task [17]. It contains selected news-wire articles from the Reuters Corpus. Each article is annotated manually. It is divided in three parts: training (*train*) and two testing sets (*testa* and *testb*). The training part contains 219,554 lines, while the test sets contain 55,044 and 50,350 lines, respectively. The task was evaluated on the *testb* file, as in the original task.

(2) WikiNER (WNER): A more recent dataset [4] of selected English Wikipedia articles, all of them annotated automatically with the author's semi-supervised method. In total, it contains 3,656,439 words.

(3) Wikigold (WGLD): Also a corpus of Wikipedia articles, from the same authors of the previous corpus. Nonetheless, this was annotated manually. This dataset is the smaller, with 41,011 words. We used this corpus to validate human-tagged Wikipedia text. These three datasets are tagged with the same four types of entities: Location, Organization, Person and Miscellaneous.

Evaluation Measures. We evaluate our NER models following the standard CoNLL-2003 evaluation script. Given the amount of experiments we carried on, and the size constraints, we report exclusively the total F-measure for the four types of entities (Location, Organization, Person, Miscellaneous). WNER and WGLD datasets are evaluated on a 5-fold cross validation.

Results. We present in this subsection the results obtained in the named entity recognition task, while employing the 4 levels of fusion proposed in the previous section.

In contrast to other related fusion works [2,5,9], we do not focus our analysis on the impact of the parameters of the fusion operators. Instead, we focus our analysis on the effect of the type of linguistic data being used and how, by transferring information from one feature type to another, they can be experimentally recombined to generate more complete representations.

Regarding the fusion operators' parameters, we empirically found the best configuration for β, from late fusion $L_\beta(A, B) = \beta \cdot A + (1 - \beta) \cdot B$, is $\beta = 0.5$. This implies that an equal combination is the best linear fusion for two different types of features.

In respect of the γ parameter, used in cross fusion $X_\gamma(A, B) = \mathbf{K}(A, \gamma) \times B$, we set $\gamma = 5$. This indicates that just few high quality similarities attain better results than utilizing a larger quantity of lower quality similarities.

Table 1. NER F-measure results using the Single Features over the three datasets. These values serve as a first set of baselines.

A	Single Features		
	CONLL	WNER	WGLD
M^{STD}	77.41	77.50	59.66
M^{LEX}	69.40	69.17	52.34
M^{SYN}	32.95	28.47	25.49

Single Features. Looking at Table 1, we see that the best independent features, in terms of F-measure come from the standard representation matrix M^{STD}. This is not surprising as these features, simple as they may be, have been used and proved extensively in the NER community. On the other hand, M^{LEX} performs relatively well, considering it only includes information contained in the dataset itself. Nevertheless, this representation that this kind of lexical context features are the foundation of most word embedding techniques used nowadays. While we expected better results from the syntactical features M^{SYN}, as they are able to provide not only general word similarity, but also functional, getting close to synonymy-level [11], we believe that the relatively small size of the datasets do not provide enough information to generalize.

First Degree Fusion. In Table 2 we present the First Degree fusion level. The best performance is obtained by trivially concatenating the representation matrices. This baseline proved to be the toughest result to beat. Late fusion does not perform well in this setting, still, we see further on that by linearly combining weighted representation matrices, we can add information to an already strong representation. Finally, regarding the cross fusion techniques, cross early and late fusion, we see that they depend directly on the information contained in the similarity matrices. We note that, as is the case on single features, the combinations with matrix S^{STD} yield almost always the best results. While these fusion techniques by themselves may not offer the best results, we see below that by recombining them with other types of fusion we can improve the general performance of a representation.

Second Degree Fusion. The second degree fusion techniques presented in Table 3 show that the recombination of cross fusion techniques gets us closer to the early fusion baseline. With the exception of cross late early fusion, the rest of the recombination schemes yield interesting results. First, in cross early fusion, the best results, for the most part, are obtained while using the S^{LEX} matrix combined with the output of $E(M^{LEX}, M^{STD})$, which is still far from the baseline values. Concerning, EXEF, we get already close to surpass the baselines with the M^{STD} matrix, with the exception of the CONLL dataset. In LXEF, even though the cross fusion $X(S^{SYN}, M^{LEX})$ is not the best performing, we found experimentally that by combining it with M^{LEX} through a late fusion, it gets a strong complementary representation. Our intuition in this case was to

Table 2. NER F-measure results using first degree fusion (1F). B is either indicated on the table or specified as follows. Looking at EF, $\hat{b}_{EF} = E(M^{SYN}, M^{STD})$. In XEF, b^*_{XEF} takes the matrix from the set $\{M^{LEX}, M^{STD}\}$ which yields the best performing result. In XLF, $\hat{b}^*_{XLF}$ corresponds to the best performing matrix in $\{S^{LEX}, S^{SYN}\}$. These configurations serve as the main set of baseline results.

Table 3. NER F-measure results using second degree fusion (2F). In XLEF, a^* corresponds to the best performing matrix in the set $\{X(S^{STD}, S^{LEX}), X(S^{LEX}, S^{STD}), X(S^{STD}, S^{SYN})\}$. For XEEF, $\hat{b}_{XEEF} = E(M^{LEX}, M^{STD})$. In EXEF, b^*_{EXEF} takes the best performing matrix from $\{X(S^{SYN}, M^{LEX}), X(S^{LEX}, M^{LEX}), X(S^{LEX}, M^{STD}), X(S^{SYN}, M^{LEX}), X(S^{SYN}, M^{STD})\}$. Finally, in LXEF, $\hat{b}_{LXEF}$ takes the best possible matrix from $\{X(S^{LEX}, M^{STD}), X(S^{SYN}, M^{STD}), X(S^{SYN}, M^{LEX})\}$.

A	B	Early Fusion		
		CONLL	WNER	WGLD
M^{LEX}	M^{SYN}	72.01	70.59	59.38
M^{LEX}	M^{STD}	78.13	79.78	61.96
M^{SYN}	M^{STD}	77.70	78.10	60.93
M^{LEX}	$\hat{b}_{EF}$	**78.90**	**80.04**	**63.20**
		Late Fusion		
		CONLL	WNER	WGLD
S^{LEX}	S^{SYN}	61.65	58.79	44.29
S^{LEX}	S^{STD}	55.64	67.70	48.00
S^{SYN}	S^{STD}	50.21	58.41	49.81
		Cross Early Fusion		
		CONLL	WNER	WGLD
S^{LEX}	M^{STD}	49.90	70.27	62.69
S^{SYN}	M^{STD}	47.27	51.38	48.53
S^{STD}	b^*_{XEF}	52.89	62.21	50.15
		Cross Late Fusion		
		CONLL	WNER	WGLD
S^{LEX}	S^{STD}	27.75	**59.12**	38.35
S^{SYN}	b^*_{XLF}	36.87	40.92	39.62
S^{STD}	b^*_{XLF}	**41.89**	52.03	**39.92**

A	B	Cross Late Early Fusion		
		CONLL	WNER	WGLD
$\hat{a}$	M^{STD}	37.69	**59.44**	**41.71**
$\hat{a}$	M^{LEX}	**38.31**	58.73	41.56
$\hat{a}$	M^{SYN}	29.31	52.06	34.91
		Cross Early Early Fusion		
		CONLL	WNER	WGLD
S^{STD}	$\hat{b}_{XEEF}$	**54.34**	64.20	39.59
S^{LEX}	$\hat{b}_{XEEF}$	49.71	**71.84**	**45.14**
S^{SYN}	$\hat{b}_{XEEF}$	47.54	53.77	43.32
		Early Cross Early Fusion		
		CONLL	WNER	WGLD
M^{STD}	b^*_{EXEF}	49.58	**77.32**	**61.69**
M^{LEX}	b^*_{EXEF}	49.79	66.22	53.54
M^{SYN}	b^*_{EXEF}	**51.53**	70.94	53.70
		Late Cross Early Fusion		
		CONLL	WNER	WGLD
M^{STD}	$\hat{b}_{LXEF}$	54.82	**75.70**	**54.73**
M^{LEX}	$\hat{b}_{LXEF}$	**56.53**	62.27	52.39

complement M^{LEX} with itself but enriched with the S^{SYN} information. In the N-degree fusion results we discover that indeed this propagation of information helps us beat the baselines we set before.

N-degree Fusion. Finally, the last set of experiments are shown in Table 4. Using a recombination of fusion techniques, a so-called hybrid approach, we finally beat the baselines (single features and early fusion) for each dataset. We note that the best configuration made use of a weighted early fusion with $\alpha = 0.95$. This indicates that the single feature matrix, M^{LEX} is enriched a small amount by the

Table 4. F-measure results using N-degree fusion (NF). In ELXEF, $\hat{b}_{ELXEF} = L(M^{LEX}, X(S^{SYN}, M^{LEX}))$. For EELXEF, $\hat{b}_{EELXEF} = E(E(M^{STD}, L(M^{LEX}, X(S^{SYN}, M^{LEX}))), L(M^{LEX}, X(S^{STD}, M^{LEX})))$ for CONLL and $\hat{b}_{EELXEF} = E(E(M^{STD}, L(M^{STD}, X(S^{SYN}, M^{STD}))), L(M^{LEX}, X(S^{SYN}, M^{LEX})))$ for WNER and WGLD. The best result is obtained in EELXEF when $\alpha = 0.95$.

A	B	Early Late Cross Early Fusion		
		CONLL	WNER	WGLD
M^{STD}	$\hat{b}_{ELXEF}$	67.16	79.45	62.37
		Early Early Late Cross Early Fusion		
		CONLL	WNER	WGLD
M^{LEX}	$\hat{b}_{EELXEF}$	65.01	78.02	62.34
$M^{LEX}_{\alpha=0.95}$	$\hat{b}_{EELXEF}$	**79.67**	**81.79**	**67.05**
EF Baseline		78.90	80.04	63.20

fusion recombination, which is enough to improve the results of said baselines. In CONLL, the early fusion (see Table 2) baseline being 78.13, we reached 78.69, the lowest improvement of the three datasets. Regarding the Wikipedia corpus, in WNER, we passed from 79.78 to 81.75; and in WGLD, from 61.96 to 67.29, the largest improvement of all. It is important that we tried the weighted Early Fusion operator with different α and the best result does not beat these fusion results.

In the next section we transfer the knowledge gained in this task to a new one, word sense induction and disambiguation.

4.2   Word Sense Induction and Disambiguation

Having learned the best fusion configuration from the previous task, in this experiments we set to test if the improvements achieved can be transfered into another NLP task, namely Word Sensed Induction and Disambiguation (WSI/WSD).

Pre-processing. We simply remove stopwords and tokens with less than three letters.

Features. We use the same set of features from the previous task, with the exception of the standard NER features, that is, those represented by M^{STD}, as they are specifically designed to tackle NER.

Test Dataset. The WSI/WSD model is tested on the dataset of the Semeval-2007 WSID task [1]. The task was based on a set of 100 target words (65 nouns and 35 verbs), each word having a set of instances, which are specific contexts where the word appear. Senses are induced from these contexts and applied to each one of the instances.

Table 5. Supervised Recall and Unsupervised F-measure for the Semeval-2007 corpus. We also display the average number of clusters found by each fusion configuration.

Method	Recall (%)			FM (%)			# cl
	All	Noun	Verb	All	Noun	Verb	
Single Features							
M^{LEX}	79.20	82.10	75.80	72.70	76.90	67.90	4.13
M^{SYN}	79.10	81.60	76.20	69.30	69.40	69.20	4.47
Early Fusion							
$E(M^{LEX}, M^{SYN})$	78.70	81.11	76.10	74.00	76.66	71.11	4.46
Cross Early Fusion							
$X(S^{LEX}, M^{LEX})$	79.20	82.30	75.70	76.20	79.60	72.50	3.63
$X(S^{LEX}, M^{SYN})$	78.30	80.90	75.30	74.60	75.10	73.90	3.08
$X(S^{SYN}, M^{LEX})$	78.60	80.90	76.10	78.90	80.70	76.90	1.08
$X(S^{SYN}, M^{SYN})$	78.90	81.40	76.10	73.70	77.70	70.00	2.72
Cross Late Fusion							
$X(S^{SYN}, S^{LEX})$	78.70	80.90	76.20	78.90	80.80	76.80	1.01
$X(S^{LEX}, S^{SYN})$	78.80	80.90	76.06	78.70	80.50	76.80	1.33
Cross Late Early Fusion							
$X(X(S^{LEX}, S^{SYN}), M^{LEX})$	78.40	80.40	76.10	70.00	68.70	71.40	3.11
$X(X(S^{LEX}, S^{SYN}), M^{SYN})$	78.90	81.80	75.60	75.20	77.40	72.80	3.16
Early Cross Early Fusion							
$E(M^{LEX}, X(S^{LEX}, M^{LEX}))$	79.20	82.40	75.70	76.00	79.50	72.10	3.57
$E(M^{SYN}, X(S^{LEX}, M^{LEX}))$	78.30	80.50	75.80	75.20	75.40	75.00	1.95
Late Cross Early Fusion							
$L(M^{SYN}, X(S^{LEX}, M^{SYN}))$	78.60	81.10	75.80	67.80	71.40	63.80	4.22
$L(M^{LEX}, X(S^{LEX}, M^{LEX}))$	79.50	82.80	75.70	76.09	79.10	72.70	3.96
Early Late Cross Early Fusion							
$E(M^{LEX}, L(M^{SYN}, X(S^{LEX}, M^{SYN})))$	78.50	81.40	75.40	74.20	78.20	69.80	4.26
$E(M^{LEX}, L(M^{LEX}, X(S^{LEX}, M^{LEX})))$	79.50	82.70	75.90	75.80	78.50	72.70	3.99

Evaluation Measures. Being an unsupervised task, the evaluation metrics of WSI/WSD are debated in terms of quality [7]. We consider supervised recall and unsupervised F-measure, as in the competition original paper [1]. The first one maps the output of a system to the true senses of the target words' instances and the second one measures the quality of the correspondence between the automatically found clusters and the senses. We consider that the number of senses found by the system is also a rather good indicator of performance: the best competition baseline assigns the most frequent sense to each target word (this baseline is called MFS), thus this baseline system would have an average of 1 sense (cluster) per word. A system that goes near this average may be indeed not resolving the task efficiently but finding the MFS trivial solution.

Consequently, to show that we do not fall in the MFS solution, we display in our results the average number of clusters.

Results. Word sense induction and disambiguation results are found in Table 5. Again, we aim to surpass the baseline of the single features and early fusion. We experimentally set $\beta = 0.90$ and $\gamma = 50$. In this task, in late fusion, when the first matrix is deemed more relevant than the second one, the performance is higher. This may be due to the fact that, in this task, the feature matrices rows contain types (that is, each line represent an unique word), and thus they are more dense, which may entail more noisy data. By reducing the relevance of the second matrix in late fusion, we are effectively attenuating the less important information. Regarding $\gamma = 50$, again due to the denser characteristic of the matrices, there is a larger quantity of true similar words that are useful to project information into another matrix, through cross fusion.

The WSI/WSD results are shown in Table 5. In the following paragraph, we will discuss these result all at once. Due to the page limit constraint, we omit certain configurations that do not yield interesting results either by converging to the MFS solution (1 sense found per target word) or because the performance shown by those configurations is not interesting.

Regarding Single Features, M^{LEX} comes on top of M^{SYN} again. Nonetheless, M^{SYN} is much closer in terms of performance, and as expected, it is actually higher with regards to verbs.

On the 1F level, we see that the early fusion technique in this task does not surpass the independent features representation. Our intuition is that the similarities of both matrices seem to be correlated. In cross early fusion, the best result is obtained by $X(S^{LEX}, M^{LEX})$, regarding the unsupervised F-measure. This configuration already beats our baselines, improving both noun and verb results on the unsupervised evaluation, improving the supervised recall of nouns, and staying on the same level considering all words. Also, it produces more senses than the MSF average number of senses (1 sense per target word), which is good but not indicative of results correctness. Regarding cross late fusion, given the average number of clusters produced, it seems that both results converge towards the MFS, therefore we do not consider these results.

Beginning with the fusion recombinations, in level 2F, both cross late early fusions yield average results. In cross early cross early fusion, the early fusion of M^{LEX} with $X(S^{LEX}, M^{LEX})$ yields very similar results than $X(S^{LEX}, M^{LEX})$. The next natural step is to test this fusion via a linear combination, with a late fusion. The result obtained confirmed the intuition of enriching a single feature matrix with another weighted-down matrix to improve the performance. Indeed, we consider that $L(M^{LEX}, X(S^{LEX}, M^{LEX}))$ gets the best results in terms of all-words supervised recall and the second best all-words unsupervised F-measure (we do not consider solutions that are too close to the MFS baseline).

We test the same configurations as in NER, within the NF level, to try and improve our results. Nonetheless, in general, they do not overcome the best result found previously.

In general, we found that the recombination fusion techniques work in terms of improving the performance of the tasks addressed. In the following, we make our final remarks and the future work to be done regarding fusion techniques on NLP tasks.

5   Conclusion and Future Work

In this paper, we presented a comparative study of multimedia fusion techniques applied to two NLP tasks: Named Entity Recognition and Word Sense Induction and Disambiguation. We also proposed new fusion recombinations in order to complement the information contained in the single representation matrices. In order to accomplish this goal, we built upon basic fusion techniques such as early and late fusion, as well as cross media fusion to transfer quality information from one set of features to another.

We found that by taking a strong feature, in our case lexical context, M^{LEX}, and enriching it with the output of rather complex fusion combinations, we can improve the performance of the tasks addressed. The enrichment has to give more relevance to the strong feature matrix, by selecting the right parameters.

While there is an improvement, we do note that fusion techniques augment the computing time and memory consumption of the tasks at hand by enlarging the feature space or by making it more dense. In that sense, more intelligent ways of finding the most appropriate fusion must be researched. This is indeed one of our future work paths: determining an optimal fusion path from single features to a N-degree fusion recombination. Coupled with this, the automatic determination of the parameters is still ongoing research in the multimedia fusion community. Consequently, we believe that efficiently determining both parameters and fusion combinations is the general domain of our future work. Another route we would like to explore is testing these techniques on other tasks and with datasets from different domains, in order to assert its effectiveness.

References

1. Agirre, E., Soroa, A.: Semeval-2007 task 02: evaluating word sense induction and discrimination systems. In: Proceedings of the 4th International Workshop on Semantic Evaluations, SemEval 2007, pp. 7–12. Association for Computational Linguistics, Stroudsburg (2007)
2. Ah-Pine, J., Csurka, G., Clinchant, S.: Unsupervised visual and textual information fusion in CBMIR using graph-based methods. ACM Trans. Inf. Syst. **33**(2), 1–31 (2015)
3. Atrey, P.K., Hossain, M.A., El-Saddik, A., Kankanhalli, M.S.: Multimodal fusion for multimedia analysis: a survey. Multimedia Syst. **16**(6), 345–379 (2010)
4. Balasuriya, D., Ringland, N., Nothman, J., Murphy, T., Curran, J.R.: Named entity recognition in wikipedia. In: Proceedings of the 2009 Workshop on The People's Web Meets NLP: Collaboratively Constructed Semantic Resources, People's Web 2009, pp. 10–18. Association for Computational Linguistics, Stroudsburg (2009)

5. Clinchant, S., Ah-Pine, J., Csurka, G.: Semantic combination of textual and visual information in multimedia retrieval. In: ICMR, p. 44. ACM (2011)
6. Collins, M.: Discriminative training methods for hidden markov models: theory and experiments with perceptron algorithms. In: Proceedings of the ACL 2002 Conference on Empirical Methods in Natural Language Processing, EMNLP 2002, vol. 10, pp. 1–8. Association for Computational Linguistics, Stroudsburg (2002)
7. de Cruys, T.V., Apidianaki, M.: Latent semantic word sense induction and disambiguation. In: ACL, pp. 1476–1485. The Association for Computer Linguistics (2011)
8. Daume III, H.C.: Practical Structured Learning Techniques for Natural Language Processing. Ph.D. Thesis, Los Angeles, CA, USA (2006). aAI3337548
9. Gialampoukidis, I., Moumtzidou, A., Liparas, D., Vrochidis, S., Kompatsiaris, I.: A hybrid graph-based and non-linear late fusion approach for multimedia retrieval. In: CBMI, pp. 1–6. IEEE (2016)
10. Goyal, K., Hovy, E.H.: Unsupervised word sense induction using distributional statistics. In: COLING, pp. 1302–1310. ACL (2014)
11. Levy, O., Goldberg, Y.: Dependency-based word embeddings. In: ACL (2), pp. 302–308. The Association for Computer Linguistics (2014)
12. Manning, C.D., Surdeanu, M., Bauer, J., Finkel, J., Bethard, S.J., McClosky, D.: The Stanford CoreNLP natural language processing toolkit. In: Association for Computational Linguistics (ACL) System Demonstrations, pp. 55–60 (2014)
13. Nadeau, D., Sekine, S.: A survey of named entity recognition and classification. Lingvisticae Investigationes 30(1), 3–26 (2007)
14. Navigli, R.: Word sense disambiguation: a survey. ACM Comput. Surv. (CSUR) 41(2), 10 (2009)
15. Panchenko, A., Faralli, S., Ponzetto, S.P., Biemann, C.: Unsupervised does not mean uninterpretable: the case for word sense induction and disambiguation. In: Proceedings of the 15th Conference of the European Chapter of the Association for Computational Linguistics (EACL 2017). The Association for Computer Linguistics (2017)
16. Ratinov, L., Roth, D.: Design challenges and misconceptions in named entity recognition. In: CoNLL, pp. 147–155. ACL (2009)
17. Sang, E., Meulder, F.D.: Introduction to the conll-2003 shared task: language-independent named entity recognition. In: CoNLL, pp. 142–147. ACL (2003)
18. Yu, S.I., Jiang, L., Mao, Z., Chang, X., Du, X., Gan, C., Lan, Z., Xu, Z., Li, X., Cai, Y., et al.: Informedia@ trecvid 2014 med and mer. In: NIST TRECVID Video Retrieval Evaluation Workshop, vol. 24 (2014)

Author Index